CONTEMPORARY AND EMERGING GLOBAL ENVIRONMENTAL CHALLENGES

Second Edition

by
JIAN PENG

Cover © Shutterstock.com

www.kendallhunt.com
Send all inquiries to:
4050 Westmark Drive
Dubuque, IA 52004-1840

Copyright © 2020, 2021 by Kendall Hunt Publishing Company

ISBN: 978-1-7924-8292-2

All rights reserved. No part of this publication may be reproduced, stored in a retrieval system, or transmitted, in any form or by any means, electronic, mechanical, photocopying, recording, or otherwise, without the prior written permission of the copyright owner.

Published in the United States of America

Dedication

This book is dedicated to the students of the environmental field and to our future generations.

Contents

Preface	xv
Acknowledgments	xix

CHAPTER 1   How to Build a Sustainable Planet — 1

- Learning Outcomes — 2
- Key Concepts — 2
- 1.1 Environmental Issues and the History of the Earth — 3
- 1.2 The Origin of the Physical Universe — 5
  - 1.2.1 The Big Bang — 5
  - 1.2.2 Synthesis of Materials in Stars — 7
  - 1.2.3 Formation of Planets — 9
  - 1.2.4 The Genesis and Evolution of the Earth — 11
- 1.3 The Origin and Evolution of Life — 12
  - 1.3.1 The Timeline — 12
  - 1.3.2 The Building Materials for Life — 13
  - 1.3.3 The Cell — 16
  - 1.3.4 The Origin of Life on Earth — 17
  - 1.3.5 DNA and the Tree of Life — 18
- 1.4 The Origin and Dominance of Humankind — 19
  - 1.4.1 The Origin of Humankind — 19
  - 1.4.2 Human Evolution and Dominance — 20
  - 1.4.3 Life Beyond Earth — 22
- Further Reading — 24
- Web Resources — 24
- Questions and Exercises — 25

CHAPTER 2   Global Biogeochemical Cycles — 27

- Learning Outcomes — 28
- Key Concepts — 28
- 2.1 The Concept of Global Biogeochemical Cycle — 28

2.2	Plate Tectonics and Rock Cycles	29
	2.2.1 Plate Tectonics	29
	2.2.2 The Rock Cycle	32
2.3	The Water Cycle	33
	2.3.1 The Erosional Power of Water	33
	2.3.2 The Water Cycle	35
2.4	Biogeochemical Cycles of Elements	38
	2.4.1 The Carbon Cycle	38
	2.4.2 The Nitrogen Cycle	45
	2.4.3 Biogeochemical Cycles of Other Elements	48
2.5	Biogeochemical Cycles and Environmental Challenges	49
Further Reading		49
Web Resources		50
Questions and Exercises		50

CHAPTER 3 Environmental Regulations — 51

Learning Outcomes		52
Key Concepts		52
3.1	Need for Environmental Regulations	52
	3.1.1 Tragedy of the Commons	52
	3.1.2 The History of Environmental Regulations	54
	3.1.3 Implementation of Environmental Regulations in the United States in Europe	54
3.2	Environmental Policy-Making Process	57
	3.2.1 Overview	57
	3.2.2 Policy Making Process	59
	3.2.3 The National Environmental Policy Act (NEPA)	60
	3.2.4 The Policy Cycle	61
3.3	Environmental Politics and Policy	62
	3.3.1 Environmental Politics	62
	3.3.2 Interest Groups in Environmental Politics and Policy	63
	3.3.3 Environmental Groups and Their Roles in Environmental Politics and Policy	64
	3.3.4 Litigation, Regulation, Legislation, and Negotiation in Environmental Politics	66
3.4	Environmental Regulations	67

	3.4.1	Environmental Regulation in the United States	67
	3.4.2	Environmental Regulation in the European Union	68
	3.4.3	Environmental Regulation in China and other Developing Countries	69
3.5		Environmental Justice	70

Further Reading — 71
Web Resources — 72
Questions and Exercises — 72

CHAPTER 4 Air Pollution and Greenhouse Gases — 75

Learning Outcomes — 76
Key Concepts — 76
- 4.1 The Earth's Atmosphere — 76
  - 4.1.1 Overview and History — 76
  - 4.1.2 Structure of the Earth's Atmosphere — 77
- 4.2 Air Pollution — 78
  - 4.2.1 Overview — 78
  - 4.2.2 Sources of Air Pollution — 78
  - 4.2.3 Main Air Pollutants — 80
- 4.3 Air Quality Issues and Regulations — 84
  - 4.3.1 Air Pollution Regulations in the United States — 84
  - 4.3.2 Air Pollution Regulations in China — 86
  - 4.3.3 Air Quality Issues and Regulations in Europe and Beyond — 88
- 4.4 Engineering Measures for Air Pollution Control — 91
  - 4.4.1 Overview — 91
  - 4.4.2 Vehicle Emission Control — 91
  - 4.4.3 Control of Other Air Pollutants — 93
- 4.5 Greenhouse Effect and Management — 94
  - 4.5.1 Greenhouse Effect — 94
  - 4.5.2 Other Greenhouse Gases — 97
  - 4.5.3 Geoengineering Options for Atmospheric Carbon Dioxide — 97
- 4.6 Other Air Quality Issues — 98
  - 4.6.1 Indoor Air Pollution — 98
  - 4.6.2 Ozone Depletion — 99
  - 4.6.3 Global Deoxygenation — 99
  - 4.6.4 Long-Range and Global Transport of Air Pollutants — 100

Further Reading — 100
Web Resources — 101
Questions and Exercises — 102

CHAPTER 5 Water Pollution — 103

   Learning Outcomes — 104
   Key Concepts — 104
   5.1 Water Quality Regulations — 105
      5.1.1 The Clean Water Act — 105
      5.1.2 The Water Framework Directive in the European Union — 108
      5.1.3 Water Quality Regulations in China and India — 109
   5.2 Common Causes of Water Pollution — 110
      5.2.1 Eutrophication — 113
      5.2.2 Pathogen Contamination — 115
      5.2.3 Toxic Contaminants — 117
      5.2.4 Trash — 124
      5.2.5 Sediment — 126
   5.3 Stormwater Issues — 127
      5.3.1 Overview — 127
      5.3.2 Urbanization and Stormwater Flow and Quality — 128
      5.3.3 Stormwater Management and Low Impact Development — 132
   5.4 Water Treatment and Reuse — 134
      5.4.1 Background and History of Water Treatment — 134
      5.4.2 Overview of Water Treatment Process — 135
      5.4.3 Water Recycling and Biosolid Management — 137
   5.5 Emerging Water Quality Issues — 139
      5.5.1 Overview — 139
      5.5.2 Microplastics — 139
      5.5.3 Per- and Polyfluoroalkyl Substances (PFAS) — 142
      5.5.4 Other Contaminants of Emerging Concern — 144
      5.5.5 Long-Range Transport of Pollutants — 145
   5.6 Water Quality Monitoring — 145
   Further Reading — 147
   Web Resources — 149
   Questions and Exercises — 150

CHAPTER 6 Water Resources — 151

   Learning Outcomes — 152
   Key Concepts — 152
   6.1 Overview — 152
   6.2 Water Quality and Water Resources — 155
      6.2.1 Nexus Between Water Quality and Water Resources — 155
      6.2.2 Novel Potable Water Production Technologies — 156
   6.3 Water Shortage and Drought — 158

	6.3.1	Global Water Shortages	158
	6.3.2	Water Security and Resilience	160
	6.3.3	Water Resource and Sustainability	161
6.4		Water Supply Infrastructure and Its Environmental Impacts	161
	6.4.1	History of Water Supply Infrastructure	161
	6.4.2	Dams around the World	162
	6.4.3	Ecological and other Impacts of Dams	163
6.5		Water Reuse and Recycling	168
	6.5.1	Overview of Water Reuse	168
	6.5.2	Types of Potable Water Reuse: Direct and Indirect	168
6.6		Desalination	169
	6.6.1	Overview of Desalination	169
	6.6.2	Desalination Process and Potential Impacts	170
6.7		Water Balance and Integrated Water Management	171
	6.7.1	Overview of Water Balance	171
	6.7.2	Case Study: Water Balance for North Orange County, California, USA	172
6.8		Water Footprint	174
	6.8.1	The Concept of Water Footprint	174
	6.8.2	Water Footprint of Common Items	174
	6.8.3	Water Footprint and Sustainability	175
Further Reading			177
Web Resources			178
Questions and Exercises			178

CHAPTER 7 Groundwater and Soil Contamination   181

Learning Outcomes			182
Key Concepts			182
7.1		Groundwater Hydrology	182
	7.1.1	Groundwater and Aquifer	182
	7.1.2	Groundwater Resources	184
7.2		Groundwater and Soil Contamination	187
	7.2.1	Groundwater and Drinking Water	187
	7.2.2	Soil and Groundwater Contaminants	187
	7.2.3	Human Health Risk Assessment	192
7.3		Groundwater and Soil Remediation	193
	7.3.1	Underground Storage Tanks	195
	7.3.2	Soil Investigation	197

	7.3.3 Groundwater Investigation and Conceptual Site Model	197
	7.3.4 Soil and Groundwater Remediation	201
Further Reading		204
Web Resources		205
Questions and Exercises		205

CHAPTER 8   Ecology and Biodiversity — 207

Learning Outcomes		208
Key Concepts		208
8.1	Principles of Ecology	208
	8.1.1 Concepts and Principles	208
	8.1.2 Population Ecology	212
	8.1.3 Ecosystem Health	212
8.2	Biodiversity	213
	8.2.1 The Significance of Biodiversity	213
	8.2.2 The Sixth Mass Extinction	214
8.3	Humans and the Ecosystem	217
	8.3.1 Humans as Part of the Global Ecosystem	217
	8.3.2 Ecosystem Services	218
	8.3.3 Human Impact on Ecosystems	219
	8.3.4 The Ecological Impact of Pollution	225
	8.3.5 The Ecological Impacts of Habitat Loss	227
	8.3.6 Ecological Impact of Hunting and Fishing	232
8.4	Genetically Modified Organisms	234
	8.4.1 The Definition of GMOs	234
	8.4.2 Benefits and Risks	236
8.5	Other Ecological Challenges	236
	8.5.1 Endangered Species	236
	8.5.2 Invasive Species	237
	8.5.3 Deep Ecology: A Scientific Perspective	239
Further Reading		240
Web Resources		241
Questions and Exercises		241

CHAPTER 9   Solid Waste — 243

Learning Outcomes	244
Key Concepts	244

9.1	Solid Waste Classification and Generation Rate		245
	9.1.1	Overview	245
	9.1.2	Solid Waste Generation in the United States	247
	9.1.3	Solid Waste Generation around the World	247
9.2	Landfill		250
	9.2.1	Overview and Waste Collection	250
	9.2.2	Landfill Operation	251
	9.2.3	Landfill Stormwater Management	252
	9.2.4	Landfill Trash Management	253
	9.2.5	Landfill Leachate Management	254
9.3	Solid Waste Recycling and Energy Recovery		257
	9.3.1	Solid Waste Recycling	257
	9.3.2	Waste to Energy	260
9.4	Hazardous Waste		263
	9.4.1	Definition	263
	9.4.2	Hazardous Waste Management	265
9.5	Plastics and Trash Pollution		266
	9.5.1	Overview	266
	9.5.2	Basel Convention	269
	9.5.3	OECD Convention	271
Further Reading			272
Web Resources			273
Questions and Exercises			273

CHAPTER 10 Energy and Sustainability — 275

Learning Outcomes			276
Key Concepts			276
10.1	Global Energy Consumption and Trends		277
	10.1.1	Energy Sources	277
	10.1.2	Trends in Global Energy Consumption	278
10.2	Conventional Energy		281
	10.2.1	Overview of Conventional Energy	281
	10.2.2	Crude Oil	282
	10.2.3	Coal	285
	10.2.4	Natural Gas	288

10.3	Renewable Energy	289
	10.3.1 Overview of Renewable Energy	289
	10.3.2 Solar Energy	291
	10.3.3 Wind Energy	294
	10.3.4 Other Renewable Energy Sources	297
10.4	Nuclear Power	300
	10.4.1 Overview	300
	10.4.2 Advantages of Nuclear Power	303
	10.4.3 Challenges of Nuclear Power	304
10.5	Energy and Sustainability	308
Further Reading		311
Web Resources		311
Questions and Exercises		312

CHAPTER 11 Climate Change, Sea Level Rise, and Ocean Acidification — 313

Learning Outcomes		314
Key Concepts		314
11.1	Climate Change	314
	11.1.1 Greta	314
	11.1.2 Global Climate System	315
	11.1.3 Earth's Energy Budget	320
	11.1.4 Greenhouse Gases and Global Warming	320
	11.1.5 Climate Variability and ENSO	327
11.2	Sea Level Rise	332
	11.2.1 The Concept	332
	11.2.2 Causes of Sea Level Rise	332
	11.2.3 Sea Level Rise in Different Scenarios	333
11.3	Ocean Acidification	335
	11.3.1 What is Ocean Acidification?	335
	11.3.2 The Mechanism of Ocean Acidification	336
	11.3.3 The Impact of Ocean Acidification	336
11.4	Mitigation Measures	338
	11.4.1 The Global Climate Feedback Loop	338
	11.4.2 The Carbon Problem	338
	11.4.3 Potential Mitigation Measures	339

Further Reading		340
Web Resources		341
Questions and Exercises		341

CHAPTER 12 Environment, Human Development, and Sustainability — 343

- Learning Outcomes — 344
- Key Concepts — 344
- **12.1** The Definition and Concept of Sustainability — 344
- **12.2** Environmental Issues through Economic Lenses — 346
    - 12.2.1 The Environmental Industry — 346
    - 12.2.2 Environmental Economics — 347
    - 12.2.3 Environmental Regulation and Economics — 350
- **12.3** Life Cycle Environmental Impacts and Costs — 351
    - 12.3.1 Life Cycle Cost — 351
    - 12.3.2 Ecological Footprint — 352
    - 12.3.3 Carbon Footprint — 352
    - 12.3.4 Life Cycle Environmental Cost — 355
- **12.4** Sustainability Issues — 359
    - 12.4.1 Environmental Mitigation — 359
    - 12.4.2 Nature's Intrinsic Value — 359
    - 12.4.3 Deep Ecology and Sustainability — 360
    - 12.4.4 Resource Depletion — 362
- **12.5** Global Sustainable Development — 366
    - 12.5.1 Global Sustainable Development Goals — 366
    - 12.5.2 The Earthrise — 370
- Further Reading — 373
- Web Resources — 374
- Questions and Exercises — 375

Bibliography — 377
Knowledge Box Entries — 389
List of Acronyms — 393

Preface

When I was wrapping up the last chapters of this book, the world was suddenly turned upside down by COVID-19, or coronavirus. The 50th anniversary of the Earth Day also came to pass with little fanfare due to lockdowns around the globe. Millions of people were infected, hundreds of thousands people died. Businesses were destroyed. The root cause of the pandemic was not exactly known yet, but most likely due to a breakage of harmony between humans and the environment.

When most of the people were locked down at home, and businesses and factories halted their operations, one striking phenomenon emerged. The water became cleaner. The air was clearer. Wild animals took over some urban landscape. When people witnessed these changes from their windows, many realized how much environmental impact humans have caused to the environment.

When I was teaching at the University of California Irvine, a classroom routine was to start each lecture with new and emerging environmental issues in the news media and internet. In the span of three months and 30 lectures, there was no shortage of new topics. Per- and polyfluoroalkyl substances (PFAS) and the movie Dark Water; Tesla's Cybertruck; extreme weather events and wildfires in California, Australia, and in the Amazon rainforest; Beijing's air pollution; presidential election issues and debates (and never-ending bickering about climate change and environmental issues); pollution incidents (as a pollution responder, I shared a handful real life stories with them); to name a few. If one needs a topic for small talk with a stranger, weather is too mundane, politics too dangerous, but environmental issues could easily start an animated conversation.

Despite the profound interest in environmental issues, many people do not have a good grasp of the fundamental science and how much they are influenced by the popular media and word-of-mouth, especially by online social media. A lot of people support or oppose climate change and global warming without a basic understanding of the underlying science. People were appalled by the fires in the Amazon rainforest, but few know that the situation was largely in line with recent history of forest clearing and reclamation (i.e. it is indeed a problem—it is just not a new problem as depicted). Politicians adeptly play environmental cards in their election campaign by inciting fear and indignity. Since environmental issues have been heavily politicized, this lack of respect for science on environmental issues could be detrimental to society in the long run.

Freshmen and sophomore students in many disciplines take a general environmental challenge course for different reasons. Some are deeply committed to the environmental field and will pursue a professional career in the field after graduation. Some are interested in environmental issues in general but may or may not be committed to a career in the environmental field. Many stumbled into this course to fulfill general education credit requirements with no particular goal or interest.

For these students, an ideal textbook should be the one that covers all major fields in environmental issues and provides concise but solid background information. It should also be up-to-date with the latest development of various environmental issues. However, I found it difficult to find such a textbook so I took on this daunting task of writing one on a wide range of global environmental challenges that are fast-changing, controversial, and sometimes difficult to explain with plain language. For example, PFAS became a major water quality issue in 2019 when USEPA and California started regulating the compounds with limitations that are difficult to achieve, resulting in closure of many drinking water wells. In the same year, a documentary film 'Dark Waters' came out and became a sensational hit that caused many people to be concerned about PFAS. Even though PFAS are not a 'new' contaminant, they have been mostly unnoticed by the general public. However, it appears that PFAS will be the pollution of concern and of attention for years to come. For textbooks that cover water quality, especially drinking water, the one that does not cover PFAS may be deemed 'outdated'. At the same time, people's opinion may change over time. Science will also advance with time. Some environmental challenges may have been largely resolved (CFC and ozone depletion); some issues may have turned out not to be as severe as people thought (Amazon fire and extreme weather events). I have written book chapters and many peer-reviewed scientific papers. This will be the first textbook for me, and it is a huge challenge due to the breadth of issues and the need to strike a balance between comprehensiveness and brevity.

This book will NOT be 'global environmental challenges for dummies' type of book. It will not introduce the basic STEM (science, technology, engineering, and mathematics) concepts that are foundational to understanding the environmental issues, since we assume the reader has acquired these understandings and capabilities through past education in high school or junior college. It will not go into extensive details on these issues, but it will clearly lay out the scientific and engineering principles of each issue and list the most authoritative reference materials and online resources to allow further reading.

This book will adopt a science-based, objective, and holistic approach to environmental issues. In many cases, both sides of the issues will be presented and analyzed without prejudice. I am deeply influenced by the book series "Taking Sides—Clashing Views on Controversial Environmental Issues" (by different authors for different editions).

Clearly, it is nearly impossible NOT to be opinionated on environmental issues. Many environmental issues hit home hard, and most of the people are affected one way or another. Politicians use environmental issues to draw voters. Businessmen often view environmental regulations as excessive and are a threat to the businesses' bottom lines. Younger generation views environmental issues as vital to their future. For the general public, not-in-my-backyard (NIMBY) principle often determines one's stance on particular local environmental issues.

The book starts with the origin of the universe, solar system, and the Earth, and covers the history of the Earth and biogeochemical cycles of rocks, water, and some key elements before the environmental issues are introduced. I believe that with this broad and fundamental background, the reader will be anchored solidly on scientific facts. He/she will have a good understanding of temporal and spatial aspects of environmental issues based on the Earth's history and various global biogeochemical cycles. Then the book elaborates on different environmental issues related to air, water, soil and groundwater, solid waste, energy, ecology, global climate change, and sustainability. Each of the topics provides sufficient materials for a week worth of teaching, including lectures, discussions, and potential project ideas.

To facilitate teaching, each chapter begins with 'learning outcomes' and 'key concepts' to summarize the gist of the chapter. At the end of each chapter, 'further reading', 'web resources' and 'questions and exercises' are provided for students to learn more about the issues and to think about some thought-provoking questions and project ideas.

Preface to the Second Edition

A year has passed since the first edition was published, and the world is still not out of the shadow of the great coronavirus pandemic. The US went through a dramatic and tumultuous presidential election, in which environmental issues and climate change were again important issues. While the global economy has mostly recovered from the pandemic low of around March-April of 2020, many other sectors including education and scientific research are still profoundly affected, with most schools open only for virtual classes. I myself have been working from home for more than a year. I was able to find some (but not enough) time to read through the book and made some edits and enhancements, as are reflected in this edition. I am committed to spending more time for a more comprehensive review and revision/update in the next edition.

In the thick of the pandemic, I traveled alone for a trip to a number of water and energy projects in California and nearby states. These projects included Ivanpah Thermosolar Plant, the world's largest; Colorado River Aqueduct near California-Arizona-Mexico, Los Angeles Aqueduct, and California's State Water Project. I drove past places such as Salton Sea, Sonoran Desert, Death Valley, and Sierra Nevada. I took photos along the way and some are included in the second edition of this book. I have been to most of these places before, but this trip (I self-dubbed it an 'environmental-energy tour') gave me a fresh perspective about issues such as water, energy, agriculture, sustainability, among many others. This second edition reflected many such perspectives, together with some editorial improvements suggested by many helpful reviewers. They include: Dr. Wang Zijian of the Center for Eco-Environmental Sciences, Chinese Academy of Science; Dr. Qian Zhang of the Chesapeake Bay Program/University of Maryland; Dr. Stuart Goong, my colleague at OC Environmental Resources; among many others. Of these reviewers, Dr. Goong provided the most detailed and comprehensive reviews for the entire book. I added more Knowledge Box entries and questions and exercises to facilitate students to learn and think.

Since this book is designed to be a college textbook on general environmental issues, it uses only well-established scientific facts and prevailing opinions. To improve the flow of information, the book does not generally indicate the sources of information where they are cited. Instead, these citations are listed in the bibliography of each chapter. Other than the water balance scheme (Chapter 6) that I developed based on my own work and photos I personally took, I do not claim credits for any factual materials in this book.

Acknowledgments

I am indebted to Dr. Sunny Jiang, professor and chair of the UCI Henry Samueli School of Engineering, Department of Civil and Environmental Engineering, for inviting me to teach the underground breadth environmental course, CEE60, Contemporary and Emerging Environmental Challenges, that inspired this book. I would also like to thank Kendall Hunt and its staff Randy Weiskettal,[1] Angela Lampe, Rachel Guhin, Natalie Digman, and Deepthi Mohan for inviting me to write this book and for their support throughout the process.

The students in the fall quarter of 2019 were overwhelmingly supportive of the book idea, and they provided the top 10 global environmental issues to be addressed in this book. The topics reflected the general flow of my lectures and students' input. The students were attentive and interested throughout the quarter and gave me and the course very positive evaluations. This gave me confidence that my way of teaching this course and developing this textbook may as well be a good one.

For the fall quarters of both 2018 and 2019, many guest speakers came and gave lectures on their respective field of expertise. I have learned a great deal from all of them. They are: Dick Zembal, Brian Glenn, Jennifer Shook, Daniel Apt, Karen McLauphlin, Roy Herndon, Sam Pascual, Isaac Novella, Robert Chang, Brian Lochrie, Colin Kelly, Tom Meregillano, Megan Plumlee, Cindy Lin, and Scarlett Zhai. My teaching assistants Derek Meinheim and Hunter Quon were great teachers themselves and managed discussion sessions very well. I would not have survived without their help.

I also want to thank my supervisors Chris Crompton, Amanda Carr, and Khalid Bazmi for allowing me to teach outside of my full-time job at Orange County Environmental Resources. My colleagues uniformly encouraged me (the guest speakers Jennifer Shook and Brian Glenn are my colleagues) and I appreciate their support and encouragement.

Dr. Adeyemi Adeleye of UC Irvine Department of Civil and Environmental Engineering wrote the sections on microplastics and PFAs. Dr. Scarlett Zhai of the California Department of Toxic Substances Control provided comments and revision on Chapter 6 (she gave guest lectures on the groundwater and soil remediation to the class I taught). Dr. Peter Bowler of UC Irvine Department of Social Ecology provided insightful comments and revisions to Chapter 8. My colleague Stuart Goong provided an editorial review. Dr. Zhen Baixin of Helmholtz Centre for Environmental Research

---
[1] Randy is no longer with Kendall Hunt.

(Lepzig, Germany) and Dr. Wang Zijian of the Research Center for Eco-Environmental Sciences, Chinese Academy of Science provided key insights and reference materials on environmental issues in the EU and in China.

Lastly, I want to thank my family for their tolerance and support during my teaching and writing when I turned into a ghost in the house. Had they not told me that it was cool for me to write a textbook on global environmental challenges, I might have had second thoughts on taking on such a daunting challenge.

Chapter 1

How to Build a Sustainable Planet

Learning Outcomes		2
Key Concepts		2
1.1	Environmental Issues and the History of the Earth	3
1.2	The Origin of the Physical Universe	5
	1.2.1 The Big Bang	5
	1.2.2 Synthesis of Materials in Stars	7
	1.2.3 Formation of Planets	9
	1.2.4 The Genesis and Evolution of the Earth	11
1.3	The Origin and Evolution of Life	12
	1.3.1 The Timeline	12
	1.3.2 The Building Materials for Life	13
	1.3.3 The Cell	16
	1.3.4 The Origin of Life on Earth	17
	1.3.5 DNA and the Tree of Life	18
1.4	The Origin and Dominance of Humankind	19
	1.4.1 The Origin of Humankind	19
	1.4.2 Human Evolution and Dominance	20
	1.4.3 Life Beyond Earth	22
Further Reading		24
Web Resources		24
Questions and Exercises		25

Learning Outcomes

- Knowledge of the origin of the universe through the Big Bang theory
- Knowledge of the formation of the Milky Way and other galaxies; Solar System; and the Earth
- Recognition that stars are the "factories of elements"
- General knowledge and understanding of Earth's history
- General understanding of the origin and evolution of life on Earth
- Recognition of the human dominance of the Earth and causes of environmental challenges

Key Concepts

Big Bang theory; redshift; Olber's Paradox; Solar System; the Great Oxygenation Event; Earth's history; human dominance; Anthropocene; nucleosynthesis; Miller–Urey Experiment; Oparin-Haldane hypothesis; DNA

Nowadays more than ever, many people around the world start to realize that humans are probably headed toward self-destruction within the next millennium due to imminent catastrophic environmental disasters caused by human activities. They would point to the rising global temperature that keeps setting new records. Sea level is rising at an alarming rate. Environmental pollution, global pandemic, resources depletion, geopolitical instabilities, and so on, point to a deeply troubled world we are in and will hand down to our future generations. Many, however, would argue for the contrary by pointing out that during the Earth's 4.6 billion year history, much more hostile conditions have existed and the Earth can survive these issues the same way it did in the past. However, scientists point out that the sheer rate of change since humans started to emerge on Earth about a hundred thousand years ago (literally a blink of an eye compared to the Earth's history), especially since the Industrial Revolution some 200 years ago, is so blazingly fast that there is little doubt that the delicate equilibrium of the Earth system will soon be irreversibly disrupted. In this chapter, we will take a broader look at this issue by focusing on the Earth's history, the origin of life, and human dominance.

The topic of this chapter is inspired by the fantastic book "How to Build a Habitable Planet—The Story of Earth from the Big Bang to Humankind," by Charles H. Langmuir and Wallace (Wally) Broecker. Langmuir is a geochemist at Harvard University, and the late Broecker was a geochemist at the Lamont-Doherty Earth Observatory at Columbia University. As a geochemist myself, I found it fitting to open a book about global environmental issues with something about the origin and evolution of the universe, the formation of the Solar System and the Earth, and the origin of life and mankind. Without a good understanding of where our planet and ourselves came from, it would be difficult to figure out how to keep our planet Earth habitable and how to live with our physical environment and other species in harmony. Therefore, Langmuir and Broecker's book fits well my vision of such a construct. Coincidentally, Professor Broecker was my academic grandfather—he was the PhD advisor of my PhD advisor (Dr. Teh-Lung Ku, professor emeritus at the University of Southern California).

The reader is also encouraged to read Stephen Hawking's "A Brief History of Time—from the Big Bang to Black Holes," a fascinating book. Without a deep-rooted understanding and appreciation of space, time, and the physical world (which are all products of the Big Bang), it may be difficult to approach some issues to be covered in this book holistically and sanguinely. The other good reference books include Charles Darwin's "The Origin of Species", and Ernst Haeckel's "The History of Creation". These two books introduce the origins and evolution of life and human race and are important for us to appreciate our role and responsibilities as part of the global ecosystem. Without putting mankind in the context of our ecosystem, human history in the context of genesis and evolution of life on Earth; or without putting our Earth in the context of the Solar System or the genesis of the universe, one might not appreciate many global environmental issues covered in this book.

1.1 Environmental Issues and the History of the Earth

Environmental issues to be covered by this book are mostly human-induced phenomena.[1] Before the advent of human civilization about 200,000 years ago,[2] the Earth underwent many drastic changes, sustained countless natural disasters, and suffered several mass extinctions. Most of the time in the **Earth's history**, the natural environment was rather hostile. However, these harsh conditions are not 'environmental issues' we will cover in this book. To a naturalist, nature has its intrinsic value and behaves according to its own set of rules. Earthquakes and hurricanes are not natural disasters. Rather, they are just part of the normal earthly occurrences. With humans being the only "intelligent" species on the planet Earth, its dominance over the other species and the natural environment is a major event in the Earth's history.

---

[1] Some natural events such as extreme weather events, sea level rise, and loss of biodiversity could be natural or man made, and the distinction may be difficult to draw. In this book these phenomena will be discussed from the angle of potential human linkage as at least a contributing factor.

[2] The limit has been shifting and consistently been pushed back further. This reflects the current general scientific consensus. For example, early human behavior, such as bipedalism (walking with hind legs), may have started as early as 10 million years ago (Leakey, 1994).

Humans lived with other species and natural environments in a largely peaceful manner until quite recently. While farming during the Neolithic era (7,000–10,000 years ago) allowed mankind to proliferate, the other watershed event was perhaps the Industrial Revolution, symbolized by the invention of steam engine in the early eighteenth century, which allowed humans to exert increasing and often damaging power over nature. The environmental issues including air, water, and soil pollution; greenhouse gas and global climate change; ecological degradation; and loss of biodiversity; etc., can all be traced back to the Industrial Revolution and associated population explosion, ever-increasing thirst for energy and products, and unfortunate damage to the environment in the process.

Therefore, when we ponder the numerous environmental challenges we are facing and wonder where humans will end up in the next few centuries or, if we are lucky enough, in the next millennium or longer, having a good understanding of the history of the Earth will put these thoughts into perspective. If we worry that the Earth will soon become uninhabitable and start to look elsewhere, a good understanding of the history of the Solar System, the Milky Way, or even the universe would put that thought in perspective as well.

Knowledge Box

Doomsday/Apocalypse—Doomsday or apocalypse is a depressing concept that humans will perish from the Earth. Without worrying too much about various religious doomsdays (there are many; we have survived several so far thanks to good luck), there could be natural disasters such as asteroid impact event; super-volcano eruption, a lethal gamma ray burst, hostile extraterrestrial life, and anthropogenic catastrophes such as global warfare, nuclear holocaust, pandemic, or global climate change (both natural or anthropogenic) that makes the Earth uninhabitable. Someday hostile artificial intelligence could take over the Earth. These science-fiction types of dire scenarios, surprisingly, are not impossible. In fact, it is highly improbable that humans can travel outside of the Solar System, and the sun will not last forever. In 5 billion years, it will turn into a red giant and will engulf Mercury and Venus and make the Earth uninhabitable. In the author's view, humans should face the doomsday scenario by doing what we can to ensure a sustainable Earth (or, in other words, to avoid self-destruction). We may even be able to mitigate or avoid some scenarios with technology advancements, such as colonizing another planet. Or we could outsmart extraterrestrial life some day.

1.2 The Origin of the Physical Universe

1.2.1 The Big Bang

The **Big Bang theory** is a revolutionary theory that could be difficult to comprehend. It is still a theory because there is no definitive evidence, but there are many supporting observations that suggest such a scenario offers the most plausible explanation. In 1927, a Belgian astronomer Georges Lamaitre proposed that a cosmic egg explosion started the universe. Hubble then observed the **redshift**, indicating that the neighboring galaxies are speeding away from us at very fast speeds. All observations point to such an event about 13.7 billion years ago, when it seemed that time, space, and all the physical matter in the universe started abruptly. Immediately after the Big Bang, the universe had such a high temperature and pressure that all matter existed in its most fundamental form, or so called "quark soup."[3] The observations supporting the Big Bang theory include a fast expanding universe observed by Hubble; a weak microwave background which could best be explained to be a remnant of such a Big Bang; the explanations to Olber's Paradox (see the Knowledge Box) and the absence of "Big Crunch," where the universe would have eventually collapsed due to gravitational pull (**Figure 1.1**).

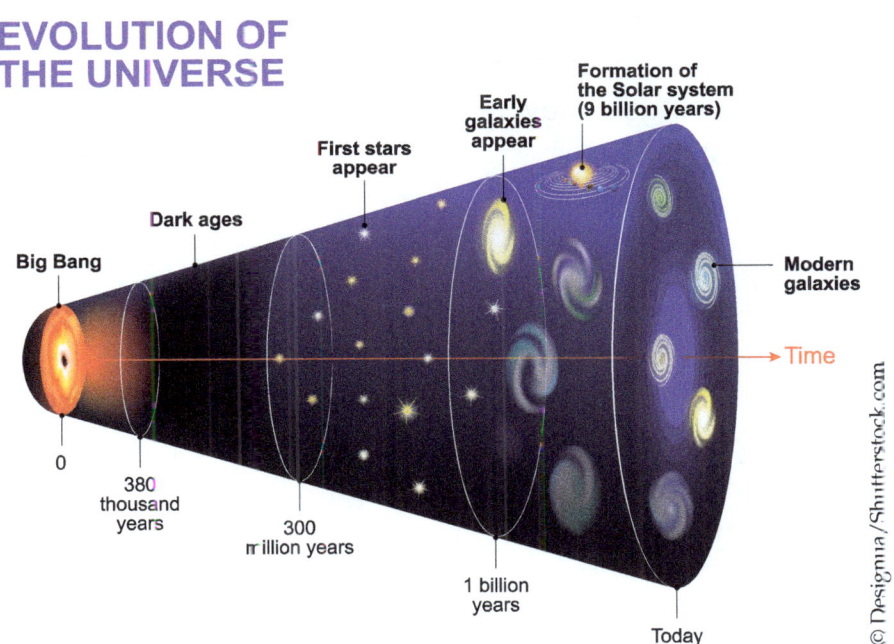

Figure 1.1 The Big Bang and subsequent expansion and evolution of the universe, including the formation of galaxies (such as the Milkey Way galaxy) and the Solar System.

---

[3] Quark is any of a number of subatomic particles carrying a fractional electric charge, postulated as building blocks of most of the basic matter such as electrons, neutrons, protons, etc.

> ## Knowledge Box
>
> **Olber's Paradox**—Named after the German scientist Heinrich Olbers (1758–1840), the **Olber's Paradox** argues that the darkness of the night sky conflicts with the assumption of an infinite, static, homogeneous, and eternal static universe, where we would have a bright night sky since there should be a star in any direction, however far it is. The darkness of the night sky suggests a dynamic universe, such as the Big Bang model. The redshift (see the knowledge box) due to the Big Bang changed most of the starlight into invisible microwave.

> ## Knowledge Box
>
> **Redshift**—Redshift is a phenomenon where physical waves (such as sound) and electromagnetic radiation such as light undergoes an increase in wavelength when the object that emits the wave or radiation is moving away from the observer. Similar to the situation where you can sense the dramatic change in the pitch of the horn from a fast train when the train is passing you, the light from a star that moves away from the Earth will have a longer wavelength than the actual light it emits. Since red light has the longest wavelength in the visible light spectrum, this phenomenon is called redshift.

How do we know when the Big Bang happened? As shown in Figure 1.1 (not proportionately), the universe started to expand rapidly after the Big Bang, and the expansion has been accelerating. Therefore, the timing of the Big Bang could be calculated by measuring how far other galaxies are from the Earth, and how fast they are moving (by measuring the redshift). If all galaxies were produced at the same time and location but travel at different velocities, their velocities and distances should be proportional and those move faster will have traveled longer distances. The relationship between the distance and velocity is determined by the point of time when the initial departure started. The fact that all galaxies can be plotted on a single velocity-distance line is itself one of the strongest evidence of the **Big Bang**. The slope of the regression line, on the other hand, provide a convenient way of calculating the time when the Big Bang happened.

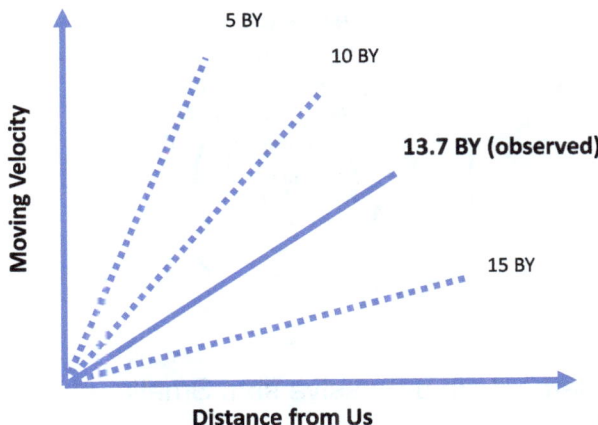

Figure 1.2 Estimating the age of the universe using the distance-velocity relationship. Based on the observations, the distances and velocities of all galaxies can be plotted on the thick line, suggesting that the galaxies in the universe originated from a single point, and that the age of the universe is 13.7 billion years.

Knowledge Box

Dark Matter and Dark Energy—Dark matter is a form of matter that cannot be observed but accounts for 85% of the matter in the universe. Dark matter is used to explain some phenomena such as how galaxies are formed and how they move around in the universe. For example, without abundant dark matter in these galaxies, they would not be spinning around their centers and would simply fly apart. Dark energy is an unknown form of energy that cannot be detected but has profound effect on the universe. Evidence of the existence of dark energy include the accelerating speed of expansion of the universe. Without dark energy, such phenomenon would be impossible to explain by the widely accepted theoretic framework such as general relativity. Together, dark matter and dark energy account for 95% of the mass-energy of the universe.

1.2.2 Synthesis of Materials in Stars

The materials from the Big Bang spun out at very high speed and cooled down quickly and, after about 70,000 to 100,000 years when the temperature of the universe dropped to about 3,000 K[4] and electrons and nucleus were stable enough to form two basic

---

[4] K here stands for Kelvin, a unit for temperature. It has the same magnitude as degree Celsius (°C) but 0°C equals to 273.15K (or, zero Kelvin is a very cold −273.15°C).

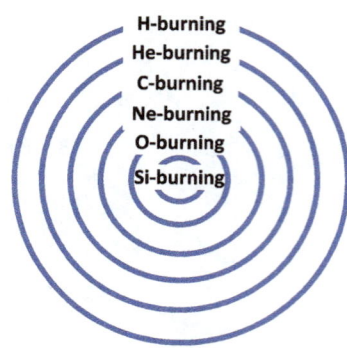

Figure 1.3 Synthesis of elements in a massive star. Smaller stars (such as the sun) only have hydrogen or helium burning.

Source: Langmuir and Broacker (2012), redrawn by Jian Peng

elements, hydrogen and helium. The amorphous hydrogen and helium "cloud" then coalesced by gravity to form galaxies and stars. It is hypothesized that the spiral form of many galaxies are simply the result of tremendous kinetic power originated from the Big Bang. Within each galaxy, the materials further coalesce and forms stars of various sizes. The high temperatures within the stars are a result of gravitational energy converted when hydrogen and helium atoms coalesce.

Stars evolve because of the **nucleosynthesis** within and resulting changes in their composition (the abundance of their constituent elements) over their lifespans, first by burning hydrogen (main sequence star), then helium (red giant star), and progressively burning heavier elements, as seen in **Figure 1.3**. However, this does not by itself significantly alter the abundances of elements in the universe as the elements are contained within the star. Later in its life, a low-mass star will slowly eject its atmosphere via stellar wind, forming a planetary nebula, while a high-mass star will eject mass via a sudden catastrophic event called a supernova. The term supernova nucleosynthesis is used to describe the creation of elements during the explosion of a massive star.

Depending on the sizes of the stars, different numbers of elements could be produced via different types of nuclear fusion, such as hydrogen burning to form helium, helium burning to form carbon and oxygen (Figure 1.3). Other elements and heavier elements are formed via other nuclear reactions, mostly involving helium nucleus. Until today, about 74% and 24% of the mass of the universe is made of hydrogen and helium, respectively. The remaining 2% of the mass is made of heavier elements, with abundances in the order of oxygen, carbon, neon, iron, nitrogen, silicon, magnesium, and sulfur. Other elements are also produced via nucleosynthesis mostly via neutron capture subsequent to the formation of the above abundant elements, as seen in **Figure 1.4**, the periodic table of the elements. It is no coincidence that most

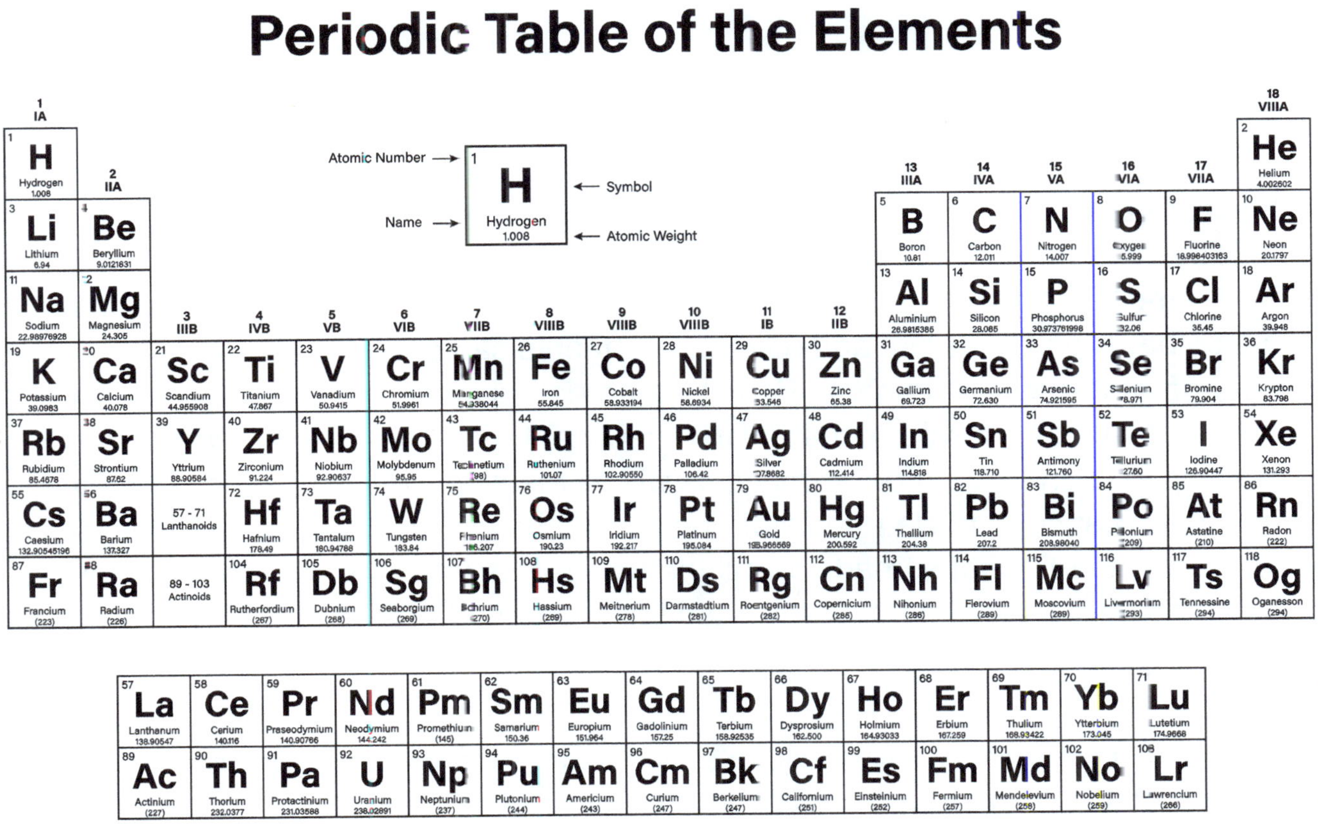

Periodic Table of the Elements

Figure 1.4 The periodic table. Elements with atomic masses divisible by 4 (i.e. carbon, oxygen, neon, magnesium, silicon, sulfur, argon, calcium, etc) have far higher abundances than their neighbors.

of the abundant elements have atomic masses of multiples of 4 (helium's atomic number), and relative stabilities of the nucleus of these elements determine their relative abundance.

1.2.3 Formation of Planets

The formation of planets can best be explained by the history of our very own **Solar System**, which formed about 4.6 billion years ago from the gravitational collapse of a giant interstellar molecular cloud composed of mostly hydrogen. The collapse of the nebula produced the sun, and conservation of angular momentum caused the Solar System to spin faster and to take the form of a disc. The planets are formed through accretion from the materials in the disc, and leftover materials formed the asteroid belt and other components. The Solar System is located in the Orion Arm, about 26,000 light years[5] from the center of the Milky Way galaxy, which contains 100 to 400 billion stars. The age of the Solar System is mostly determined by radioisotope dating (see the Knowledge Box).

---

[5] A light year is a measure of distance that light travels in a year, which is 9.5×10^{12} km.

> ## Knowledge Box
>
> **Radioisotope Dating**—Radioisotope dating is a method to determine the age of an object, such as a rock or a mineral. When a rock or mineral is initially formed, the natural radionuclides begin to decay to form "daughter elements." If both nuclides are conserved (i.e., have not escaped from the media where they are measured), and since the half-life of the radionuclide is known, the relative abundance between the parent and daughter elements can be measured to determine the age of the rock or mineral. Radioisotope dating could also be done on other materials such as fossils, ocean water, sediment, or living things using different assumptions and logic. For example, as will be mentioned later in this chapter, the global atmospheric nuclear tests during the 1960s produced many radionuclides (such as Cs-137) that spread around the world. Therefore, recent sediment could be dated by the 'event horizon' representing that period of time.

More than 99% of the Solar System's mass is in the Sun. The four terrestrial inner planets, Mercury, Venus, Earth, and Mars are composed mostly of rock and metal. The four outer planets Jupiter, Saturn (both gas planets made of hydrogen and helium), Uranus and Neptune (both ice planets made of heavier volatiles such as water, ammonia and methane) are giant gas planets. The planets orbit the sun in a nearly flat plane called the ecliptic (**Figure 1.5**). From the composition and nature of these planets, the influence from the sun is fairly obvious, with rocky inner planets and gaseous/icy outer stars because of solar wind, which is a continuous stream of charged particles traveling at a very high speed (about 1.5 million km/h). The solar wind strips off the

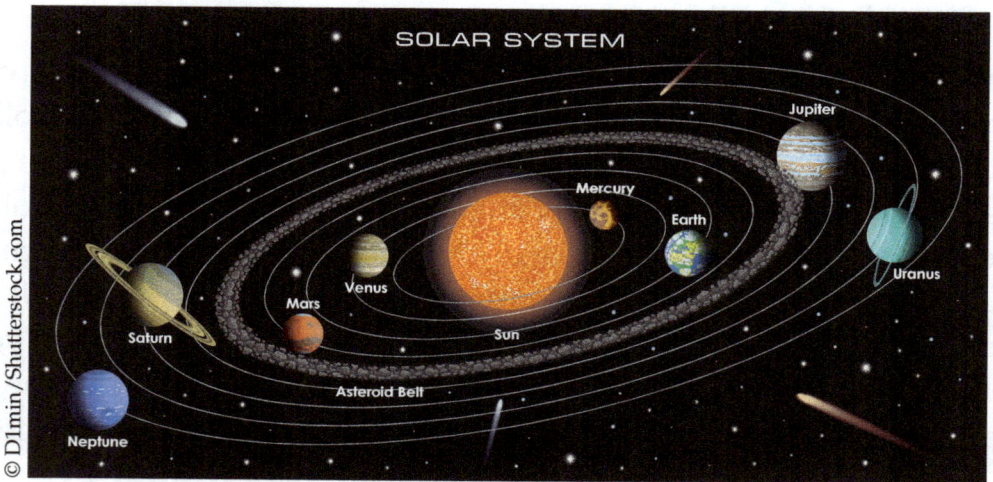

Figure 1.5 Solar system with orbits of all planets and the asteroid belt.

atmospheres of the inner planet except for the Earth[6] and pushes the gases and other volatile material to the outer planets.

1.2.4 The Genesis and Evolution of the Earth

The Earth is a terrestrial inner planet and consists of mostly oxygen, magnesium, silicon, and iron, which account for about 90% of the Earth's mass. The rest are led by calcium, aluminum, nickel, and sulfur. The abundances of the rest of the elements in the Earth are mostly controlled by their volatilities, with more volatile elements being "blown away" by solar wind during the initial period of Solar System formation.

When the Earth was first formed about 4.6 billion years ago, the abundant short-lived radionuclides such as potassium kept the Earth very hot. Gradually and along with the cooling down of the Earth, gravitational force caused the heavier materials (iron and nickel) to sink to the center to form the Earth's core, and lighter materials such as silicates "float" to the surface, forming the mantle. Yet even lighter materials on the very surface formed the crust, which are most silicates of calcium, magnesium, and potassium (**Figure 1.6**).

Aside from the Earth's solid part (lithosphere), the water (hydrosphere) and air (atmosphere) are likely formed by degassing of the Earth's interior and volcanic activities. Some of them, as well as other volatile materials, could be supplied by meteorites that impacted the Earth. As will be seen later in the book, the hydrosphere and atmosphere are both critical to the origin and proliferation of life on Earth. The early

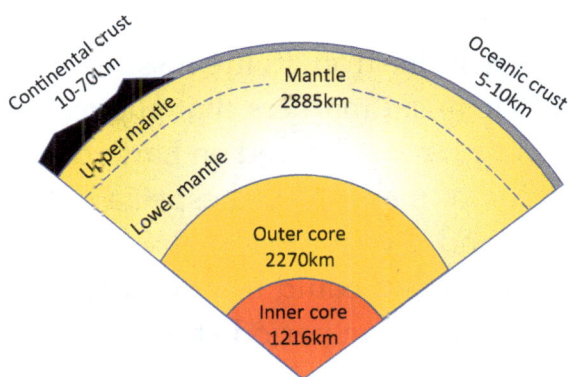

Figure 1.6 Earth's structure with numbers indicating thickness. The boundary between upper and lower mantle is not definitive. Continental crust is lighter and thicker and oceanic crust is heavier and thinner.
Source: Created by Jian Peng

---

[6] The Earth's atmosphere is mostly protected from the solar wind thanks to its magnetic field, which deflects the charged particles of the solar wind to the North Pole (and producing the spectacular aurora along the way).

atmosphere likely provided enough greenhouse effect to keep the Earth warm despite a weaker Sun (30% less energy than it is today). The atmosphere also evolved in a way that allowed increasingly complex life forms on Earth, as will be discussed later. Conversely, the life on Earth changed the composition of the atmosphere as well.

The current structure of the Earth, as indicated by the formation of the crust, formed around 4 billion years ago.[7] As will be discussed later in this book,[8] the Earth has been far from static or stable. Rather, volcanic activities, plate tectonics, and other processes have been continuously shaping and changing the Earth's surface and interior. Deep inside the Earth, the liquid core undergoes some fascinating processes that remain largely unknown to us[9] and produces the Earth's magnetic field. Amazingly, the magnetic field has reversed many times in the Earth's history, a phenomenon that first perplexed scientist but later provided them a useful tool to study Earth's history.[10] With the formation of the Earth's oceans and atmosphere that moderate the temperature and protect the Earth from harmful ultraviolet light, the Earth created a rather favorable environment for the advent of life.

1.3 The Origin and Evolution of Life

1.3.1 The Timeline

The earliest life on Earth might have occurred between 3 and 3.5 billion years ago, as evidenced by fossils. For example, using state of the art electron microscopes, probable single-cell fossils were discovered from 3.2 billion-year-old rocks. Similar fossils with intricate external cell structures were identified in 1.6 billion-year-old rocks in China. Other evidence of early life include stromatolite fossils, which have modern day analogues; stable isotope; and chemical markers that suggested that oxygen-producing, photosynthetic single cells may have appeared on Earth during this time frame. Due to the difficulties of preservation of many fossils, especially for single-celled organisms, which could be rather simple and not easy to identify, the actual time when the first life emerged on Earth was uncertain and could have occurred earlier. The earliest undisputable fossils were from the Cambrian period about 540 million years ago.

---

[7] This could be deduced by the age of the oldest rock on Earth, which was dated to be formed 4.4 billion years ago (Simon A. Wilde, et al.: Evidence from detrital zircons for the existence of continental crust and oceans on the Earth 4.4 Gyr ago, *Nature Geoscience*, 2001).

[8] See Chapter 2.2.

[9] One such theories is the so-called geodynamo theory, where the Earth's outer core of liquid metal undergoes rotation and produces electromagnetic field.

[10] In fact, there is a field of science called paleomagnetism just to study this phenomenon.

1.3.2 The Building Materials for Life

The nucleosynthesis as described in Section 1.3 produced essential elements for organic life such as carbon, hydrogen, oxygen, sulfur, and other trace elements. The atmosphere of the early Earth perhaps was filled with nitrogen, methane, carbon dioxide, water vapor, and hydrogen. These gases were not necessarily conducive to modern life, but **Miller–Urey experiment** showed that they could produce essential organic molecules under certain conditions such as lightning (**Figure 1.7**). It is quite possible that carbohydrates, lipids, amino acids (**Figure 1.8**) could be synthesized through similar processes, especially if enough time and iterations are given.

A critical step beyond these molecules is nucleic acids, the building block of deoxyribonucleic acid or **DNA** and ribonucleic acid or **RNA** (**Figure 1.9**), the genetic material that can be passed along from generation to generation. However, to make even the

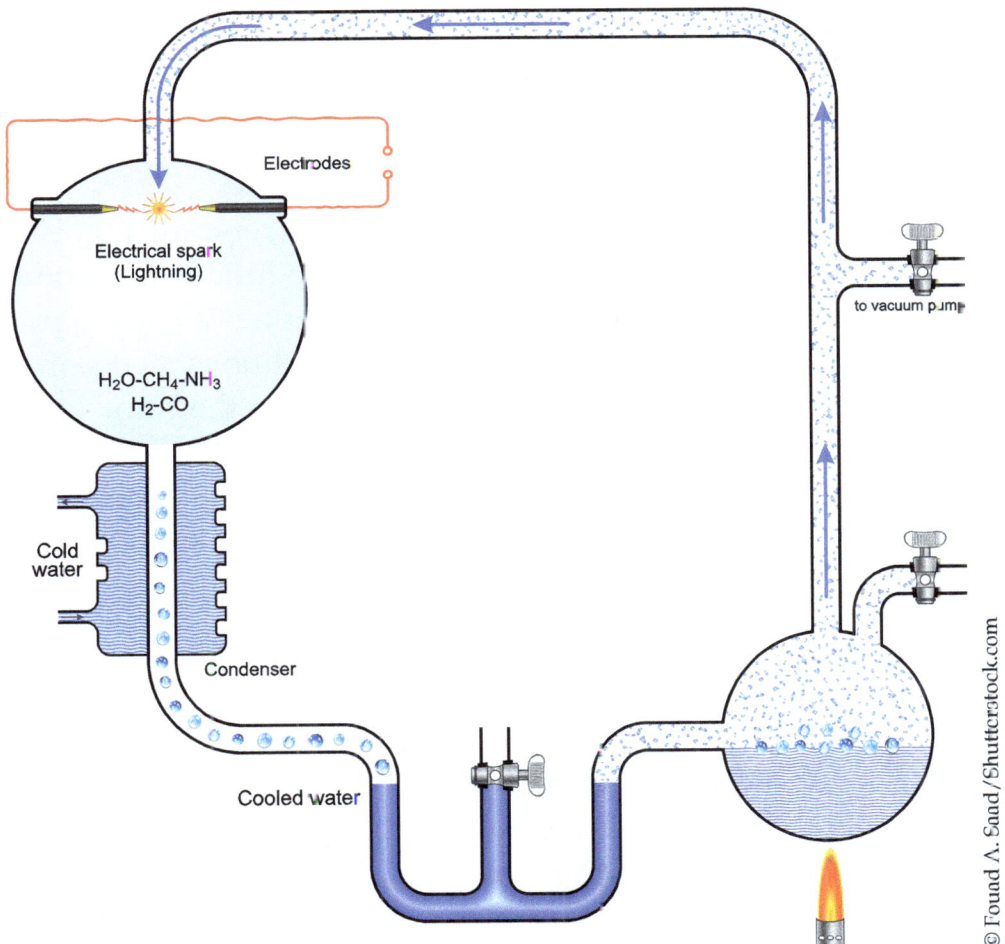

Figure 1.7 The Miller–Urey Experiment where the flask on the left contains water, nitrogen, ammonia, hydrogen, carbon dioxide, and methane—the gases in the atmosphere of the early Earth.

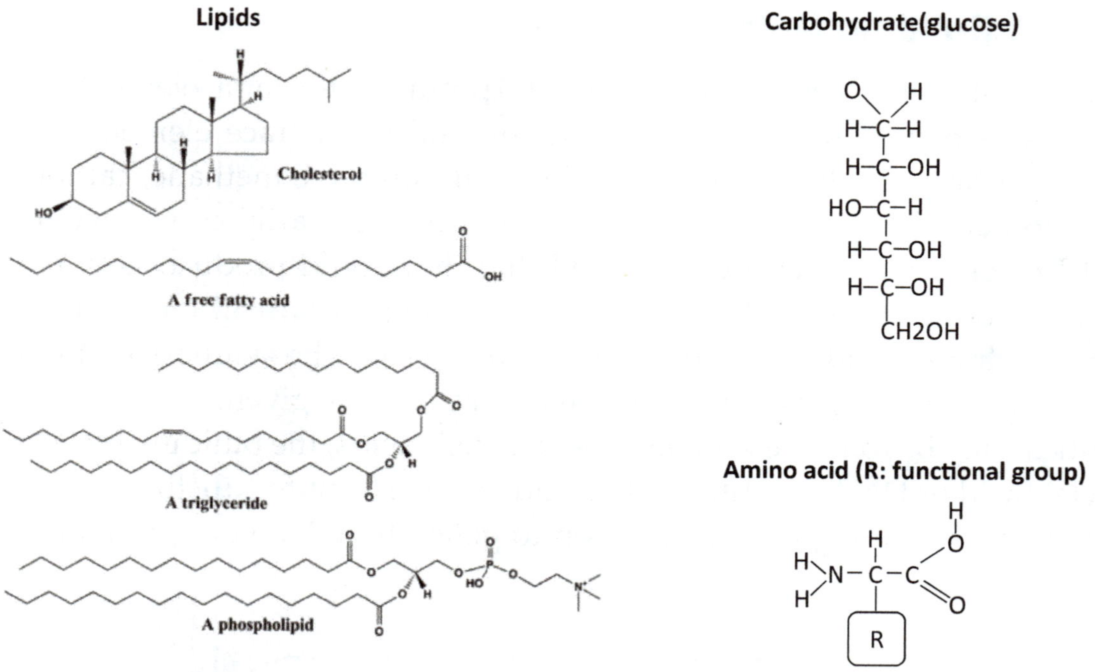

Figure 1.8 Chemical structures of lipids (cholesterol, fatty acid, glycerine, phospholipid); a carbohydrate (glucose); and amino acid. They are the basic building blocks of life on Earth.
Source: Created by Jian Peng

simplest form of life, a unicellular cell that is capable of reproducing, is much more difficult than it appears. After all, Darwin's theory of "a little warm pond" might as well work, where favorable physical and chemical conditions created a prototype cell by chance, after millions of trials and permutation of conditions, that started the earliest form of life. One possible evidence is stromatolites (see the Knowledge Box) that can still be found in the modern world. Another possibility is undersea vents where potentially favorable, albeit drastically different environment from "a little warm pond," and a similar process could happen that resulted in the first living cell.

Knowledge Box

Stromatolite—Stromatolite or stromatoliths (literally meaning "layered rocks") are sedimentary rocks with layered structure that are thought to be indicative of early life forms on Earth. This is evidenced by modern stromatolites, which are shown to form due to the interaction of microorganisms such as cyanobacteria and their inorganic environment, forming biofilms and accretion of inorganic materials and layered structure. The finding of very old stromatolite fossils in ancient rock formations probably indicates that life originated early in the Earth's history. See Figure 1.10 for more details.

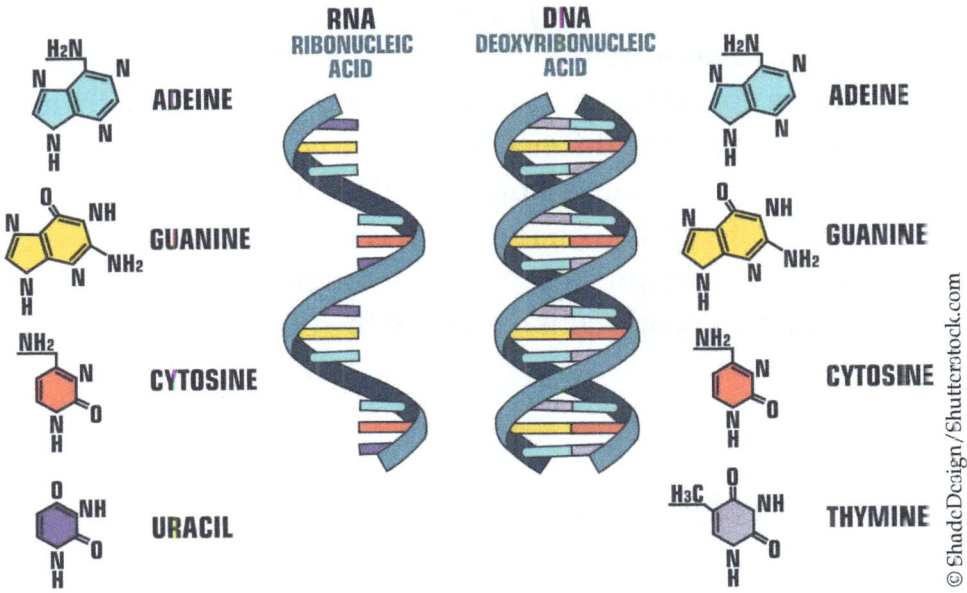

Figure 1.9 DNA chemical structure with adenine, thymine, guanine, and cytosine with phosphate-deoxyribose backbone. The structure will form a three-dimensional "double helix" to be functional. A ribonucleic acid, or RNA section, is shown on the left for comparison. Compared to DNA, RNA has uracil instead of thymine.

Figure 1.10 Early forms of life? Left: Hamelin Pool Stromatolites, Australia. These stromatolites are calcareous mounds built up of layers of lime-secreting cyanobacteria and trapped sediment. Right: a 650 million years old fossil from a pre-Cambrian geological formation in Bolivia showing stromatolite-like structure, suggesting that life (such as cyanobacteria) emerged on Earth at least as early as 650 million years ago.

1.3.3 The Cell

The strongest evidence that all life on Earth originated from simple life forms is the fact that cell is the basic unit for nearly all life (viruses may be an exception but they may not be considered "life" themselves). Life on Earth most likely originated from single-celled organisms. There are two types of such cells, prokaryotic and eukaryotic cells (Figure 1.11). The former have no nucleus and less structured interior and cell membrane performs many critical functions. Eukaryotic cells, on the other hand, are much more complex with a nucleus housing DNA and different organelles, such as mitochondria, chloroplasts, and so on (Figure 1.11).

Eukaryotic cells are larger and reproduce in longer intervals (every 24 hours vs. every 20 minutes for prokaryotic cells). Most of the cells have similar structures and

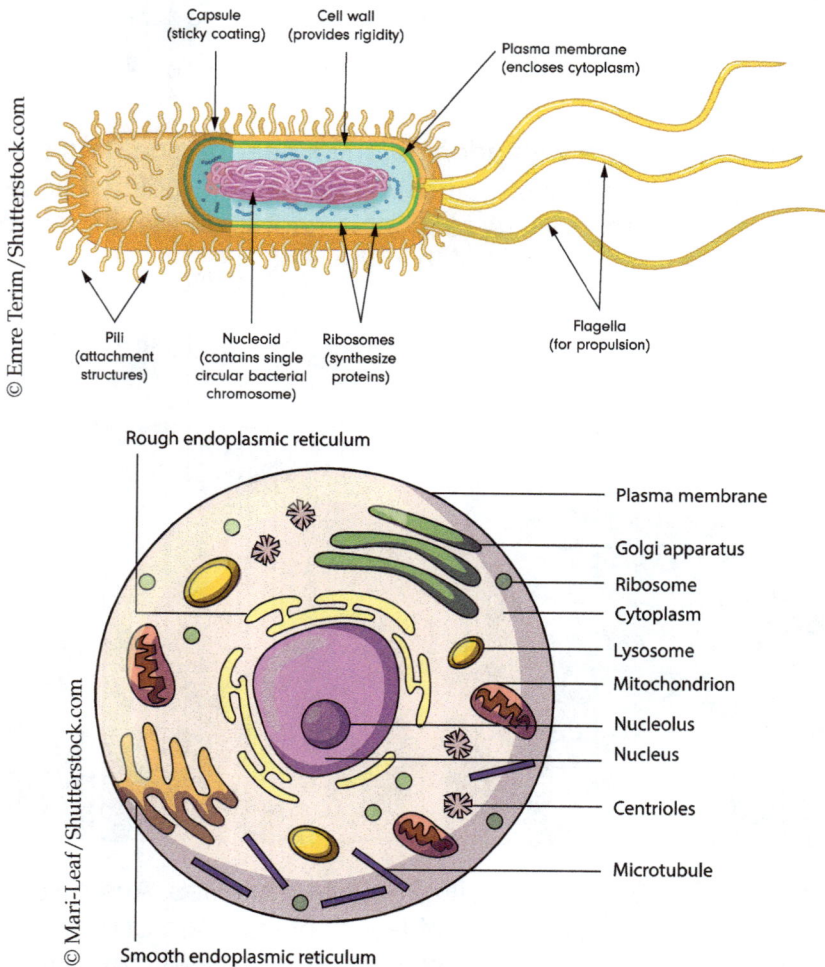

Figure 1.11 Prokaryotic (top) and Eukaryotic cells (bottom).

appearance, such as cell walls, nucleus, organelles (mitochondria, vacuole, endoplasmic reticulum, Golgi body, etc.), and protoplasm. They use similar groups of molecules to build themselves, and similar mechanisms for energy via adenosine triphosphate (ATP), and use DNA for genetic materials, which uses RNA to pass along the genetic information to make proteins, the building materials of most of the cells.

1.3.4 The Origin of Life on Earth

The question about the origin of life on Earth is the one about how single-celled organisms emerged on Earth. While the formation of double-walled cell membranes was possible and replicated *in vitro*, how the leap toward self-sustaining and reproducing cells took place is incredibly complex and still largely unknown. Sidney Fox, inspired by the Miller–Urey experiment as well as by the fact that Apollo 11 brought back soils from the moon and trace amounts of amino acids were found, confirmed that under certain conditions, these amino acids could coagulate spontaneously and form tiny microspheres that Fox found "lifelike." Even though this theory was largely discredited later, the quest for the understanding on how life began on Earth is still ongoing. Therefore, it is not surprising that some scientists, including Crick himself, hypothesized that the earliest bacteria might come from outer space. However, even for the outer space, the same question of the original of life remains unsolved.

Perhaps the most plausible theory about the original of life on Earth, in addition to Darwin's famous "little warm pond," is the one discovered independently by a Soviet scientist Alexander Oparin and Scottish scientist John Haldane, who postulated that, even before the Miller–Urey experiment, primitive life could emerge on the young Earth. The so-called **Oparin-Haldane hypothesis** envisioned that the primitive life probably originated when the Earth had no free oxygen. Instead, methane and ammonia filled the atmosphere. The lack of ozone allowed UV light to reach the Earth's surface unchecked. There were also frequent volcanic activities. This high-energy environment with abundant UV light may have enabled production and increasingly complex organic compounds and structures. So, in such a "hot little soup," organic compounds could form pre-cellular structures that are between nonlife and life. These molecular aggregates are simpler than the single-celled organisms, but complex enough to eventually carry out metabolism and reproduction. What is truly remarkable is that Haldane realized that virus, being "half-living" organisms themselves, might hold the key to cell reproduction, which Haldane believed to be the hallmark of true life.

1.3.5 DNA and the Tree of Life

No long after Fox's theory and Oparin-Haldane hypothesis about possible mechanisms that the early cells could be formed, Watson and Crick took advantage of the state of the art X-ray crystallography capabilities at the famed Cavendish Laboratory of the Cambridge University and identified the double helix structure of DNA. Through the structure of the DNA, Watson and Crick hypothesized that DNA could represent a copying mechanism. However, a complete understanding of DNA as the key to reproduction and protein synthesis did not come about until many years later. The variety of species and complexities of many life forms often make people wonder how nature could produce such intricate life forms without some divine assistance. Nonetheless, the favorable environment of the earth, the eons of time (3–4 billion years) that allowed things to happen, and fossil records clearly point to progressively more complex life forms.

While the Earth provided an ideal environment for the origin of life, early life on Earth also shaped the Earth profoundly and induced more and more complex life forms. Around 2–3 billion years ago, the Earth's atmosphere changed from a highly toxic mixture of hydrogen, methane, ammonia, carbon dioxide, and nitrogen, to mostly nitrogen and increasing levels of oxygen, presumably as a result of photosynthesis by the early microbes such as cyanobacteria. The so-called "**the Great Oxygenation Event**"[11] around 2.5 billion years ago, supported by carbon isotope data, is widely considered to be induced by biological activities. In turn, the rising oxygen level made it possible for more complex life forms such as animals to appear.

Darwin's phylogenetic "tree of life," as improved by Haeckel (**Figure 1.12**) provided strong evidence of not only the origin of species, but also the origin and evolution of life. At the center of Darwin's theory of biological evolution is natural selection. In this process, random genetic variations occur and are passed on to the subsequent generations. The survival of the fittest of these individuals allows the species evolve and differentiate over a sufficiently long period of time. This notion was further strengthened by the American scientist Carl Woese, who established the phylogenetic tree of life based on the genomic structure of various microbes. Strikingly but not surprisingly, it was quite similar to the tree of life that Darwin drew for a wider range of species. It is so important that it was dubbed "the Woesian Revolution" that changed people's thoughts on the origin of life.

---

[11] It is not until around 1 billion years ago when the oxygen level started to increase to the current level of 19.5%.

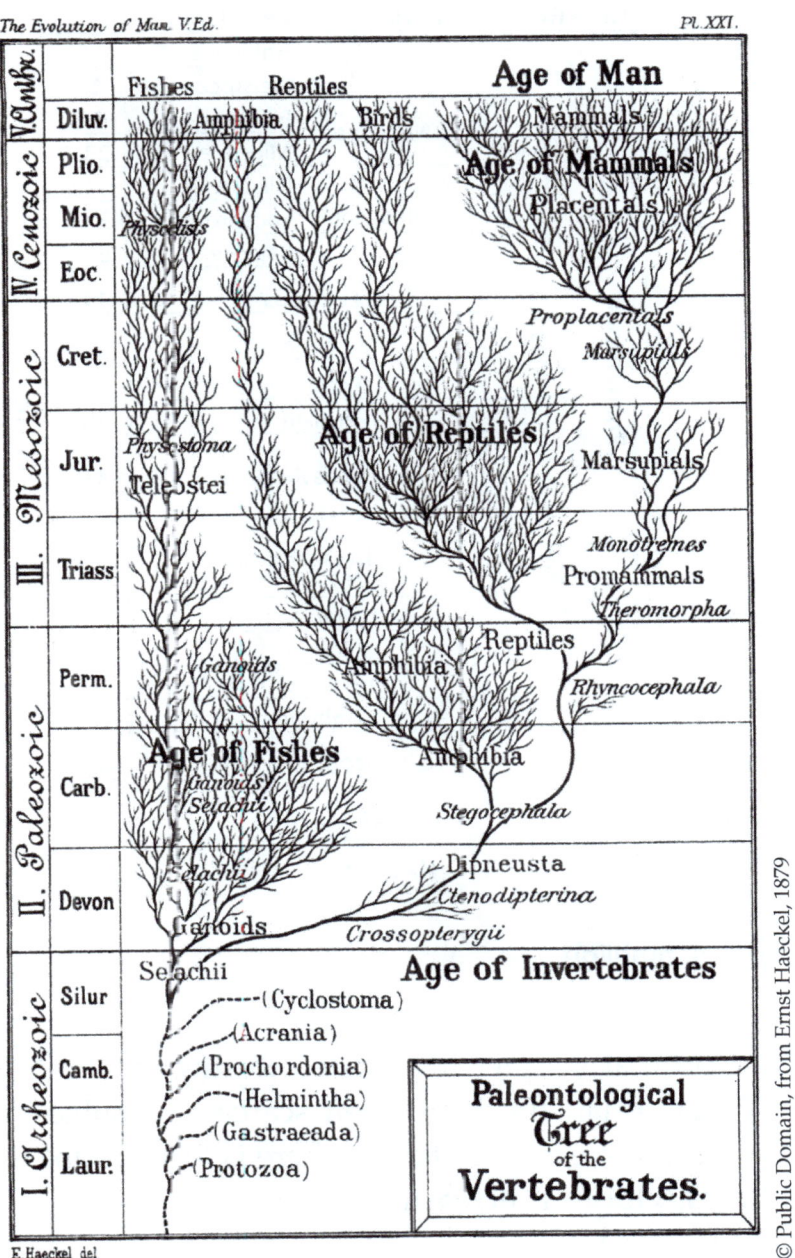

Figure 1.12 Evolution of life on Earth.

1.4 The Origin and Dominance of Humankind

1.4.1 The Origin of Humankind

The evolution of life on Earth has been recorded by fossils in sedimentary rocks and helped establish the Earth's geological timeline. As shown in **Figure 1.13**, from Paleozoic era, especially during the Cambrian period, life started to be abundant on Earth.

EON	ERA		Period	EPOCH	
Phanerozoic	Cenozoic		Quaternary	Holocene	0.01
				Pleistocene	1.6
		Tertiary	Neogene	Pliocene	5.3
				Miocene	23.7
			Paleogene	Oligocene	36.6
				Eocene	57.6
				Paleocene	66.4
	Mesozoic		Cretaceous		144
			Jurassic		208
			Triassic		245
	Paleozoic		Permian		266
		Carboniferous	Pennsylvanian		320
			Mississippian		360
			Devonian		406
			Silurian		438
			Ordovician		505
			Cambrian		570
Precambrian			Proterozoic		2500
			Archean		3800
			Hadean		4550

(Age in millions of years before present)

© Public Domain, from Ernst Haeckel, 1879

Figure 1.13 Geological timeline.

There is little question that simpler life existed during the Precambrian period, but fossil records of Precambrian life have been scarce and subject to strong scrutiny. Despite uncertainty in these early life forms, there is little doubt that the evolution of life has followed a trajectory of increasing complexity and larger brains from fishes, amphibians, to reptiles, birds, and mammals.

Given the trend that increasingly complex life forms started to colonize the Earth, especially when many species of primates started to emerge since the Tertiary–Quaternary period, it is perhaps inevitable that intelligent species such as humans would appear eventually.

1.4.2 Human Evolution and Dominance

Humans share a common ancestor with chimpanzees until roughly 7 million years ago (with large uncertainties), when early humans perhaps originated in Africa. The famed paleontologist Richard Leakey found skeletal remains of an early human that lived more than 1.5 million years ago. Earlier stone tools were found in Africa about 2.5 million years ago, when fossil records indicated a significant increase in the average brain size of early humans. Around or less than 2 million years ago, *Homo erectus* perhaps wandered out of Africa and started to colonize an increasing expanse of the world. Since then, tool-making, meat-eating, use of fire, until the emergence of "modern" humans around 200,000 years ago, showed a clear progress toward increasing sophistication in tools, arts, and language. We are now officially in the Holocene (meaning "entirely new"), which began 11,700 years ago since the last ice age. Since humans have changed the Earth so much, there has been some intense debate whether the current period should be instead called "**Anthropocene**" ("anthropo-" means "human"). This term received objection from traditional strategraphers, who argued that the classification of geological periods is based on fossil records or other clear divides between two periods. The question about when humans emerged as a distinct species is still largely unanswered, with new findings pushing the limit further and further.

Knowledge Box

Human Dominance or Self-Destruction?—One depressing fact about **human dominance** of the Earth is that we still own enough nuclear weapons that could wipe out humans around the globe many times. In the United States alone, there were more than 32,000 nuclear warheads during the peak time of 1996 before the significant reduction thanks to the end of the Cold War. During the peak of the Cold War, the United States and the Soviet Union had enough nuclear weapons to destroy the human race on Earth 40 times. Other nations such as Russia, China, United Kingdom, France, India, Pakistan, even North Korea now own nuclear weapons. During the height of atmospheric nuclear testing in the early 1960s, so much man-made radionuclides (carbon-14; cesium-137 (Cs-137); tritium (hydrogen-3), etc.) were released that they were spreaded around the globe and become part of our natural environment and embedded in our body. The only silver lining of this is that these radionuclides become a useful tool in environmental research.[12]

---

[12] For example, cesium-137 is a useful tool to calculate sediment accumulation rate because the 1963 time horizon in the sediment profile is often marked by a sharp peak in Cs137 abundance.

> **Knowledge Box**
>
> **Anthropocene**—The term Anthropocene is coined by Paul Crutzen, a Dutch scientist who also discovered ozone-depletion compounds such as chlorofluorocarbons (CFCs). Troubled by human's dominance and its markings on Earth that are distinct from other geological era including the Holocene, he invented this term in a paper on the prominent journal Nature to point out several distinctive changes that human has caused globally, including transformation of more than a third of the surface of the Earth; damming most of the major rivers and changing the global sedimentary processes; disturbance of nitrogen cycles by producing more synthetic nitrogen than natural processes; global fisheries; and unsustainable consumption of more than half of the global freshwater resources. These profound changes could be assessed using the traditional stratigraphic criteria because much of the above will be recorded in the sediment rocks.

Anthropocene or not, there is no question that humans are dominating the Earth. **Figure 1.14**, by way of nightlight around the globe, shows how humans have occupied nearly all of the inhabitable places of the Earth. While humans often seem powerless in the face of natural disasters, we nonetheless have impacted the Earth in many fundamental ways. Many global changes such as the Great Oxygenation Event or other dramatic changes in the past often took millions of years. However, human history is virtually a blink of an eye compared to the Earth's history.[13] With this blink of an eye, population explosion spurred increasing and often unsustainable consumption of food and energy and brought about a host of environmental issues such as pollution of air, water, soil; climate change; habitat loss; loss of biodiversity, and many other emerging global issues. There are indications that if these trends are not curbed in the next few centuries, human existence may become questionable.

1.4.3 Life Beyond Earth

Are there other Earth-like planets in the universe? Is human civilization unique or inevitable? These questions have fascinated astronomers, other scientists, and many curious minds alike for many years. With environmental issues and global sustainability moving increasingly to the front and center of people's attention, the search of habitable planets outside of the solar system and for extraterrestrial civilizations is getting more attention recently. This issue will be addressed in more detail at the end of the book.

---

[13] To put this in perspective, if the Earth history were 1 day (or 24 hours), the length of modern human history is about one single second.

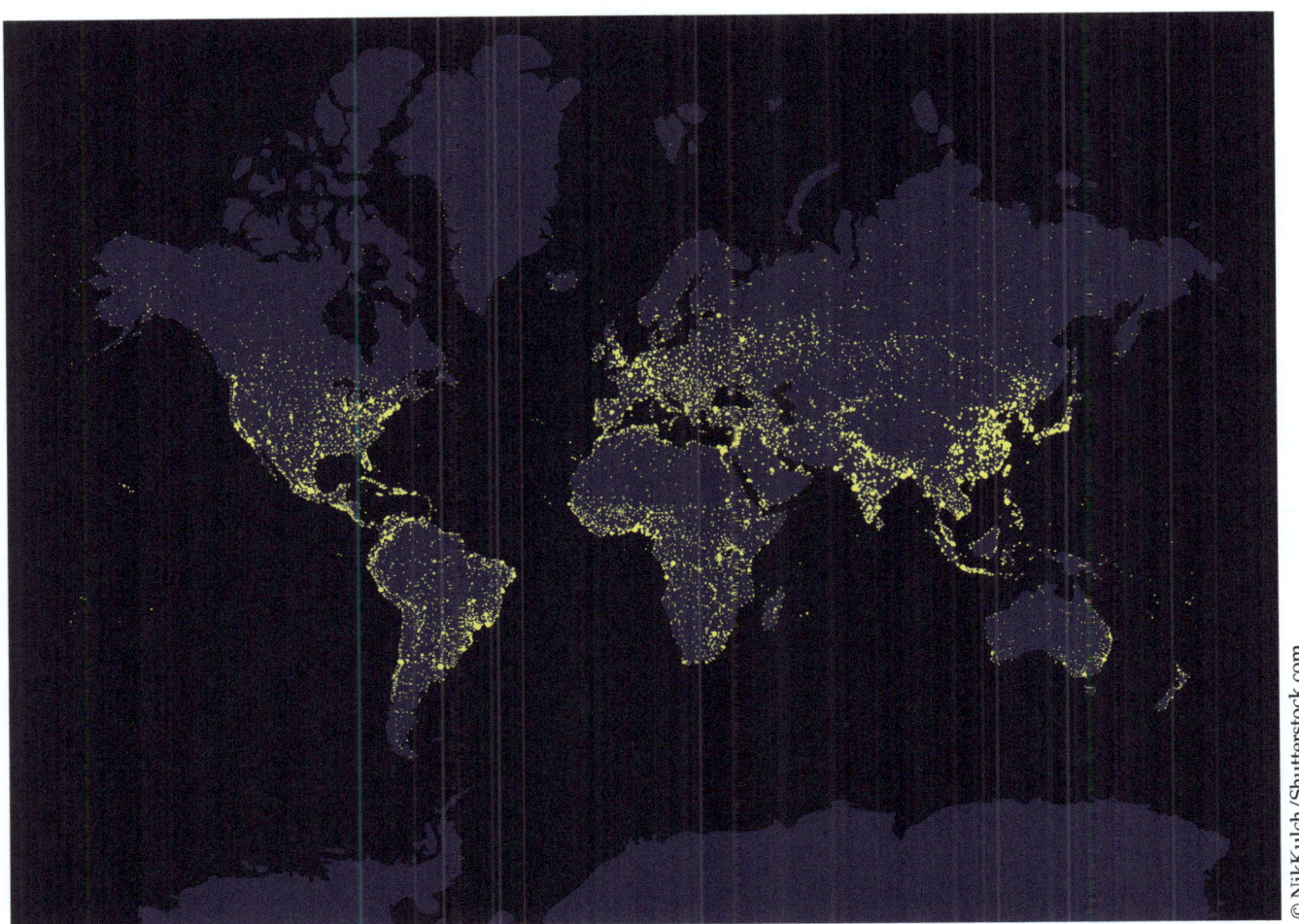

Figure 1.14 Human dominance of the Earth: the night light.

> ### Knowledge Box
>
> **Gaia Hypothesis**—The Gaia Hypothesis (Gaia is the goddess in Greek mythology that personifies the Earth) is a theory that proposes that the Earth's biosphere and the rest of its abiotic environment interact and self-regulate in such a way that life on Earth can survive and prosper. On the surface, the Gaia Hypothesis is evidenced by the near optimal conditions of the Earth for life, such as the stabilities in the global temperature, ocean salinity, oxygen level in the atmosphere, and global ecosystem, which all have intricate interactive and feedback mechanisms. This hypothesis has drawn criticism for lack of scientific evidence and for its implication that the Earth can self-regulate and optimize itself when in fact the Earth system is in a delicate and precarious state.

Further Reading

Leakey, Richard E. The Origin of Humankind. New York, NY: Basic Books, 1994.

Mesler, Bill and H. James Cleaves II, A Brief History of Creation, New York, NY: W.W. Norton and Company, 2016.

Langmuir, Charles H. and Wallace S. Broecker. How to Build a Habitable Planet—The Story of Earth from the Big Bang to Humankind. Princeton, NJ: Princeton University Press, 2012.

Stephen W. Hawking. A Brief History of Time—from the Big Bang to Black Holes. London, UK: Bantam Dell Publishing Group, 1988.

Martin, Daniel, Helen McKenna, and Valerie Livina. "The Human Physiological Impact of Global Deoxygenation." *Journal of Physiological Science*, 67 (2017): 97–106.

Ernst Haeckel. The History of Creation. New York, NY: D. Appleton and Company, 1880 (this book is freely available at Project Gutenberg at: http://www.gutenberg.org/files/40472/40472-h/40472-h.htm).

Woese, Carl R., O. Kandler, and M. L. Wheelis. "Towards a Natural System of Organisms: Proposal for the Domains Archaea, Bacteria, and Eucarya." *Proceedings of the National Academy of Sciences of the United States of America*, 87, no. 12 (1990): 4576–79.

Zalasiewicz et al, 2011. Stratigraphy of the Anthropocene. Philosophical Transactions of the Royal Society A, 369//1938. 13 March 2011

Web Resources

Big Bang theory explained: https://www.youtube.com/watch?v=wNDGgL73ihY

Ted Talks—The history of our world in 18 minutes, by David Christian: https://www.youtube.com/watch?v=yqc9zX04DXs

A fascinating video about our universe in the eye of the Hubble Telescope: https://www.youtube.com/watch?v=cHSftdHskxs

Wikipedia: the Oldest dated rocks, found in Canada, about 4.03 billions years old: https://en.wikipedia.org/wiki/Oldest_dated_rocks

Wikipedia: Solar System: https://en.wikipedia.org/wiki/Solar_System

The history of the Earth from 4.5 billion years ago to present, with temperature, atmosphere composition, length of day, and ocean-continent distribution: https://www.youtube.com/watch?v=Q1OreyX0-fw

Youtube: from atom to whole universe—Amazing scale video: https://www.youtube.com/watch?v=DWouwx3Hxmk

Questions and Exercises

1. On a mental note, imagine the world on different scales—from quarks to the whole universe. Which scale do you think is the most fascinating?
2. On a mental note, go through the entire Earth history from Precambrian to the present. Which period do you prefer living in?
3. From what you have learned from this chapter, do you believe there will be another planet just like Earth? With life? With intelligent life?
4. Conduct a brief research on the Drake's Equation. Consider how it apples to the Earth, and how it applies to the other planet in the universe.
5. Considering the origins of time and space, the Universe, Milky Way, solar system, the Earth, life on Earth, and finally humans—why is it important that we lead a sustainable lifestyle?
6. Imagine that you were a super intelligent extraterrestrial life. You came cross the Earth on a UFO. Using a scoring system of 0-10 on sustainability, what score you would give to the earthlings once you have learned the past and present of the Earth?
7. In your opinion, which step in the pathway of the origin of life is the most critical?
    a. creation of organic molecules
    b. creation of DNA/RNA
    c. formation of first viable cells
    d. formation of multicellular organisms

Chapter 2

Global Biogeochemical Cycles

Learning Outcomes		28
Key Concepts		28
2.1	The Concept of Global Biogeochemical Cycle	28
2.2	Plate Tectonics and Rock Cycles	29
	2.2.1 Plate Tectonics	29
	2.2.2 The Rock Cycle	32
2.3	The Water Cycle	33
	2.3.1 The Erosional Power of Water	33
	2.3.2 The Water Cycle	35
2.4	Biogeochemical Cycles of Elements	38
	2.4.1 The Carbon Cycle	38
	2.4.2 The Nitrogen Cycle	45
	2.4.3 Biogeochemical Cycles of Other Elements	48
2.5	Biogeochemical Cycles and Environmental Challenges	49
Further Reading		49
Web Resources		50
Questions and Exercises		50

Learning Outcomes

- Knowledge and understanding of the concept of global biogeochemical cycles
- Knowledge and understanding of the concept of rock cycles and water cycles
- Understanding of the global biogeochemical cycles of important elements, including the most important reservoir of each element, and the most important processes that drive the cycling of each element
    - Carbon
    - Nitrogen
- Understanding of the linkage between biogeochemical cycling of rocks, water, and different elements
- Knowledge of how to approach environmental issues from the perspective of biogeochemical cycling

Key Concepts

Global biogeochemical cycle; geochemistry; biogeochemistry; reservoir; plate tectonics; rock cycle; paleomagnetism; Redfield ratio; water cycle; carbon cycle; inorganic carbon cycle; nitrogen cycle

As we learned from the last chapter, the Earth has changed dramatically over time. The Earth is 'living' not only because of life it supports, but also because it undergoes constant changes. During the early years when the Solar System itself was initially formed, the Earth had a hostile environment that was not conducive to life. Later on, when early life started to emerge, they also gradually changed the Earth's environment by providing oxygen through photosynthesis. The geological and fossil records show that the Earth underwent continuous changes, such as **plate tectonics**, volcanic activities, catastrophic events such as meteorite impacts; dramatic climate changes, and so on, and the Earth's surface have constantly been carved by hydrological processes. Many of such processes are affected by the biosphere through complex interaction between the living organisms and their inorganic environment.

2.1 The Concept of Global Biogeochemical Cycle

Geochemistry is a field of natural science that uses the principle of chemistry to study geological, hydrological, and other processes on Earth.[1] **Biogeochemistry** is a subfield of geochemistry to study biology-related geochemistry phenomena, such as

---

[1] The "cool factor" of geochemistry can be illustrated by Antonio Lasaga's quote "Old geochemists never die, they merely reach equilibrium."

spatial and temporal distribution and variation of oxygen in different parts of the ocean, lakes, and sediments, diagenesis of organic materials in the sediment, how biological processes have affected these processes, among others. **Global biogeochemical cycle** is a process where, mediated by biological processes, the materials on Earth cycle within and among different compartments including lithosphere, hydrosphere, biosphere, and atmosphere. It is a concept that is better understood with the global environment in mind. Similar to the issue of the genesis of the universe and evolution of the Earth and the life thereon, an understanding of global biogeochemical cycles is beneficial to environmental professionals and should be essential to environmental scientists.

Global biogeochemical cycles are often studied by observing the cycling of certain elements or chemical molecules, such as water, oxygen, carbon, nitrogen, phosphorus, etc. With the advancement of science, the cycling of some rare elements has also been studied recently and these studies provided important insights to our understanding of the natural world. In this book, the **rock cycle** and cycling of water, nutrients (exemplified by nitrogen), and carbon will be discussed as examples.

To understand global biogeochemical cycles, the concept of mass and energy conservation is needed. The law of conservation of matter states that matter can neither be created or destroyed, or they should always be conserved. The law of conservation of energy, on the other hand, states that energy will always be conserved. The biogeochemical processes also obey chemical, physical, and biological principles and one needs to be knowledgeable on atmospheric science, oceanography/limnology/hydrology, ecology, geology, and many other fields of science. Therefore, biogeochemistry is a highly interdisciplinary science.

2.2 Plate Tectonics and Rock Cycles

2.2.1 Plate Tectonics

Throughout most of human history, the Earth was thought to be static and firm. There are legends about drastic changes in land and sea,[2] but it is unlikely that even the wisest early men understood the inner workings of the Earth. Early geologists and geochemists, by studying fossils in geological formations and using tools such as radioisotope dating and paleomagnetism (see the knowledge box about this concept), gradually realized that the Earth had undergone drastic changes in the past millions

---

[2] For example, the Chinese idiom "the ocean turning into cropland"; and the legend about Atlantis, which, according to Plato, sank under the Atlantic Ocean (and gave the Atlantic Ocean its name).

and billions of years, and figured out that the land could subside and turn into ocean; and ocean floor could uplift and turn into land and even mountains, and the process could repeat itself.

> ### Knowledge Box
>
> **Paleomagnetism** is a branch in geology that uses magnetism of rocks to study the movement of rocks after its formation. The Earth's magnetic field will magnetize an igneous or metamorphic rock (which always contains trace amounts of magnetic minerals that can be magnetized by the Earth's magnetic field) when it is cooling down. The orientation of the Earth's magnetic field will then be recorded by the rock. If the rock is then moved due to geological processes (such as faults, subsidence, lifting, or plate tectonics), the movement could be deciphered by measuring the orientation of its magnetic field. Interestingly enough, as mentioned in Chapter 1, the Earth's magnetic field reversed many times in the past, and these inversions were also recorded in the rocks. The study of paleomagnetism has provided strong evidence for the theory of plate tectonics and many geological phenomena.

It was not until the development of the theory of plate tectonics when we started to link large scale geological phenomena together. Plate tectonics describes large-scale motion of seven large plates and many smaller plates of the Earth's crust (**Figure 2.1**; also refer to Figure 1.5). These plates were "floating" on the upper mantle (asthenosphere), which is semi-fluidic. It is hypothesized that the asthenosphere is convective, and the convection and perhaps other forces drive the movement of the plates above the asthenosphere.[3] The lateral movements of these plates (sliding; diverging, or converging/colliding, see Figure 2.1) cause a number of geological processes, including mountain-making; movement of continents; earthquakes; volcanic eruptions; and many other processes that shape the surface of the Earth. Plate tectonics likely started between 3.3 and 3.5 billion years ago when the Earth's crust was completely formed.

Plate tectonics takes place in a time scale of millions of years or longer. For example, the plates move about a few centimeters each year, and it took about 115 million years for the Atlantic Ocean to become one of the major global oceans when the

---

[3] It remains unclear, however, what is the driving force for the movement of these plates.

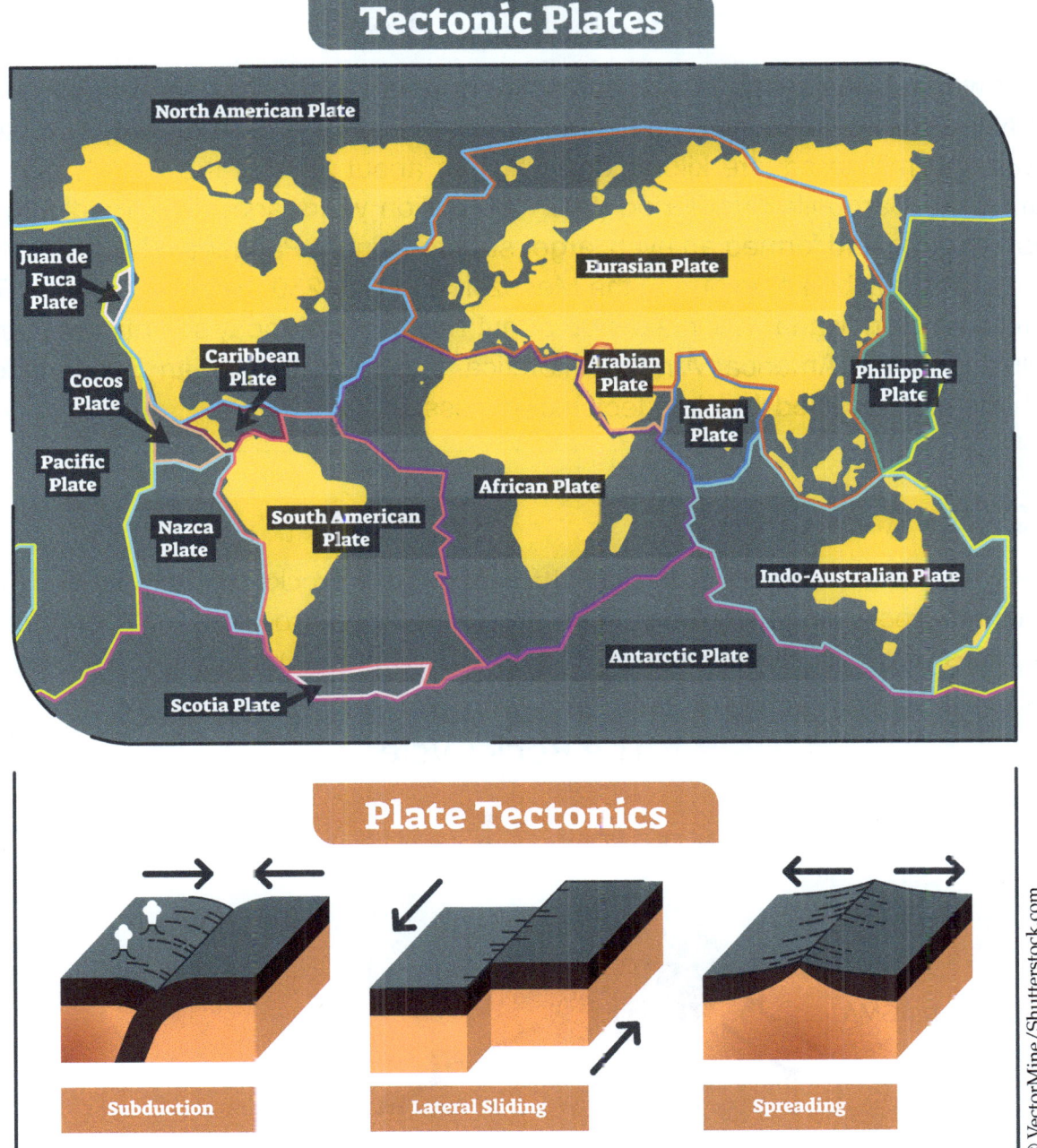

Figure 2.1 Plate tectonics.

great continent Gondwana (see knowledge box) broke up to form South America and Africa. With time, the Atlantic Ocean will keep growing, and the Pacific Ocean will be shrinking. Plate boundaries are geologically active areas where new rocks are formed and older rocks or sediments are recycled. Other activities such as earthquakes and volcanic eruptions tend to take place at plate boundaries as well. These processes shape the surface of the Earth profoundly but patiently.

> **Knowledge Box**
>
> **Gondwana**—Gondwana is a supercontinent about 550 million years ago during the Neoproterozoic period (see Figure 1.13 for the geological timescale). It covered about 100 million square kilometers, which is about 20% of the Earth's surface. During the Carboniferous Period (300–400 million years ago), it merged with yet another plate and formed an even larger supercontinent Pangaea. The supercontinent eventually broke up during the Mesozoic Era (70–250 million years ago). The remaining Gondwana now makes up about two thirds of today's continental area, including South America, Africa, Antarctica, Australia, the Indian Subcontinent, with the rest 'recycled' by plate tectonic processes.

2.2.2 The Rock Cycle

It was well known to geologists that different types of rocks, namely igneous rocks, sedimentary rocks, and metamorphic rocks, can undergo processes that turn them into other types of rocks, as indicated in **Figure 2.2**. The aforementioned plate tectonics

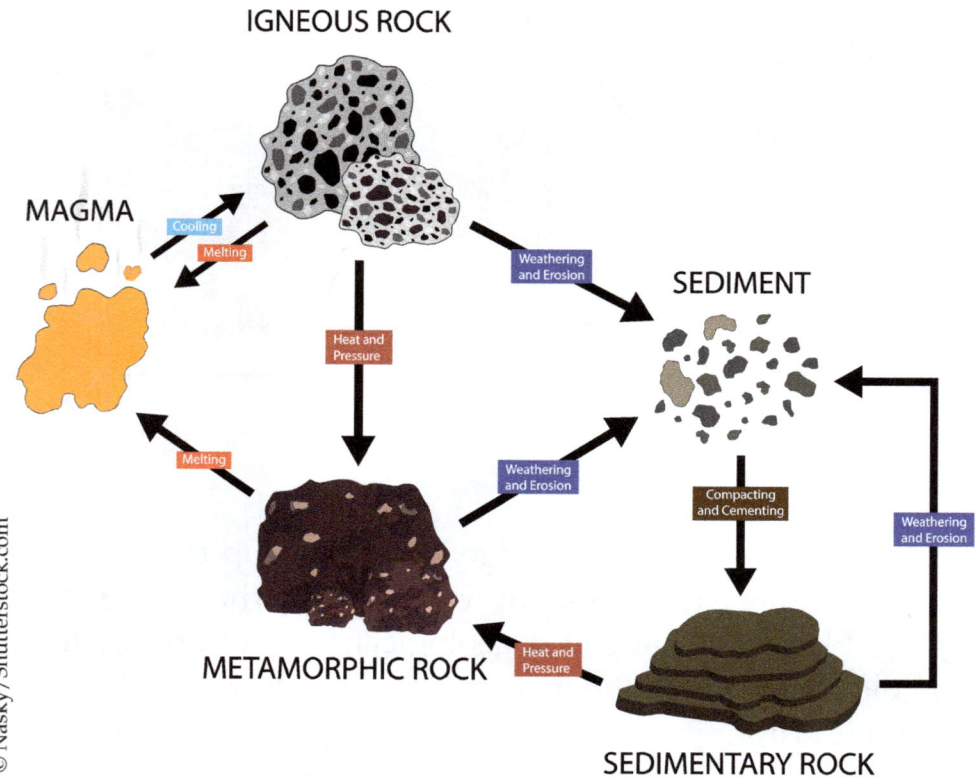

Figure 2.2 The rock cycle.

plays a major role in driving the rock cycle. Igneous rocks are formed by the cooling of magma or lava. The raw materials for magma or lava could come directly from the crust or upper mantle, or from complete or partial melting of existing solid rocks. Igneous rocks can be classified into many different subcategories based on the chemical composition, such as felsic (high silica content; >63%), intermediate (52–63% silica), mafic (<52% silica), and ultramafic rock (<45% silica). The rate of cooling affects the shape and size of various mineral crystals, and igneous rocks can be classified into phaneritic (with observable crystals) or aphanitic rocks.

Sedimentary rocks are formed by weathering, aggregation, and cementation of other rocks or minerals. In most cases, sedimentary rock materials undergo physical and chemical weathering and transportation before the rock-forming process. Common examples of sedimentary rocks are sandstone, shale, limestone, and many other types.

The correlation between plate tectonics and the rock cycle can be easily established, especially by observing what is taking place at a convergent plate boundary, such as the west coast of the Americas. When the Pacific Plate plunges underneath the North American Plate, causing the uplift of the American Plate and forming the Andes, the oceanic crust and sediment rocks dive under the continental crust, into the upper mantle and melt, and are thus regenerated. At the divergent plate boundary such as mid ocean ridge, fresh magma from the upper mantle produces fresh igneous rocks that will be subject to erosion (to produce sedimentary rocks) and metamorphism (to produce metamorphic rocks). As mentioned above about plate tectonics, it could take millions of years for the Earth's crust to be regenerated through plate tectonics. So-called "fresh" magma at the mid-ocean ridges and volcanoes is likely "recycled" material that has undergone prior rock cycles. As the rock cycle provides raw materials for the other processes (see the following sections), it is critical to the global biogeochemical cycles.

As will be seen in the subsequent sections, rock cycles are closely linked to the biogeochemical cycles of other materials and elements. To study the linkage, it is beneficial to introduce the terms of endogenic cycle and exogenic cycle. An endogenic cycle is the rock cycle itself, and an exogenic cycle involves atmosphere, biosphere, hydrosphere, and pedosphere (i.e., sediment and soil). The biogeochemical cycles for water, carbon, nitrogen, and other elements are all exogenic cycles. They actually interact with the endogenic cycle quite closely, as will be discussed later.

2.3 The Water Cycle

2.3.1 The Erosional Power of Water

Water is the most important feature on the surface of the Earth, covering 71% of the Earth's surface. However, the volume of the Earths' water is rather small compared to that of the Earth. If we put all water on Earth in a bubble, the diameter of the bubble

Figure 2.3 The volume of the Earth's water compared to the volume of the Earth.

would be 1,385 km as compared to 12,472 km for the Earth (Figure 2.3). The Oceans hold about 96.5% of the Earth's water. The rest of the water exists in glaciers, soil, rivers, and lakes. Later in this book (Chapter 6), another important aspect of global **water cycle** will be discussed.

As discussed in Chapter 1, nearly all of the water in the oceans are from degassing of the interior of the Earth and some may come from extraterrestrial sources such as meteorites. Together with other gases in the atmosphere during the early years in the Earth's history, water vapor in the atmosphere helped keep the planet warm through the greenhouse effect. Water was also essential in facilitating the genesis of life on Earth (see Figure 1.6). Therefore, the reader should be really thankful that the Earth is gifted with the abundance of water. In fact, as will be discussed in Chapter 12, when we look for possible extraterrestrial life beyond Earth (for curiosity or for finding other planets for potential human colonization), it is essential to assess whether there is water on that planet. So far, scientists have been disappointed by the findings, including those about the Mars (Figure 2.4), the Earth's closest neighbor and the single most promising candidate as a potential future colony for the earthlings.

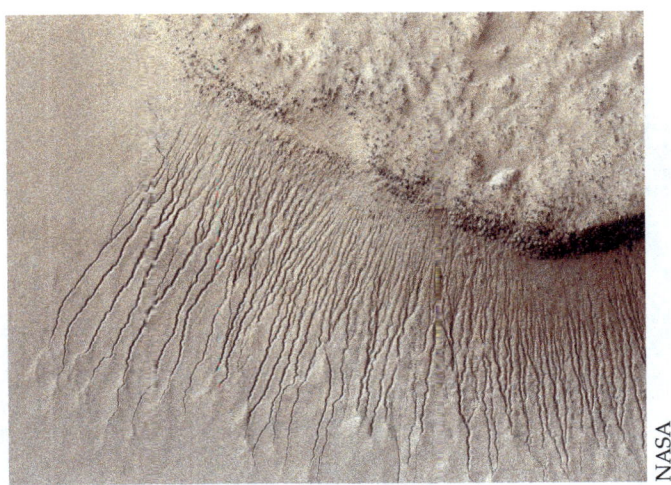

Figure 2.4 The surface feature on a scarp in the Hellas impact Basin on Mars taken by the High Resolution Imaging Science Experiment (HiRISE) camera on NASA's Mars Reconnaissance Orbiter in 2011. This caused much excitement in the scientific community. However, it is not yet clear whether this feature is caused by water or liquid carbon dioxide.

The water is an excellent solvent that can host many gases and leach minerals from soils, rocks, and sediments and bring them to the oceans or other receiving waters. It produces Karst geomorphology by dissolving carbonate rocks. Water also expands upon freezing, making it a powerful tool to break apart rocks from the cracks. Without this process, it would need a lot longer for these rocks to weather and disintegrate. Water's specific gravity renders sufficient power to runoff to erode rock and soils and carry the sediment downstream. It dissolves and precipitates different minerals at different places depending on chemical equilibrium of ions. Glaciers on high-altitude mountains can flow, albeit at a very slow rate, and grind down the mountain ranges. During ice ages, glaciers can produce many significant geomorphological features on continental Earth, including the Great Lakes. Under favorable conditions, rivers can cut into rock formations deeply and form spectacular geomorphological features such as the Grand Canyon in Arizona, USA, or other deep-cut canyons such as the Red Mountain Grand Canyon in Xinjiang, China (**Figure 2.5**).

2.3.2 The Water Cycle

Water appeared on Earth around 4 billion years ago, when geological records suggested the formation of first sedimentary rocks, including both erosional sedimentary rocks as well as carbonate rocks. Just as Ovid said, "dripping water hollows out stone, not through force but through persistence," the power of water that has chiseled the surface of the Earth lies also in the constant cycling of water. Rivers and glaciers are replenished by rain or snow, which are produced from clouds, and clouds are the result

Figure 2.5 The erosional feature at the Red Mountain Grand Canyon, Xinjiang, China.

of evaporation from any moist surfaces such as the ocean, lake, forest, and so on. The continuous cycling of water keeps the rain falling, the rivers flowing, and the glaciers creeping downhill. All of these processes continuously shape the surface of the Earth.

Compared to the rock cycle, the water cycle goes on at a much faster rate. Evaporation and precipitation of water takes days to weeks. Rain or snow melt flow to a lake or ocean in weeks or a few months. Groundwater and the water in the deep ocean take much longer (up to a few thousand years), but the time scale is still orders of magnitude shorter than the rock cycle, which takes millions of years. A comparison could be made to help the reader appreciate the difference. For an average rock to go through a complete cycle through plate tectonics, it would take 50 million years. During this period of time, an exposed rock would have been subject to the erosion power of a typical river 300 million times,[4] or it would have been grinded by a typical glacier 500,000 times.[5]

These factors make water cycle the most important biogeochemical cycles on Earth. Similar to other biogeochemical cycles, the water cycle is driven by solar energy,[6] which amounts to about 174 petawatts (10^{15} watts) for the entire surface of

---

[4] Assuming a typical large river takes 2 months to run its course.

[5] Assuming a typical mountain glacier takes 100 years to run its course.

[6] To be exact, about 99.98% of the energy that drives hydrological cycle is from solar energy, with the rest 0.02% by the heat from the interior of the Earth (Berner and Berner, 1987).

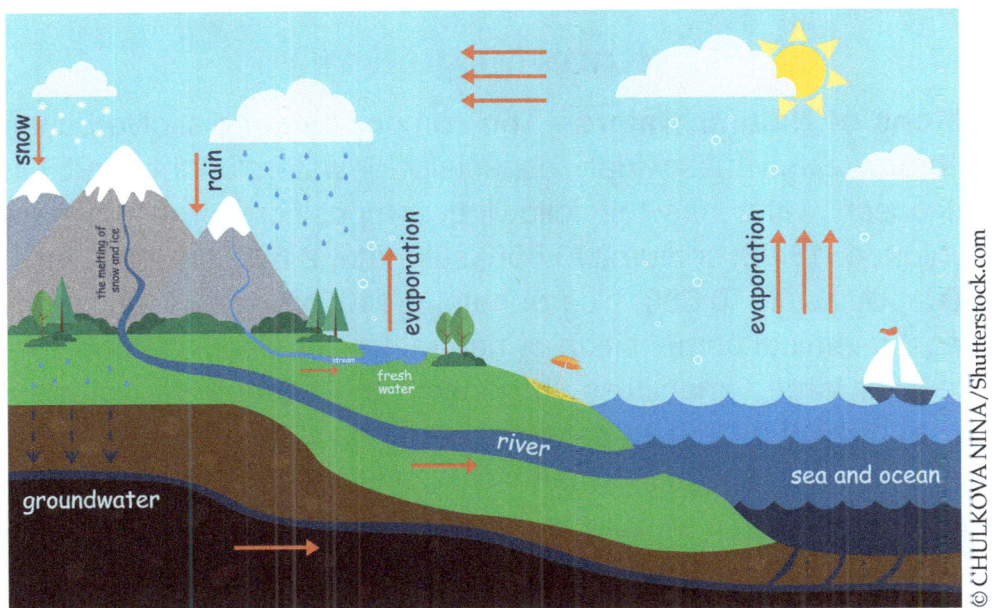

Figure 2.6 The water cycle, including evaporation, precipitation, surface runoff, groundwater flow.

the Earth. The sun drives the weather system (evaporation, heating of the Earth's surface, production of winds, etc.[7]) and ocean circulation. The precipitation supports much of the plant life on earth and drives the hydrological cycles that shape the Earth's surface. As will be discussed in the next sections, water cycle also drives the biogeochemical cycles of other elements through a range of processes such as physical erosion and transportation, chemical dissolution, precipitation, and water-mediated biological processes.

Water cycle also has profound impact on both water quality and water resources. Its linkage to water resource is obvious, because the amount of water that is available for human consumption (mostly freshwater in rivers and lakes, as well as groundwater) in a water cycle, as shown in **Figure 2.6**, is a small portion of the total amount of water on Earth. This topic will be discussed in more detail in Chapter 6. Along the water cycling pathways, it could be contaminated by a number of natural and anthropogenic processes. Rivers, lakes, groundwater, and oceans could all be polluted by various contaminants such as nutrients, bacteria, trace metals, and pesticides, that will affect the beneficial uses of the water. The water quality issues will be discussed in more detail in Chapter 5. Understanding water cycle will help put both water quality and water resource issues in a wider context.

---

[7] While solar energy starts the weather system, the gravity will take care of the rest.

> **Knowledge Box**
>
> **Compositions of Natural Waters**—The composition of dissolved chemical species in natural waters varies widely, especially for waters on the land. On average, the global ocean water has the following composition (order by per thousand weight): Chloride, 18.98; sodium, 10.556; sulphate, 2.649; magnesium, 1.272; calcium, 0.400; potassium, 0.380; bicarbonate, 0.140; bromide, 0.065; borate, 0.026. The 'average' chemical composition of natural freshwaters is impossible to calculate due to their large variabilities. In general, the abundances of these species could rank from high to low as follows: bicarbonate, sulphate, chloride, sodium, calcium, silicate, magnesium, and potassium. Also see Chapter 5.2.

2.4 Biogeochemical Cycles of Elements

2.4.1 The Carbon Cycle

The element carbon is important to us for two main reasons. First, it is the building block of all living organisms on Earth (perhaps for extraterrestrial life as well). Second, we rely on carbon as our main energy source (fossil fuels), and this reliance has caused many environmental issues. Carbon is also one of the most abundant elements in the universe and Solar System other than hydrogen and helium.[8] However, the Earth may have lost much of its carbon (one of the "volatile" elements) to the outer planets. The remaining carbon probably existed in lithosphere and was released to the atmosphere via volcanic activities.

The reason why carbon is the building block of all living organisms is its special chemical property. With an atomic number of 6, a carbon atom has two layers of electrons (**Figure 2.7**). The inner two electrons are not available for chemical reactions, but the outer four can readily form covalent bonds with other elements, including carbon itself. In fact, the ability of building covalent bonds with other carbon atoms make it a versatile building block, because one-, two-, or three-dimensional structures can be readily formed using carbon's four covalent bonds. As shown in Figure 1.7, carbon can form virtually unlimited number of organic molecules, including carbohydrate (the energy source of many biological processes and building blocks of most of plants); amino acids (the components of proteins, which are the building block of most animals); cholesterols, fatty acids, triglyceride, phospholipids, etc. The genetic material

---

[8] This is because of the nature of nucleosynthesis, see Figure 1.2. Within a massive star, hydrogen burning will be followed by helium burning, which generates carbon.

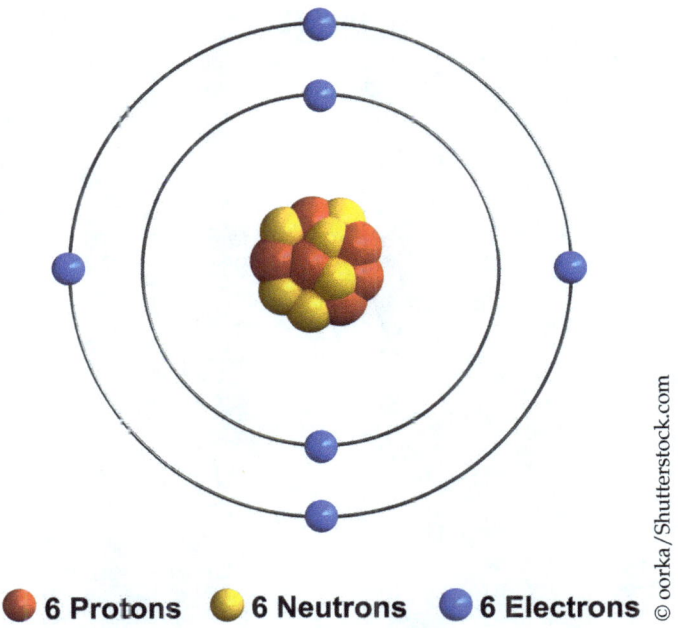

Figure 2.7 Illustration of a carbon atom.

DNA (Figure 1.8) is a large, complex, and three-dimensional molecule built on an intricate double helix formed by smaller organic functional groups. Simply put without carbon, there will be no biosphere on Earth.

The **carbon cycle** consists of many compartments, or **reservoirs**, such as atmosphere, plants, animals, detrital, and fossil fuels. The carbon cycling is the combination of many processes that transport or convert carbon in between different reservoirs (**Figure 2.8**). For example, plants and other primary producers on land or in the ocean produces carbohydrates and other organic materials from water and carbon dioxide from the air or from the ocean.[9] Nearly all of this is done through photosynthesis propelled by solar energy,[10] and oxygen is produced as a by-product,[11] as shown in the chemical reaction (1) below. This first step, by converting inorganic carbon to organic carbon that can be used by biota on different trophic levels, is the most critical step for the biosphere.

$$6CO_2 + 6H_2O \rightarrow C_6H_{12}O_6 + 6O_2 \tag{1}$$

---

[9] The dissolved carbon dioxide in the surface ocean can largely be deemed as originating from atmosphere through air–water interaction.

[10] The exception is chemical autotrophs that rely on chemical energy instead of solar energy.

[11] As discussed in Chapter 1, the advent of early life (likely one that was capable of photosynthesis) on Earth produced oxygen in the atmosphere, enabling the emergence of other species, especially animals that require oxygen to survive.

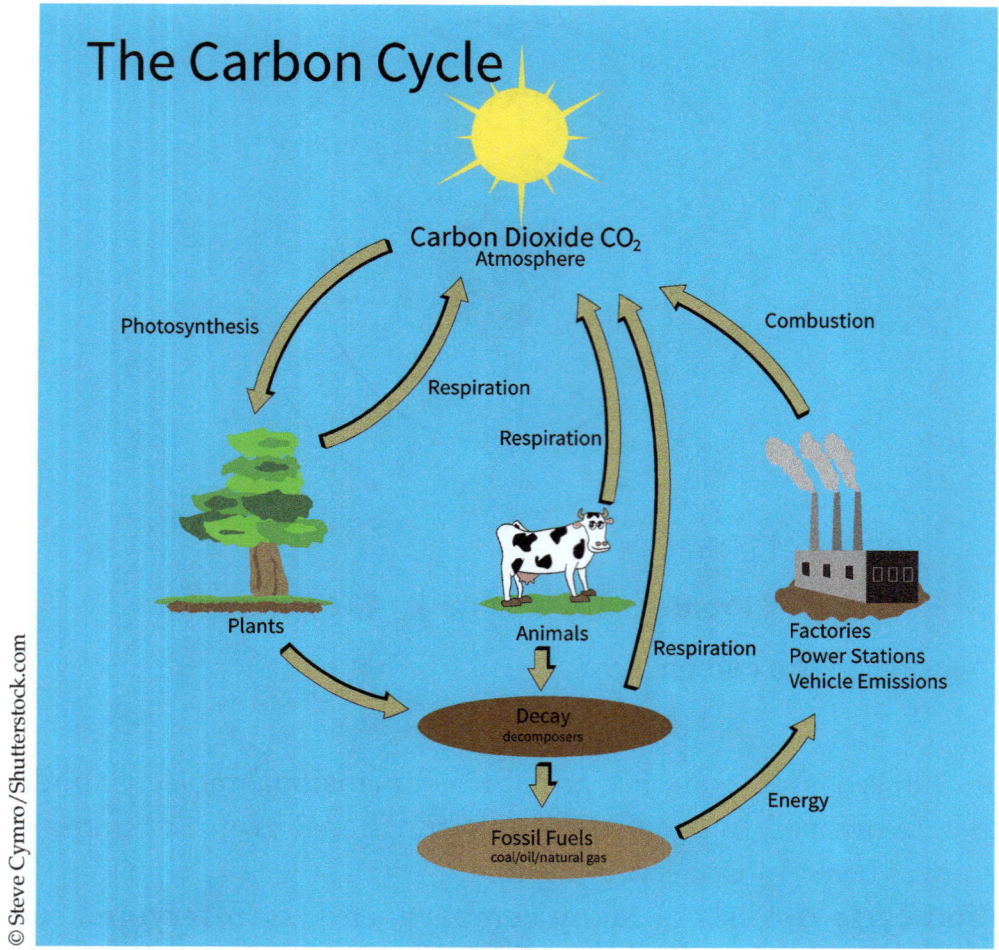

Figure 2.8 The nitrogen cycle.

After photosynthesis, the organic materials will be cycled through different trophic levels, losing materials and energy along the way (please refer to Chapter 8 for more information) to the decomposers to return the organic carbon back to carbon dioxide. A portion of organic carbon may be buried to form sedimentary organic carbon. Under favorable conditions and if accumulated in sufficient purity and quantity, fossil fuels such as coal, crude oil, or natural gas may be formed after millions of years. Some of these fossil fuels will be extracted by human for a wide range of uses. The consumption of fossil fuels again returns the carbon back to the atmosphere, completing the carbon cycle, as shown in Figure 2.8.

Many marine organisms use carbonate to build their shells. When they die, the shells may be preserved and form carbonate rocks. Carbonate ions in the oceans and some terrestrial waters could precipitate and form carbonate rocks as well. The sheer amount of carbonate rocks (about 10% of all sediment rocks on Earth) is a manifestation of the importance of such a process.

A lesser-known, but important carbon cycle is the inorganic carbon cycling, or carbonate-silicate cycle. This process includes long-term transformation of silicate rocks (the most common rocks) to carbonate rocks by weathering and sedimentation, as seen in the reaction (2) below, and the transformation of carbonate rocks back into silicate rocks by metamorphism and volcanism, as seen in the reaction (3), which is essentially the reverse of reaction (2). These processes are both part of the rock cycle as shown in Figure 2.2. In the first process, carbon dioxide is removed from the atmosphere during burial of weathered minerals. In the second process, carbon dioxide is returned to the atmosphere through volcanism.

$$2CO_2 + H_2O + CaSiO_3 \rightarrow Ca^{2+} + 2HCO_3^- + SiO_2 \qquad (2)$$

$$CaCO_3 + SiO_2 \rightarrow CaSiO_3 + CO_2 \qquad (3)$$

On million-year time scales, the carbonate-silicate cycle is a key factor in controlling Earth's climate because it regulates carbon dioxide levels and therefore global temperature. Over the past 40 million years, the Earth's temperature has been on a mostly cooling trend in part because of the uplift of the Himalayas that absorbed carbon from the atmosphere. This **inorganic carbon cycle** also involves the water cycle, as discussed in Section 2.3, because the reaction (2) requires water to break down silicate rocks to allow reaction with carbon dioxide. After reaction, the rock will gradually break down and some portions will be dissolved and transported to the ocean. Dissolved silica could be captured by organisms forming silica-based shells such as coccoliths. Under favorable conditions, they will form new silica-rich sediment rocks on the ocean floor.

Knowledge Box

Weathering process: Silicate rocks do not break down easily due to its structural stability and chemical inertness. However, with time, physical disintegration and chemical weathering will take place. The first step of chemical weathering after physical breakdown is to turn into soil minerals, as can often be seen in nature where soil overlays subsoil, substratum or parent rock, and bedrock or weathered parent materials are at the bottom, as shown in Figure 2.9. The chemical reaction through which bedrock turns into clay mineral can be exemplified by the one where plagioclase (one of the most common rock minerals) turns into kaolinite (one of the most common clay minerals). The reaction would look like: $177Na_{0.62}Ca_{0.38}Al_{1.38}Si_{2.62}O_8 + 246CO_2 + 367H_2O = 123Al_2Si_2O_5(OH)_4 + 110Na^+ + 68Ca^{2+} + 246HCO_3^- + 220SiO_2$. Subsequently if given enough time, kaolinite will further weather according to a reaction quite similar to (2) above, completing a cycle that involves the rock cycle, carbon cycle, and water cycle.

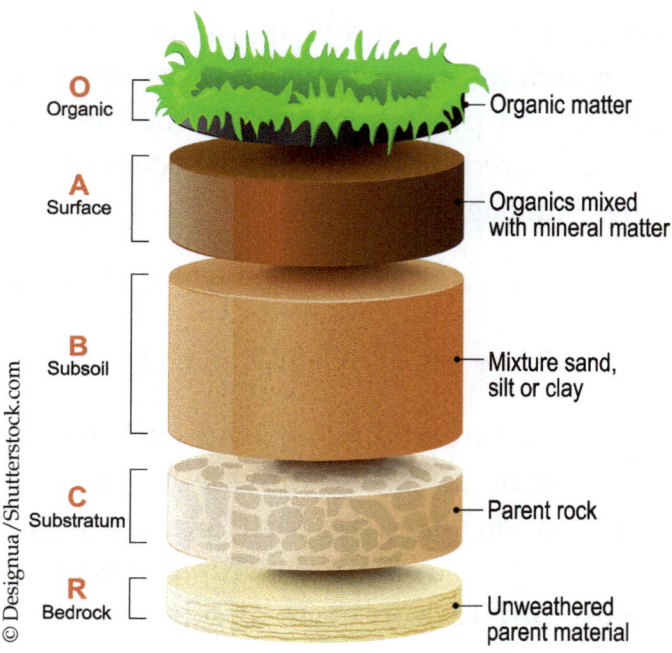

Figure 2.9 Typical soil vertical profile.

The ocean and many surface waters are heavily buffered, slightly basic solutions mostly dominated by bicarbonates, with other ions such as carbonate, boron hydroxide, silicate acid, chloride, sulfide, organic anions, hydrogen ions and hydroxyl ions. Due to the buffering effect, the pH values of these waters rarely change, which is important to the organisms living in these waters. With few exceptions,[12] bicarbonate ion is the most abundant ion among all dissolved species. Therefore, water chemistry is an important component of the carbon cycle. These ions control the pH of the water and together with biological processes, dictate the formation and dissolution of carbonate and silicate shells and minerals.

As can be seen in the reaction (3), the chemical weathering process is a significant sink for carbon in the atmosphere. After the dissolved species are carried to the ocean and under favorable conditions (when the ions are supersaturated) and mediated by shell-forming organisms such as bivalves and foraminifera, the calcium and bicarbonate react and form carbonate, releasing carbon dioxide in the process, as shown below in reaction (4):

$$Ca^{2+} + 2HCO_3^- \rightarrow CaCO_3 + CO_2 + H_2O \qquad (4)$$

---

[12] For example, the Colorado River and Rio Grande are perhaps the only two major rivers around the world that bicarbonate is not the most abundant ion. For these two rivers, sulfate is more abundant. This is due to the unique geology as well as the arid climate of the watershed.

> ### Knowledge Box
>
> **Shellfish farming as a carbon sink:** When global warming became a hot topic and people were looking for possible ways to reduce carbon from the atmosphere and the ocean, some proposed that shellfish farming in the ocean and burying the shells in landfills could help reduce the carbon dioxide in the ocean and the atmosphere. Contrary to common sense, this will not help reduce either. In fact, it may even make matter worse, because the ocean is such an overwhelmingly large reservoir of supersaturating bicarbonates, reducing bicarbonate by making shells will have little impact on the buffer. Instead, forming new shells will generate CO_2 as indicated by reaction (4), potentially contributing to ocean acidification.

A look at the relative sizes of carbon reservoirs in different geological compartments will also yield some important insights. As shown in **Figure 2.10** (note the graph is in logarithmic scale), sedimentary rocks (including fossil fuels) hold the overwhelming portion (>95%) of the carbon, with most of the rest residing in the ocean, mostly in the deep ocean. The Earth's interior likely holds a much larger carbon reservoir. However, since most of it does not participate in the carbon cycle (unless through degassing of the mantle through plate tectonics/volcanic activities), they are not discussed here.

Humans have caused significant disturbance on carbon reservoir and biogeochemical cycles. The overall mass of carbon will always be conserved (i.e., no new carbon

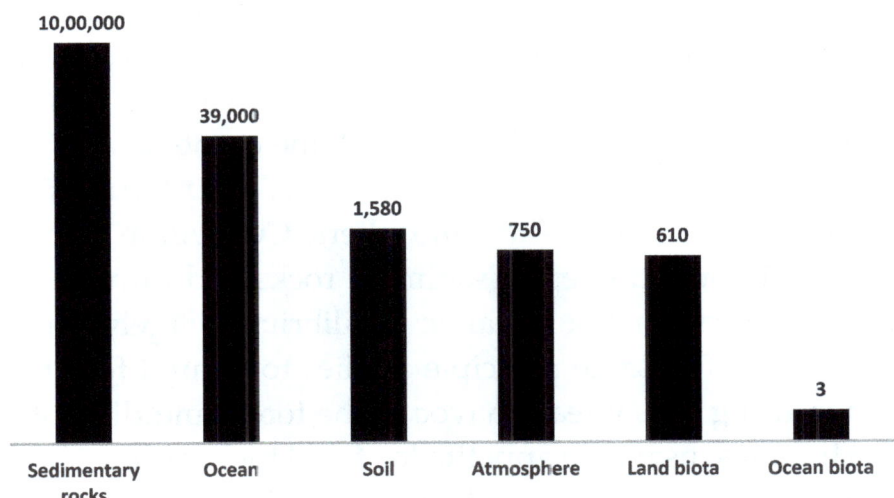

Figure 2.10 Carbon reservoir sizes.
Data source: Bar-On et al., 2018. Redrawn by Jian Peng

Figure 2.11 Tapo Canyon Tar Pit in Ventura, California, USA. The tar pit is among many in California and is associated with the oil-bearing Monterey Formation, which accounts for the bulk of California's crude oil production.

will be brought in or taken out). The issue is the rate of transportation and transformation between the reservoirs causing the changes in the reservoir sizes and subsequent systematic changes. For example, naturally-occurring fossil fuels such as crude oil and coal are formed in deep geological formations. When these fossil fuels are exposed due to geological processes (for example, Figure 2.11), the crude oil and coal will oxidize (sometimes burn) naturally over geological time scale. After burning/oxidation, they put CO_2 back into the atmosphere. The atmospheric CO_2 will in turn be absorbed by biota, and some will be buried deep in sediment rocks and turned into coal and oil. Over time, these processes reach a dynamic equilibrium, largely stabilizing the CO_2 level in the atmosphere. The same principle applies to natural forest fires. However, active deforestation using fires wreaks havoc to the forest and disrupts global carbon balance as well. Human's insatiable appetite for fossil fuels results in a fast increase in the exploration and burning of fossil fuels in a very short period of time, drastically increasing the carbon reservoir in both atmosphere and ocean. This topic will be revisited in latter parts of the book, including Chapters 10 to 12.

2.4.2 The Nitrogen Cycle

Nitrogen is one of the essential elements that plants and animals need in order to grow. As shown in Figure 1.8, nitrogen is part of biologically important organic molecules such as amino acid, cholesterol, phospholipid, as well as the DNA itself, as shown in Figure 1.9; where all four main bases cytosine, adenine, guanine, and thymine have nitrogen in their molecular framework. For all plants (i.e., the primary producers), it is an indispensable, and often limiting nutrients.[13] Therefore, nitrogen cycling is an important process that has implications on other biogeochemical cycles, especially carbon. It also has significant environmental impact because too much nitrogen, especially in slow-moving waters such as lakes, bays, and coastal oceans, could spur excessive algae growth and cause a host of environmental and ecological issues.

Nitrogen biogeochemical cycling includes the following processes (**Figure 2.12**). The atmospheric nitrogen (as the nitrogen gas) is the largest nitrogen reservoir but nitrogen in this inert form is not available for biological processes. Through lightning and more importantly through biologically mediated nitrogen fixation, the atmospheric nitrogen is transformed either to nitrate or organic nitrogen and becomes biologically available. This process is represented by the reaction below:

$$2N_2 + 6H_2O \rightarrow 4NH_3 + 3O_2 \tag{5}$$

As an essential nutrient, this nitrogen drives primary productivity and enters the biosphere and different trophic levels through various biological processes. Most of the organic nitrogen eventually turns into ammonia and is released back to the atmosphere, or is oxidized into nitrate following the following two processes:

$$4NH_4^- + 6O_2 \rightarrow 4NO_2^- + 8H^+ + 4H_2O \tag{6}$$

$$4NO_2^- + 2O_2 \rightarrow 4NO_3^- \tag{7}$$

Under anaerobic conditions, denitrification microbes will use nitrate as its energy source and reduce them into nitrogen gas and release it back to the atmosphere, following the reaction below:

$$6NO_3^- + 2H_2O \rightarrow 2N_2 + 5O_2 + 4OH^- \tag{8}$$

---

[13] A limiting nutrient is one that is at the shortest supply. If such a nutrient is added to a system, a proportionate increase in primary productivity can be observed. By comparison, adding a nonlimiting nutrient (i.e., a nutrient that is already in excess) will have no effect on productivity. If enough limiting nutrient is added to a system, another nutrient will become the 'new' limiting nutrient. The most common limiting nutrients are nitrogen and phosphorus. In some conditions, other elements such as iron and silicate could become limiting nutrients too.

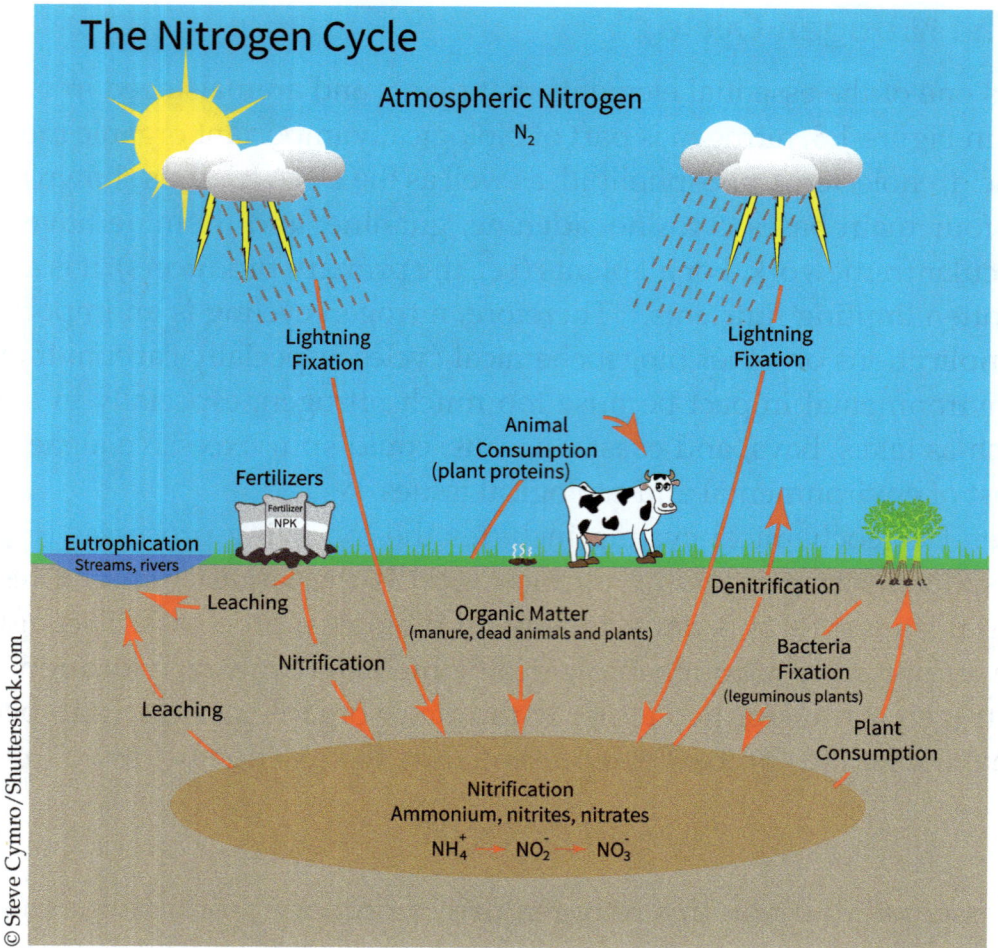

Figure 2.12 The nitrogen cycle.

Modern agriculture relies heavily on nitrogen-based chemical fertilizers containing high level of nitrogen compounds (ammonium or nitrates), which are produced by chemically fixing the nitrogen gas in the atmosphere through the Haber process:

$$N_2 + 3H_2 = 3NH_3 \tag{9}$$

Modern agriculture uses a large amount of nitrogen fertilizers to maintain productivity. A whopping 144 gigatons of nitrogen is produced annually as ammonia through the Haber process, mainly to produce fertilizer. As a result, the fertilizer production process is an important component of the global **nitrogen cycle**, especially the environmentally meaningful part of the nitrogen cycle. Agricultural-related nutrient pollution is also one of the most common water pollution issues, as will be discussed in Chapter 5.

In terms of the reservoir sizes, the predominant (~78%) global nitrogen reservoir is the atmospheric nitrogen, followed by continental crust and the ocean, as shown in **Figure 2.13**. The annual industrial ammonia production is in the same order of magnitude with major global reservoirs such as marine biota, suggesting a large

Knowledge Box

Redfield Ratio—The Redfield ratio is the ratio between three most important elements carbon, nitrogen, and phosphorus in typical marine phytoplankton (one main global carbon reservoirs, see Figure 2.10). The American oceanographer Alfred Redfield first described the relatively consistent ratios in marine biomass between these three elements to be C:N:P = 106:16:1. For terrestrial biomass, the Redfield ratio is different. Similar work later added iron as well, but the amount of iron could vary, and the Redfield Ratio is C:N:P:Fe = 106:16:1:0.1–0.01. It is clear from the Redfield Ratio that nitrogen, phosphorus, and iron are essential nutrients that plants must have to grow. Carbon is essentially limitless from the atmosphere. Iron is abundant in most places.[14]

anthropogenic disturbance of the global biogeochemical cycle of nitrogen. This is similar to the situation where human activities perturbed the carbon cycling by accelerating the exploration and burning of fossil fuels. These two processes are also connected because much of the anthropogenic nitrogen production is for food production (i.e. to provide carbon source for human consumption). Moreover, the resulting loss of excess nitrogen from fertilizers into the environment causes eutrophication and other

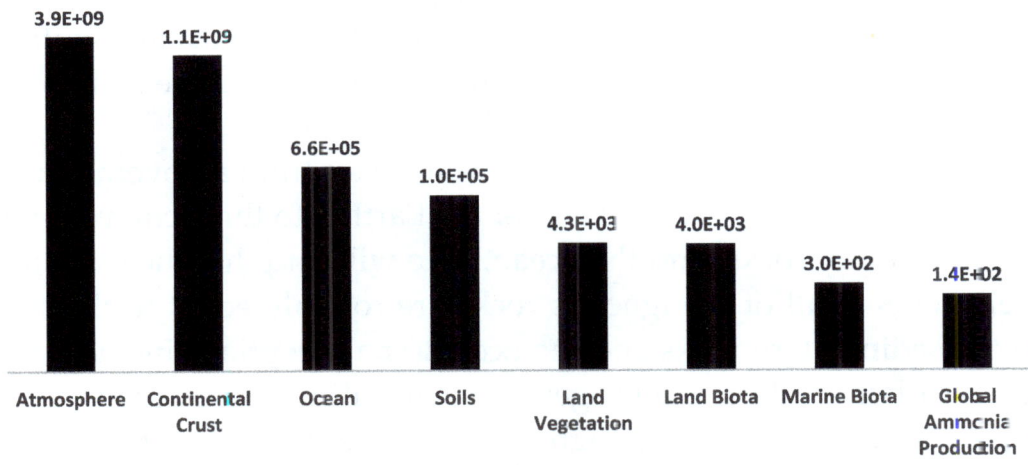

Figure 2.13 The nitrogen reservoir.
Data source: Palya et al., 2011

---

[14] With the exception of most part of the open ocean, where scientists found that low productivity there was often due to lack of iron. They found that by adding a small amount of iron, the productivities in these areas can increase dramatically. Therefore, these areas are called "high nitrogen, low chlorophyll (HNLC) ocean. In fact, one of the ingenious ways to reduce atmospheric carbon dioxide and to control global warming is to fertilize the ocean with iron.

environmental issues, many of which manifest themselves as perturbations in carbon cycling as well.

2.4.3 Biogeochemical Cycles of Other Elements

There have been many studies on the global biogeochemical cycles of other important elements, such as phosphorus, oxygen, silicon, sulfur, iron, and some trace elements. The cycling of these elements is more or less related to the biogeochemical cycles that have been discussed here, such as rock cycle, water cycle, and carbon and nitrogen cycles. For example, the global biogeochemical cycle for phosphorus is intricately related to the rock cycle. Phosphorus is an essential nutrient (see Section 2.4.2 and the Knowledge Box about the Redfield Ratio. For every 16 units of nitrogen, a primary producer would need one unit of phosphorus to grow) and is part of an extremely important compound called adenosine triphosphate or ATP, which is the intermediary of nearly all energy transfer in organisms. Phosphorus is also part of the backbone of DNA double helix (Figure 1.8, where phosphate molecules form the "ribbon" of both DNA and RNA). Phosphorus is mined extensively, perturbing its natural biogeochemical process. Excessive phosphorus usage as chemical fertilizers has impacted water quality and ecosystems in many places, especially in lakes.

The biogeochemical cycle of sulfur is also environmentally relevant in that it participates in rock cycle and water cycle. Pyrite (FeS_2) is a common mineral and sulfur dioxide is one of the main volcanic gases. Sulfate is one of the most abundant ions in the aqueous environment. Under anaerobic condition, a water body may give out pungent rotten egg smell attributable to hydrogen sulfide, most likely produced by reduction of sulfate in the water or in the sediment. Sulfur also occurs in coal at significant levels. Burning high-sulfur coal without proper treatment is the one of the main sources of air pollution and acid rain, as will be discussed in Chapter 4.

It is also intriguing to look at biogeochemical cycles from an overall mass balance perspective. If we divide the surface layer of the Earth into three compartments, igneous rocks, sedimentary rocks, and the ocean,[15] we will find that the overall masses of nearly all elements in all of the igneous rocks are roughly equal to the sum of each element in the sedimentary rocks and the ocean. The exceptions include gaseous elements such as carbon, hydrogen, nitrogen, oxygen, sulfur, and chlorine; calcium, mercury, and some others which have higher abundance in sedimentary rocks and the ocean. This may indicate that these elements are being lost from the lithosphere in an irreversible process as part of the evolution of the Earth itself. If we apply the same mass balance approach to the ocean and examine what goes in from rivers and what goes out (as marine sediment rocks), we could see a general mass balance as well, with small imbalance of the same few elements as mentioned above, in addition to sodium.

---

[15] We ignore metamorphic rocks and the atmosphere here for simplicity.

2.5 Biogeochemical Cycles and Environmental Challenges

The concept of global biogeochemical cycles puts many environmental challenges in perspective. Many global environmental challenges are due to perturbation of natural biogeochemical cycles. As mentioned previously, global warming as a result of excessive fossil fuel burning is a perturbation of the carbon cycle, in which humans put enormous amounts of carbon dioxide to the atmosphere in a century that would have taken the nature millions of years to do. Increasing carbon dioxide not only causes global warming, but also results in ocean acidification that interferes with formation of carbonate-based shells. This phenomenon is causing profound ecological impact and has the potential to bring about more disruptions to the oceanic environment. Humans also alter the water cycle by pumping groundwater, building dams and other structures to meet the ever-increasing demand for water resources and hydropower. Intensive agriculture puts excessive amounts of chemical fertilizers, much of which enters the aqueous environment, causing eutrophication and ecosystem disruptions. Large-scale animal feeding operations cause deviation of the global carbon cycling by significantly altering the natural trophic transfer of materials and energy. These animals, especially cattle, also release large amounts of methane in the air, exacerbating the greenhouse effect. Human-induced climate change have disturbed the heat balance of the planet and in turn impacted the water balance, which is sensitive to the Earth's energy balance. Later in this book, many of such environmental impacts will be discussed.

Further Reading

Bar-On, Y.M., R. Phillips, and R. Milo. 2018. The biomass distribution on Earth. PNAS. 115/25, pp. 6506–6511. www.pnas.org/cgi/doi/10.1073/pnas.1711842115

Berner, Elizabeth Kay and Robert A. Berner. The Global Water Cycle. New Jersey, USA: Prentice-Hall, 1987.

Drever, James I. The Geochemistry of Natural Waters: Surface and Groundwater Environments, 3rd Edition. New Jersey, USA: Prentice-Hall, 1997.

Intergovernmental Panel on Climate Change (IPCC). Climate Change 1994, Cambridge, UK: Cambridge University Press, 1995.

Palya, Annie P., Ian S. Buick, and Gray E. Bebout. "Storage and mobility of nitrogen in the continental crust: Evidence from partially melted metasedimentary rocks, Mt. Stafford, Australia." *Chemical Geology*, 281, no. 3–4 (February 2011) 211–26.

Faure, Gunter. Principles and Applications of Geochemistry, 2nd Edition, New Jersey, USA: Prentice Hall, 1998.

The Open University. 1989. Seawater: its composition, properties and behavior. Butterworth Heinemann. Oxford OX2 8DP. UK.

Schnoor, Jerald L., and Werner Stumm. "Acidification of aquatic and terrestrial systems." In *Chemical Processes in Lakes*, edited by Werner Stumm, pp. 311–38. New York, USA: Wiley Interscience, 1985.

Harvey Blatt, 1997. Our Geologic Environment. Prentice Hall, Upper Saddle River, NJ 07458

Web Resources

Redfield Ratio Wikipedia page: https://en.wikipedia.org/wiki/Redfield_ratio

Biogeochemical cycle Wikipedia page: https://en.wikipedia.org/wiki/Biogeochemical_cycle

Biogeochemical cycle at Khan Academy (it's a little rudimentary but is a good introductory video): https://www.youtube.com/watch?v=ccWUDlKC3dE

Biogeochemical cycle at Bozeman Science (focused mostly on nutrients): https://www.youtube.com/watch?v=Bn41lXKyVWQ

Questions and Exercises

1. Check the mass balance of the reaction equations in this chapter.
2. Check the mass balance of the plagioclase weathering formula: $177Na_{0.62}Ca_{0.38}Al_{1.38}Si_{2.62}O_8 + 246CO_2 + 367H_2O = 123Al_2Si_2O_5(OH)_4 + 110Na^+ + 68Ca^{2+} + 246HCO_3^- + 220SiO_2$. How could this highly complex formula be figured out?
3. The "real" process of photosynthesis could be more complicated than the reaction equation (1). It looks something like this (Schnoor and Stumm, 1985): $800CO_2 + 6NH_4^+ + 4Ca^{2+} + 1Mg^{2+} + 2K^+ + 1Al(OH)^{2+} + 1Fe^{2+} + 2NO_3^- + 1H_2PO_4^- + 1SO_4^{2-} + H_2O \rightarrow biomass + 16H^+ + 804O_2$. Can you derive the "average" chemical composition of biomass? What is the ratio of C:N:P? How does it compare to the Redfield Ratio? Why?
4. Some elements such as sodium, chloride, and sulfur are being accumulated in the ocean, apparently violating the assumption that our natural world is in equilibrium and steady state. Could you postulate the reason for this phenomenon? Could this trend continue to eternity?
5. Read the knowledge box entry on the chemical composition of natural waters and compare the average chemical compositions of freshwater and seawater. Can you make a clear conclusion whether the seawater is simply the accumulation and evaporation of its freshwater sources? Why or why not?

Chapter 3

Environmental Regulations

Learning Outcomes		52
Key Concepts		52
3.1	Need for Environmental Regulations	52
	3.1.1 Tragedy of the Commons	52
	3.1.2 The History of Environmental Regulations	54
	3.1.3 Implementation of Environmental Regulations in the United States in Europe	54
3.2	Environmental Policy-Making Process	57
	3.2.1 Overview	57
	3.2.2 Policy Making Process	59
	3.2.3 The National Environmental Policy Act (NEPA)	60
	3.2.4 The Policy Cycle	61
3.3	Environmental Politics and Policy	62
	3.3.1 Environmental Politics	62
	3.3.2 Interest Groups in Environmental Politics and Policy	63
	3.3.3 Environmental Groups and Their Roles in Environmental Politics and Policy	64
	3.3.4 Litigation, Regulation, Legislation, and Negotiation in Environmental Politics	66
3.4	Environmental Regulations	67
	3.4.1 Environmental Regulation in the United States	67
	3.4.2 Environmental Regulation in the European Union	68
	3.4.3 Environmental Regulation in China and other Developing Countries	69
3.5	Environmental Justice	70
Further Reading		71
Web Resources		72
Questions and Exercises		72

Learning Outcomes

- Knowledge of major environmental laws and regulations
  - Air quality laws
  - Water quality laws
- Understanding of the intent and origin of environmental policies and regulations
  - Understand the concept of "the tragedy of the commons"
- Knowledge of the environmental policy making process (6 steps);
- Knowledge of the concept of "policy cycle" (i.e., life cycle of a policy)
- Understanding of the differences and linkage between environmental policy and politics; regulations, and examples of major environmental policies in the world.

Key Concepts

Tragedy of the Commons; game theory; law; regulation; policy; politics; policy making; policy cycle; National Environmental Policy Act (NEPA); California Environmental Quality Act (CEQA); (environmental) politics; interest groups; United States Environmental Protection Agency (USEPA); European Environmental Agency (EEA); environmental justice

Environmental regulation is an important part of the environmental field. This chapter will start with discussing the issue of "**Tragedy of the Commons**" that, without environmental regulations, will result in disastrous damage of our environment. The billion-dollar environmental industry and millions of jobs in the field of wastewater treatment, remediation, environmental consulting, and so on, would be drastically different or even be nonexistent if there were no environmental regulations that prescribe requirements for environmental protection. In this chapter, we will examine the need for environmental regulations, the process and practice of developing and implementing environmental regulations, **environmental policies** and **politics**, and associated issues and challenges. Important environmental regulations in the United States, European Union, and China will be introduced, and the issue of environmental justice, which is a socioeconomic issue with environmental nexus, will be briefly discussed.

3.1 Need for Environmental Regulations

3.1.1 Tragedy of the Commons

"Tragedy of the commons" is a concept brought up by Hardin in 1968, in which individuals, acting independently (not necessarily selfishly) according to their best interest, behave contrary to the common good of the others by depleting or spoiling the shared resource through their collective actions. A good example is grazing lands, where an individual cattle owner will likely use the land as much as possible for grazing for his cattle. If there are too many of such users and there is no regulation limiting the use, the grazing land will eventually be overexploited and cannot be used for grazing any more, hurting everyone involved. There are many other examples such as water resources, water quality, air quality, forest, and other publicly owned properties. Everyone suffers when a common resource is depleted or harmed because when a resource belongs to everybody, it also belongs to nobody.

In an ideal world, a nation would be governed by a philosopher king, and its citizens would behave rationally.[1] The economy would be strong, its citizens would enjoy high quality of life, and the environment would be clean and sustainable. However, there is no philosopher king in this world. There are some rational people, but collective rationality is hard to achieve when people act at his/her best interest even without intentionally harming other people's interest. This can best be explained by **game theory**, which is the study of strategic interactions among rational decision-makers in a zero-sum game that each participant tries to maximize his or her gain at the expense of others. In the scenario of the tragedy of the commons of the grazing lands, the participants would consider how other cattle owners would do in order to figure out the best strategy of his/her own. One likely outcome, if the participants are cooperative and rational, is that they probably will establish a set of rules to enforce the sharing of common resources without hurting everyone. Each environmental legislation has a different story, but the overall rationale has been consistent and can all be traced back to the tragedy of the commons. The most eminent examples are probably international treaties in the protection of common resources, such as chlorofluorocarbons (CFCs), greenhouse gases, protection of endangered species, conservation of wetlands, etc. Before the treaties were established, every nation, as a member of the "commons," tried to develop its economy at the expense of global environmental resources. As a result, global warming, ozone depletion,[2] crashing of ocean fisheries, rapid loss of global biodiversity, and other global environmental issues are becoming increasingly acute. Some nations may have realized the issues earlier than others, but without binding international treaties and enforcement, these Tragedy of the Commons issues cannot be effectively addressed.

---

[1] According to Plato, a philosopher king is a ruler who possesses both a love of wisdom, as well as intelligence, reliability, and a willingness to live a simple life.

[2] Ozone depletion is caused by many chemicals, especially CFC because one single CFC molecule can destroy on average 100,000 ozone molecules before it is removed from the atmosphere. Frank Rowland of UC Irvine won the Nobel Prize for his pioneering work on this issue. The 1987 Montreal Protocol is credited for the global effort on CFC emission that has resulted in significant reduction of the ozone hole over Antarctica.

3.1.2 The History of Environmental Regulations

The same tragedy has happened to every nation and around the world along with population growth and economic development. The earliest "environmental regulation" perhaps took place in the 1300s, when the English King Edward I briefly banned coal fires in London due to extremely poor air quality. In the United States, one such rule is the Rivers and Harbors Appropriation Act of 1899, the oldest federal environmental law in the New World. This Act set a number of rules to protect the nation's navigable waters against dumping of waste; dredge/fill/dam, or otherwise alter such navigable water without a permit. This Act is the precursor of the monumental Clean Water Act, arguably the most successful and influential environmental law in the history of the United States. In European Union (EU), the Paris Summit meeting of heads of state and government of the European Economic Community (EEC) in October 1972 marked the beginning of the EU's environmental policy. China's first environmental protection law was passed in 1979, but significant effort on environmental protection did not start until the 2000s. In many cases, environmental regulations were spurred by worsening environmental quality, sometimes by environmental disasters. Environmental groups have also been important in raising the public awareness of these issues. Examples will be given in the discussion later in this chapter when specific regulations are described.

> ### Knowledge Box
>
> **Laws** and **Regulations:** In the United States, Federal laws are bills that have passed both houses of Congress, been signed by the president, passed over the president's veto, or allowed to become law without the president's signature. Individual laws, also called acts, are arranged by subject in the United States Code (U.S.C.). Regulations are rules made by executive departments and agencies, and are arranged by subject in the Code of Federal Regulations (C.F.R.). Regulations are the ongoing process of monitoring and enforcing the law. Essentially, regulations are the extension of the law. For example, the U.S. Congress sets the laws such as the Clean Water Act, and federal agencies such as **United States Environmental Protection Agency (USEPA)** sets environmental regulations to implement the Act by adopting water quality standards and issuing discharge permits in the spirit of the Clean Water Act. In this book, these two terms are not always distinguished from each other and they are collectively termed "policies."

3.1.3 Implementation of Environmental Regulations in the United States in Europe

It takes a legislative body such as a congress to enact environmental laws, and it takes large agencies or bureaucracies to implement environmental regulations.

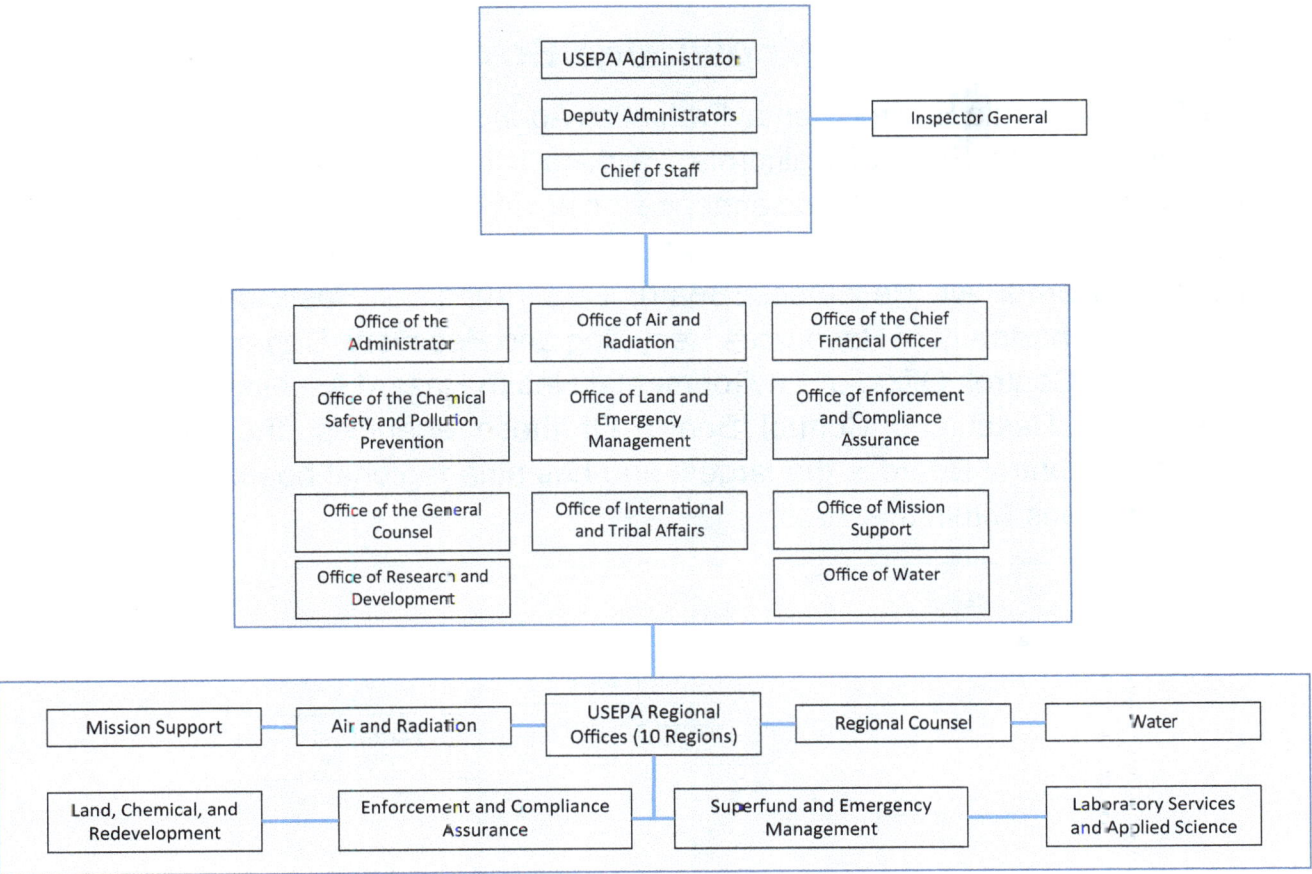

Figure 3.1 The organization chart of USEPA.

Source: Created by Jian Peng based on information from USEPA website www.epa.gov.

In the United states, USEPA is one of the major federal agencies that, though not a cabinet department (such as the Department of Agriculture or the Department of the Interior), enjoys tremendous clout and the administrator is routinely treated as a cabinet member. With a large staff of more than 14,000 full-time employees as of 2018 and a budget of more than $8 billion, USEPA is one of the most influential federal agencies, and certainly is the governing agency that oversee nearly all environmental issues. **Figure 3.1** shows the organization chart of USEPA. The 10 regional offices of USEPA enjoy significant autonomy and can implement environmental policies more tailored to each specific region. Different regional offices may differ significantly in terms of progressiveness in how they manage environmental issues. For example, USEPA Regional 9 oversees the states of Arizona, California, Hawaii, Nevada, Guam, Samoa, and Navajo Nation (a Native American tribe) and is perhaps the most progressive region among the 10 regions. In the EU, the governing environmental agency is the **European Environmental Agency (EEA)** established in 1990. EEA's task is to provide sound, independent information on the environment and to support sustainable development by helping to achieve

> **Knowledge Box**
>
> **CalEPA**—California Environmental Protection Agency (CalEPA) is the environmental agency for the State of California. Working with USEPA Region 9, CalEPA's mission is to restore, protect, and enhance the environment, to ensure public health, environmental quality and economic vitality. It has the following six major departments: California Air Resources Board; Department of Pesticide Regulation; California Department of Resources Recycling and Recovery; Department of Toxic Substances Control; Office of Environmental Health Hazard Assessment; and the State Water Resources Control Board. Of these agencies, the State Water Resources Control Board is the largest and has nine regional boards that govern different regions within the State.

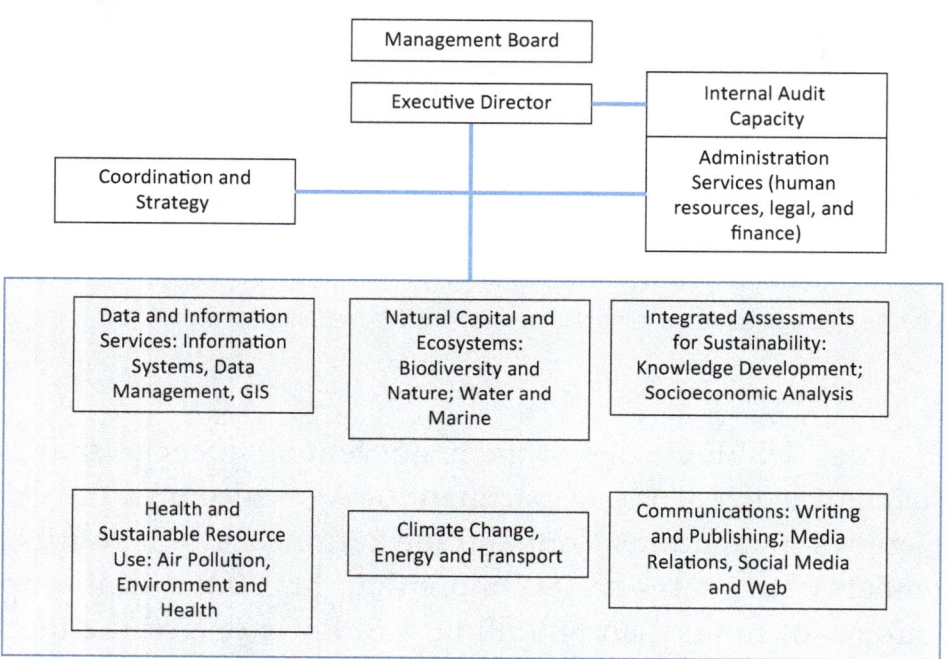

Figure 3.2 European Environmental Agency (EEA) Organization Chart.
Source: Created by Jian Peng based on the information from EEA website: eea.europa.eu.

significant and measurable improvement in Europe's environment. The organization chart for EEA is shown in **Figure 3.2.** Compared to USEPA, EEA put more emphasis on sustainability, biodiversity, and climate change. This reflects the increasing push of these global environmental issues by EU's 32 member nations and 6 cooperating nations.

3.2 Environmental Policy-Making Process

3.2.1 Overview

Similar to other public policies, environmental **policy making** process could be modeled in several different ways.[3] These models, unfortunately, largely assume that the policy-makers are fully informed, rational, and intelligent, which obviously is not the case. The more pessimistic view of the policy making process is the so-called "garbage can model."[4] Somewhere in the middle, the most plausible is the model proposed by John Kingdom (1984), where policy making is made up of three streams: problem, political, and policy. In the problem stream, environmental issues are defined as problems and receive regulatory attention. In the political stream, events, trends, institutions, and interest groups determine which problems will receive attention on the governmental agenda. In the policy stream, the agenda turns into an environmental policy after exchange and collisions of ideas, analyses, arguments, and participation of bureaucrats, congressional staff, think tanks, etc.

In terms of the administrative process, environmental policy making takes the following steps: start-up and planning; issue development, options analysis, and drafting; review and drafting of the proposal; public comment and evaluation; drafting and review of the final rule; and promulgation and implementation (**Figure 3.3**). Usually, the entire process for a typical environmental policy takes around 2 years in the US, but it could be much longer if the policy is controversial or complicated. Federal laws usually take longer than state laws. This is not surprising—there are at least 20 senate and house committees that have oversight on environmental issues. Because federal laws cover national issues, they usually generate more widespread public and institutional attention and need more time for public participation and to address the comments.[5] Emergency rule-making could take much less time.

---

[3] They include institutional model (laws are made by institutions as their formal responsibilities); system model (laws are set up to regulate entities coexisting in equilibrium in a system); group process model (laws are products of struggles and collaborations among different interest groups, and the most powerful groups win out); and net-benefit model (laws are set up to achieve the best overall benefit to the society).

[4] According to a poignant paper by Cohen et al. (1972), a "garbage can" organization is a "collection of choices looking for problems, issues and feelings looking for decision situations in which they might be aired, solutions looking for issues to which they might be the answer, and decision makers looking for work."

[5] One good example is USEPA's Clean Water Rule that aimed at revising the definition of navigable water in 2015. One of the versions of the draft law received more than a million public comments. By law, all of these comments need to be addressed. Therefore, it is not surprising that it took a very long time to move forward (only to be a victim of partisan politics 4 years later, in 2019).

Figure 3.3 Policy making process.
Source: Created by Jian Peng based on Fiorino, 1995.

> ## Knowledge Box
>
> **Policy Making Process:** Based on Fiorino (1995; see Figure 3.3), the policy making process has the following steps:
> - Start-up and planning: work on a rule is initiated in response to a legal requirement (for example, a law requires that certain regulations be established by a set deadline) or a program need (for example, an emerging environmental issue that needs to be addressed by new regulation but under the current legal framework).
> - Issue development, Options analysis, and Drafting: the issue to be addressed is defined and organized, analyzed, including the cost benefit analysis and economic analysis.
> - Review and publication of the proposal: the draft proposed rule is published for public comments.
> - Public comments and evaluation: public comments are received and addressed.
> - Drafting and review of the final rule: public comments are addressed and the rule revised and republished for further comments. This process could repeat several times depending on level and significance of the comments.
> - Promulgation and implementation: the rule is finalized, officially adopted, and implemented.

3.2.2 Policy Making Process

Figure 3.4 shows a typical policy making process for USEPA. If we apply the above model for policy-making process, the start-up and planning does not show in Figure 3.4 because this process could take many years, and many issues will not make it through the policy making process. Steps 1 through 9 will be the second bullet above (issue development, options analysis, and drafting) where a workgroup, usually consisting of technical experts in the specific field of the issue, USEPA officials, and sometimes environmental group members, takes the bulk of the work on issue and policy analysis, options evaluation, and drafting of the document. The drafted policy will undergo agency review before it is externalized to other federal government agencies such as the Office of Management and Budget (OMB)[6] or the United States Fish and Wildlife Services (USFWS)[7] before being published for public commenting. The comments are then compiled and addressed in the final rule. The promulgation and implementation of the rule usually take some additional administrative steps. On the state level, this could involve review and approval by federal agencies. On the federal level, the policy making process could take on either legislative or agency pathways. Interestingly, both federal and state environmental laws or policies, before they can be developed, must go through environmental review process. This will be addressed in the next section.

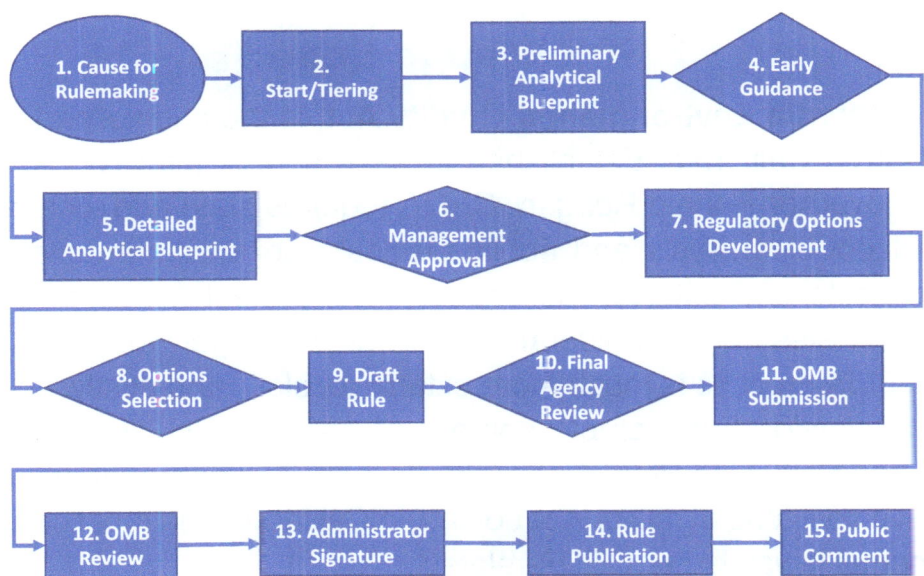

Figure 3.4 The USEPA's rule making process.
Source: Created by Jian Peng based on www.epa.gov

---

[6] OMB review is required if the rule is deemed "significant" due to its budget impact (>$100 million), significant inconsistency with other federal policies or agencies; or it raises new legal or policy issues.

[7] Pursuant to Endangered Species Act (ESA), USFWS and/or National Marine Fisheries Services review or consultation will be required if the rule may impact endangered or threatened species.

3.2.3 The National Environmental Policy Act (NEPA)

The **National Environmental Policy Act**, or NEPA, was established in 1970 as the first of a series of significant environmental laws including the Clean Air Act, Clean Water Act, Endangered Species Act, Safe Drinking Water Act, and many other environmental laws that completely changed the environmental landscape in the United States. NEPA is a foundational environmental law that requires environmental reviews for all federal projects and federal policies[8] very early in the process so that their environmental impacts can be assessed and the public be notified of these impacts. A 2015 court document has the following elegant description of NEPA (Byron, 2015):

"Following nearly a century of rapid economic expansion, population growth, industrialization, and urbanization, it had become clear by the late 1960s that American progress had an environmental cost. A congressional investigation into the matter yielded myriad evidence indicating a gross mismanagement of the country's environment and resources, most notably at the hands of the federal government. As a result, lawmakers and the general public alike called for an urgent and sweeping policy of environmental protection."

> ### Knowledge Box
>
> **CEQA**, or **California Environmental Quality Act**, is the counterpart of NEPA on the state level in California. Established in 1970 shortly after NEPA and signed by the then-governor Ronald Reagan, CEQA requires state and local agencies to follow a protocol of analysis and public disclosure of environmental impacts of proposed projects and mitigate the impacts. The project mitigation requirement significantly deviates from the federal NEPA, which is largely for public disclosure. Therefore, CEQA process is much more extensive, slow, and costly, and is subject to some controversies. For example, some large projects such as California's High-Speed Rail and Bay-Delta Project can take decades to go through the CEQA process and have cost tens of millions of dollars. In California, the most common types of environmental lawsuits are CEQA-related.

---

[8] Note that NEPA applies only to federal projects and policies. For state and local projects and policies, local NEPA-like policies apply. For example, CEQA applies to projects in California. Some projects may have both federal and state interests. In that case, both NEPA and CEQA documentations need to be developed and obligations fulfilled, making the project planning more challenging.

There are three levels of analysis for a project's environmental impacts:

- **Categorical Exclusion determination**—this category requires no additional actions or analysis if a project can be categorically excluded due to lack of environmental impact or lack of applicability.
- **Environmental Assessment/Finding of No Significant Impact**—this category includes projects that cannot be categorically excluded but it does not cause significant environmental impacts.
- **Environmental Impact Statement**—this category applies to projects that have significant environmental impact; therefore the project will be subject to more detailed and rigorous requirements, including alternative consideration and plans for mitigation and monitoring if necessary. This information will be included in the Environmental Impact Statement document that is subject to public review, commenting, and official adoption. It is also a legal document that the project owner has to abide by to avoid legal challenges.

3.2.4 The Policy Cycle

Based on Rosenbaum (2005), who took on a more holistic view of the entire policy making process and life cycle of a policy, the "**Policy Cycle**" includes the following steps: agenda setting, formulation and legitimation, implementation, assessment and reformulation, and termination. Of these processes, the first two are part of the policy making process, and the last three steps are part of the entire life cycle of a policy. The following Knowledge Box explained these steps.

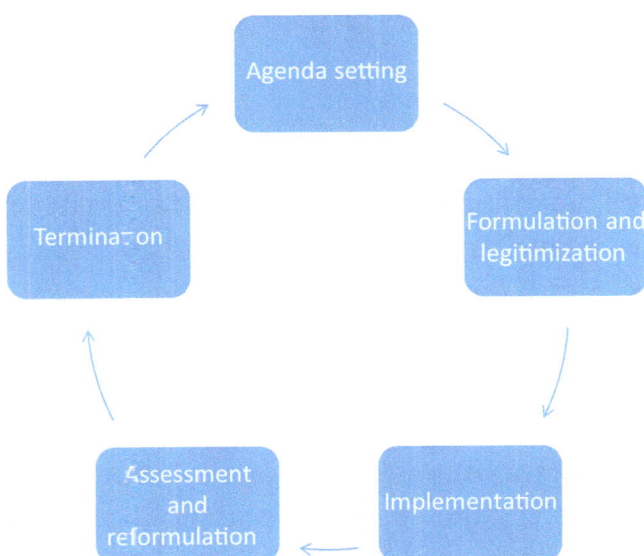

Figure 3.5 The policy cycle.
Source: Created by Jian Peng based on Rosenbaum (2005)

> ## Knowledge Box
>
> **The Policy Cycle**—Based on Rosenbaum (2005), the Policy Cycle has the following steps:
>
> - Agenda setting: the process of getting an issue urgent and significant enough to the official government agenda to be considered. Not all of the issues will be taken up, or be agendized, for further policy making.
> - Formulation and legitimization: for those issues that are agendized, policy goals will be set, plans will be created, and plans and proposals be prepared to implement the plans. Policies, once created, will then be legitimized to become effective through voting, public hearing, presidential orders, etc.
> - Implementation: this is the step where a series of actions are required to be taken in order to realize the intent of the policy. (For example, a discharge permit is one of the mechanisms to implement Clean Water Act and related policies to improve water quality.)
> - Assessment and reformulation: the impact of a policy will be monitored and assessed by the court system, media, and governmental agencies. A policy may be reformulated to address emerging issues.
> - Termination: deliberate conclusion or succession of a policy after its intent has been fulfilled.

3.3 Environmental Politics and Policy

3.3.1 Environmental Politics

Politics is a set of activities associated with the governance of a country, state or an area. It could be a neutral word depicting these activities, but often it has negative connotations because it often involves negotiation, compromise, policy making, and exercising force. Politics, interest groups, and their agendas affect policy and the environment through nearly every step in the policy making process and the "Policy Cycle" (Figures 3.3 and 3.5). For example, agenda setting is a process where a social issue becomes important enough that it is written up and agendized (i.e., getting onto a legislative calendar) to be presented before a legislative committee or authority. Within each agency, such as USEPA (Figure 3.1) and EEA (Figure 3.2), they need to "play politics" with other agencies within the U.S. government or the EU to implement their agenda. Within these agencies themselves, different departments/divisions may need to play politics to exert influence and fight for resources.

Figure 3.6 Cuyahoga River in Cleveland, Ohio. During and before the 1960s, it caught fire at least 13 times. The fire on June 22 was caught on camera and created a national sensation and boosted the momentum of the environmental movement in the United States.

On the society level as a whole, the dynamics among different interest groups is critical to the way how environmental issues are handled. Most environmental issues need a political propellent to survive the early stages of agenda setting. Had Rachel Carson not published her seminal book Silent Spring, had the Cuyahoga River in Ohio (**Figure 3.6**) not caught fire on June 22, 1969 and the shocking photo captured the nation's attention, the environmental movement in the late 1960s might not have been strong enough to prompt a slurry of environmental legislations. Many environmental initiatives and legislative agendas may be adverse to many other interest groups, and they cannot be easily agendized due to lack of support or petition by these interest groups.

3.3.2 Interest Groups in Environmental Politics and Policy

Interest groups play an increasing role in both environmental policy making and politics. On environmental issues, four main interest groups can be called out (**Figure 3.7**): legislature (congress), government, businesses/industries, and environment group. The general public is often an important interest group, but due to its heterogeneity, their positions can often be reflected in either the environmental group or the businesses/industries group. Business groups are often viewed as the most influential interest group with good reason, because their leaders manage much of a nation's economy. Business groups can have their interests represented early and forcefully in the policy making process. But they are far from the only interest group, and sometimes

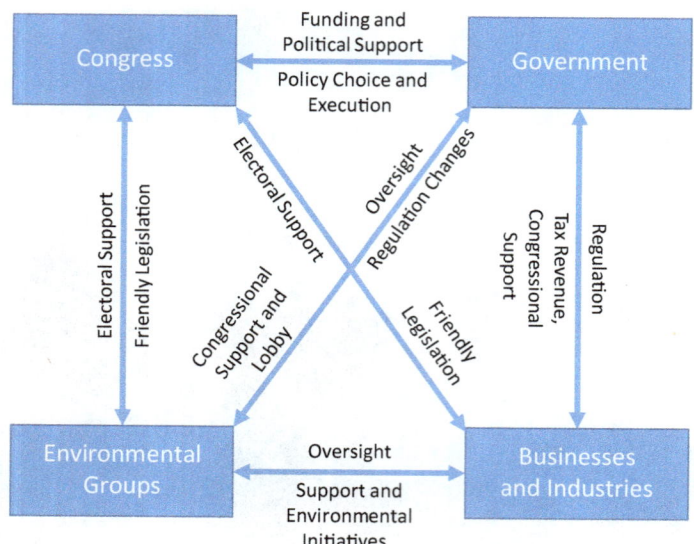

Figure 3.7 The Iron Quadrangle of Environmental Politics.
Source: Created by Jian Peng

not the dominant one on some policy issues. The business group sometimes are differentiated into smaller sub-groups, each with different or even competing interests. For example, in the automaker industry, electrical carmakers such as Tesla are a disruptive force and they often have different priorities and political agendas (especially those on environmental issues) than the traditional carmakers. Some technology firms such as Apple, Google can be very environment-friendly. Even for companies such as Monsanto (the world's largest pesticide manufacturer, now part of Bayer), ExxonMobil (one of the largest oil companies in the world), and Rio Tinto (one of the world's largest mining companies), their website and public messaging could also appear to be environment-friendly. However, they are often subject to severe environmental regulations and have to treat the other interest groups carefully to maintain profitability.

3.3.3 Environmental Groups and Their Roles in Environmental Politics and Policy

On the other end of the spectrum is the environmental groups (**Figure 3.8**), which had little political influence until the 1970s. Compared to the business groups, environmental groups have little financial means to support their political agenda. However, they have become more and more proficient in fundraising, grassroot mobilization using media and social media, as well as maximizing their influence on environmental policy making process via public participation process and legal challenges. As shown in Figure 3.7, environmental groups exert tremendous influence on all other interest groups by electoral support to the congress, lobbying and oversight to various

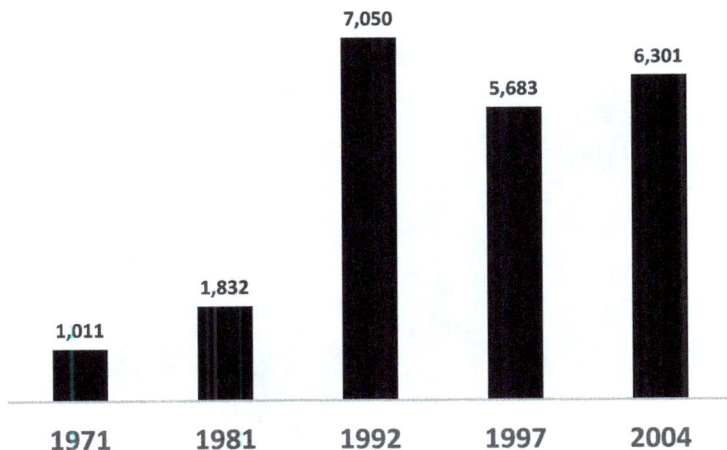

Figure 3.8 Memberships in 13 major environmental groups (the World Wildlife Fund; the Nature Conservancy; Sierra Club; National Wildlife Federation; National Audubon Society; Natural Resources Defense Council; Environmental Defense Fund; Greenpiece; Friends of the Earth; Wilderness Society; National Parks Conservation Association; Izaak Walton League).
Source: Created by Jian Peng

governmental agencies, and very strong oversight to the business groups. Without these environmental groups, the world would have been quite different, and most likely for the worse.

Environmental groups themselves have evolved and differentiated over the years. Environmentalism has recurred throughout human history in different forms and places around the world.[9] However, it was largely believed that environmentalism and environmental activism did not start until after the industrial revolution. The first environmental organization is probably the Commons Preservation Society in Britain in 1865, which promoted rural preservation against industrialization and development. In the United States, the earliest such organization is Sierra Club organized by John Muir in 1892 together with establishing the Yosemite National Park in northern California. Muir believed in nature's inherent right, especially after spending time hiking in Yosemite Valley and studying both ecology and geology. Today the Sierra Club remains to be one of the largest and most influential environmental groups in the United States, and its conservationist principles are part of the bedrock of modern environmentalism.

Along with the mainstream environmental groups such as Sierra Club, there are also other types of environmentalism (for example, School Strike 4 Climate rally led by the Norwegian teen Greta Thunberg, as shown in **Figure 3.9**). One such type of groups include those who espouse deep ecology or lifestyle transformation. Deep ecologists

---

[9] Such as India's Jainism (sixth century BC) and China's Taoism (fourth century BC), both promoting harmony between man and nature.

Figure 3.9 Sydney, Australia—March 15, 2019—20,000 Australian students gather in climate change protest rally, School Strike 4 Climate, and demand urgent action on climate change.

believe that humans are only part of the nature[10] and all forms of life have equal claim on existence. They believe that the institutional structures and socioeconomic values should all be fundamentally changed to promote global ecological integrity. However, deep ecologists lack numbers, political leverage or influence that other mainstream environmental organizations, but they make up by being more active and vocal, sometimes disruptive. Some environmental groups became radical when they are frustrated by the mainstream. They general promote nonviolence and tolerance, but some may resort to "ecotage" or "monkey-wrenching." Chapter 8 will have more discussions on deep ecology and other ecological issues.

3.3.4 Litigation, Regulation, Legislation, and Negotiation in Environmental Politics

In environmental politics, there are several levels of engagement/interactions among different interest groups. These levels of engagement/interactions include, from the degree of hostility, litigation; regulation; legislation; and negotiation. Litigation takes place when two sides have different opinions/interpretations of applicable environmental laws and regulations, and the differences have to be resolved in the court of law. This process is usually hostile, expensive, and can be time-consuming as well. However, due to the complexity of the regulatory framework, environmental litigation is fairly common in the US, and decisions of important cases often send shock waves across the environmental field. On the other end of the spectrum is negotiation, in

---

[10] See Chapter 12 of this book.

which two or more interest groups discuss the issues and try to resolve the differences without formal regulatory actions or litigation. However, negotiation could be a significant part of the regulation/legislation and even litigation process. In the middle of the spectrum, regulation is the process where regulatory agencies impose regulatory requirements on regulated entities, and legislation is the process where environmental laws are set up and promulgated, as discussed previously in this chapter. In short, environmental politics is a mixed bag of various processes, strategies, and power plays that permeate throughout every aspect of the environmental field.

3.4 Environmental Regulations

The tragedy of the commons makes it necessary to protect the shared environmental resources such as air, water, natural resources, and biodiversity, through environmental regulation. There are three main mechanisms of environmental regulation. The first method is to impose controls such as technology-based regulation, which focuses on the pollution-control technologies, and environmental quality-based regulation, which focus more on the environmental impact. Environmental regulation could also be in the form of market incentives, such as effluent fees, marketable pollution rights, subsidies, and so on. Thirdly, the publication of environmental impacts such as through the aforementioned NEPA and CEQA processes could encourage or force a business or a facility to consider alternatives to mitigate these impacts. Environmental regulations are often based on ethical considerations aside from economic ones. This is often due to the difficulties in quantifying monetary values of many factors, such as protection of nature; protection of future generations; and environmental justice issues.

In addition to the above, environmental regulations often need to consider economic burdens that it will bring to the regulated parties and the general public, and ethics issues, such as human health protection (especially the protection of future generations) and conservation of nonrenewable natural resources.

3.4.1 Environmental Regulation in the United States

In the United States, the Congress sets the environmental laws, and the USEPA is the federal agency to interpret and implement these laws. USEPA was established by Congress by passing NEPA in 1969. Then in 1970, Congress passed the Clean Air Act (CAA) and in 1972, the Clean Water Act[11] and the Federal Insecticides, Fungicides, and Rodenticides

---

[11] CWA was called the Federal Water Pollution Control Act (FWPCA) when it was first adopted in 1972. Its amendment in 1977 was commonly referred to as the Clean Water Act. Since FWPCA and CWA have the same lineage, people often call both (and their subsequent amendments) Clean Water Act.

Act (FIFRA).[12] In 1973, the Endangered Species Act (ESA) was established. In the late 1970s, Toxic Substances Control Act (TSCA) and Resources Conservation and Recovery Act (RCRA) were passed. In 1980, Congress passed the Comprehensive Environmental Response, Compensation, and Liability Act (CERCLA), also known as the Superfund Act. There are also many other lesser known environmental laws that USEPA oversees, but they are not mentioned here. It is worth noting that few of these laws were brand new when they were established in the 1970s. Their enactment during this time frame coincided with the environmental movement in the 1960s marked by Rachel Carson's Silent Spring and several notable environmental incidents, including the Cuyahoga River fire.

These environmental laws created a complex network of legal settings for policy making on both federal and state levels. Over time, these laws underwent further amendments, notably the CAA Amendment of 1990; CWA amendments of 1977 and 1987; Superfund Act Reauthorization Act of 1986, and so on. With time, these environmental laws became increasingly prescriptive and, in general, required more actions and costs to comply. Of these environmental laws, CAA, CWA, ESA, TSCA, and CERCLA will be discussed in more detail later in this book when air pollution, water pollution (and related endangered species protection), soil and ground water pollution, and remediation issues are discussed.

USEPA, as many other federal administrative agencies such as the Food and Drug Administration, was set up by Congress to regulate complex issues that Congress cannot do by itself. USEPA is a nonpartisan agency, but its head administrator and nine regional administrators are all appointed by the President of the United States. Therefore, USEPA is inevitably and profoundly influenced by politics and political partisanship.

For the most part, USEPA delegates its power to state environmental agencies and provides oversight on significant issues or issues affecting federal interests. For example, most of the states have been delegated authority to issue pollutant discharge permits under the federal Clean Water Act. However, these permits still need to be approved by the USEPA before they become effective. All state regulations that establish or change water quality standards also need USEPA review and approval.

3.4.2 Environmental Regulation in the European Union

EU is a political and economic union of 28 member states. With 6.9% of the world's population but 24.6% of the global gross domestic product (GDP), the EU enjoys very high human development index.[13] The governance of the EU is based on a series of treaties, and

---

[12] FIFRA was first passed in 1947 but was significantly amended in 1972; and the authority of implementing was passed from the United States Department of Agriculture (USDA) to USEPA.

[13] The Human Development Index (HDI) is a statistical composite index of life expectancy, education, and per capita income indicators.

there are three levels of competencies: exclusive competence (i.e., the power sits exclusively within EU); shared competency (i.e., the power is shared between the EU and individual nations), and supporting competence, where EU provides only supporting roles. EU's environmental policy is an example of shared competence. Similar to the United States, environmental issues were not given sufficient attention until the 1970s in the EU.

EU has extensive environmental laws and its environmental policy is significantly intertwined with other international and national environmental policies, especially for its member states. The body of EU environmental law amounts to well over 500 Directives, Regulations and Decisions. EU's environmental legislation addresses issues such as air (including ozone and acid rain), water, noise, waste management, and sustainable energy. Due to the various levels of competence as mentioned above, the governance of these environmental policies and regulation is quite complex, and implementation of these policies is often slow and controversial.

The Commissioner for the Environment is a member of the European Commission responsible for EU environmental policy. The Directorate-General for Environment (DG ENV) is responsible for the EU policy area of the environment. The European Environment Agency (EEA) is an agency of the EU which provides independent information on the environment. Its goal is to help those involved in developing, implementing and evaluating environmental policy, and to inform the general public. The agency is governed by a management board composed of representatives of the governments of its member states, a European Commission representative and two scientists.

In EU, the law governing water pollution control is the EU Water Framework Directive. There are also Bird Directive and Habitat Directive to protect birds, natural habitat and biodiversity. The Water Framework Directive is one of the most successful and influential environmental regulations. For air quality, the EU has the 2005 Thematic Strategy on Air Pollution (TSAP). This strategy established interim objectives for air quality and also established measures to ensure progress toward the air quality goals. Under EEA leadership, the air pollution issue is the second most important environmental issues, with climate change being the most important. In general, the EU has a more progressive stance on environmental issues and environmental regulation compared to the United States and the rest of the world.

3.4.3 Environmental Regulation in China and other Developing Countries

China provides a good example of environmental regulation in a developing country. Environmental policy in China is set by the National People's Congress and managed by the Ministry of Ecology and Environment (MEE; formerly the Ministry of Environmental Protection or MEP) of the People's Republic of China. The central government

Table 3.1 Environmental Regulations

Category	US	European Union	China	United Nations
General	National Environmental Policy Act	European Commissioner for the Environment	Environmental Pollution Control Act	United Nations Environment Programme
Air	Clean Air Act	Airborne Pollution Prevention and Control Action Plan		
Water	Clean Water Act	Water Framework Directive	Water Pollution Control Act	
Soil and Groundwater	CIRCLA; RCRA		Soil Pollution Prevention Act	
Wildlife and Conservation	Endangered Species Act	Birds Directive; Habitat Directive	Rare and Endangered Species Protection Act	Convention on Biodiversity; RAMSAR Convention
Climate Change	Section 202 of Clean Air Act	European Climate Change Programme	Paris Agreement	Paris Agreement
Ozone	Montreal Protocol on Substances that Deplete the Ozone			

issues fairly strict regulations, but the actual monitoring and enforcement is largely undertaken by local governments that have greater interest in economic growth. The environmental work of nongovernmental forces, such as lawyers, journalists, and nongovernmental organizations, is limited by government regulations. China's rapid economic expansion combined with the country's relaxed environmental oversight has caused a number of ecological problems. In response to public pressure, the national government has undertaken a number of measures to curb pollution in China and improve the country's environmental situation. **Table 3.1** summarizes major environmental laws and regulations in the US, European Union, China, and the United Nations for comparison. Some of the regulations will be discussed in more detail in subsequent chapters.

3.5 Environmental Justice

According to USEPA, **environmental justice** is the fair treatment and meaningful involvement of all people regardless of race, color, national origin, or income, with respect to the development, implementation, and enforcement of environmental laws, regulations, and policies. After observing the environmental policy making process and how environmental politics works, one can easily understand why minorities, low income citizens, and other disadvantaged or underrepresented groups are often on the short end of environmental issues. Environmental justice ensures that everyone enjoys the same degree of protection and have the same access to decision-making process and environmental policy making process as described above. Environmental justice issues are largely linked to the siting locally undesirable land uses such as highways, landfills, hazardous material storage facilities, or even new developments.

It can also be more insidious issues caused by other socioeconomic or political factors such as income inequality, property values, and political representation, therefore the solution to this issue is often difficult and controversial. Common solution to environmental justice is to compensate the disadvantaged communities impacted. Environmental justice issues could be tackled by the equal protection clause in many environmental laws. In the presidential order #12898, President Clinton ordered that all federal agencies should make environmental justice one of their missions.

In the United States, environmental justice issues are partly addressed by the NEPA process as described in Section 3.2.3. NEPA requires federal agencies to include minority and low-income populations in their NEPA-mandated environmental analysis. The president's Executive Order #12898 specifically requires that NEPA consider environmental effects on human health and economic and social effects, especially within minority and low-income populations, which are disproportionately impacted by environmental detriment.

Environmental justice could also be an international issue and is often tied to other socioeconomic issues. For example, China used to import a large amount of recyclable materials (plastic, paper products, electronic wastes) from other countries. Some of the recycling operations, especially those for electronic wastes, caused severe environmental impacts to some communities. The issue is complicated because it involves environmental responsibilities of the foreign waste-generating country and China's national or local environmental regulations.[14]

Further Reading

Carson, Rachel. Silent Spring. New York: Houghton Mifflin Books, 1962.
Hague, Rod and Martin Harrop. Comparative Government and Politics: An Introduction. London, UK: Macmillan International Higher Education, 2013.
Fiorino, Daniel J. Making Environmental Policy. Berkeley and Los Angeles, CA: University of California Press, 1995.
Rosenbaum, Walter A. Environmental Politics and Policy. Washington, DC: CQ Press, 2005.
King, Anthony. "Review of Agendas, Alternatives and Public Policies," *Journal of Public Policy*, 5, no. 2 (1985): 281–83.
Cohen, Michael D., James G. March, and Johan P. Olson. "A Garbage Can Model of Organizational Choice." *Administrative Science Quarterly*, 17 (1992): 1–25.

---

[14] Nowadays nearly all imports of such wastes/recyclables to China have stopped.

Hayden, F. Gregory. "Policymaking Network of the Iron-Triangle Subgovernment for Licensing Hazardous Waste Facilities." *Journal of Economic Issues*, 36, no. 2 (June 2002): 479.

Rowland, F. Sherwood. "Stratospheric Ozone Depletion." *Philosophical Transactions of the Royal Society B*, 361, no. 1469 (2006): 769–90.

Molina, Mario J. and F. Sherwood Rowland. "Chlorine Atom Catalysed Destruction of Ozone." *Nature*, 249 (1974): 810–12.

Byron, P.G., 2015. Legal opinion on the court case: RB Jai Alai, LLC v. Secretary of The Florida Department of Transportation, 112 F.Supp.3d 1301, 1307–1308 (M. D. Fla. 2015).

Malone, Linda A. Emanuel Law Outlines: Environmental Law. New York: Aspen Publishers, 2007.

Web Resources

USEPA website: https://www.epa.gov/
European Environmental Agency website: https://www.eea.europa.eu/
CalEPA website: https://calepa.ca.gov/
Natural Resources Defence Council (NRDC): https://www.nrdc.org/
USEPA Organization Chart: https://www.epa.gov/sites/production/files/2020-01/documents/usepa_orgchart_11x17.pdf. Access date: April 23, 2020
European Environmental Agency organization chart: https://www.eea.europa.eu/about-us/who/staff/chart
https://www.senate.gov/reference/reference_index_subjects/Laws_and_Regulations_vrd.htm (about differences between laws and regulations)
President Clinton Executive Order #12898 on Environmental Justice: https://www.epa.gov/fedfac/epa-insight-policy-paper-executive-order-12898-environmental-justice#memo1

Questions and Exercises

1. List a few possible reasons why an environmental policy could have negative environmental impacts.
2. Analyze a major project that you know and list at least five environmental impacts that it may have.
3. Use the web resource, compare the similarities and difference between USEPA and EEA.

4. One of the most influential environmental groups is the Natural Resources Defence Council (NRDC—please visit its website: https://www.nrdc.org/). List at least three main reason why NRDC is influential and effective.
5. For a disadvantaged community (e.g., minority; low-income; tribe) that is established AFTER known possible environmental pollution sources have been in place, do they have a valid argument on the environmental justice issue? Why or why not?
6. Why is it more challenging for a disadvantaged community to participating in environmental policy making process?
7. Visit the website of your home state's environmental agency and see how you can participate in environmental policy making processes on issues you are concerned about.

Chapter 4

Air Pollution and Greenhouse Gases

Learning Outcomes	76
Key Concepts	76
4.1 The Earth's Atmosphere	76
4.1.1 Overview and History	76
4.1.2 Structure of the Earth's Atmosphere	77
4.2 Air Pollution	78
4.2.1 Overview	78
4.2.2 Sources of Air Pollution	78
4.2.3 Main Air Pollutants	80
4.3 Air Quality Issues and Regulations	84
4.3.1 Air Pollution Regulations in the United States	84
4.3.2 Air Pollution Regulations in China	86
4.3.3 Air Quality Issues and Regulations in Europe and Beyond	88
4.4 Engineering Measures for Air Pollution Control	91
4.4.1 Overview	91
4.4.2 Vehicle Emission Control	91
4.4.3 Control of Other Air Pollutants	93
4.5 Greenhouse Effect and Management	94
4.5.1 Greenhouse Effect	94
4.5.2 Other Greenhouse Gases	97
4.5.3 Geoengineering Options for Atmospheric Carbon Dioxide	97
4.6 Other Air Quality Issues	98
4.6.1 Indoor Air Pollution	98
4.6.2 Ozone Depletion	99
4.6.3 Global Deoxygenation	99
4.6.4 Long-Range and Global Transport of Air Pollutants	100
Further Reading	100
Web Resources	101
Questions and Exercises	102

Learning Outcomes

- Knowledge about the Earth's atmosphere, including its structure and evolution over the geological time scale
- Knowledge of the type and health impacts of main air pollutants
- Understanding of the mechanisms of air pollution
- Knowledge of main technology and nonengineering control measures of common air pollutants
- Knowledge of the origin and structure of Clean Air Act
- Grasp of the nature of greenhouse effect and potential options to manage it
- Ability to list other major greenhouse gases

Key Concepts

Criteria Air Pollutants; Smog; Clean Air Act; Air Quality Index; PM2.5; greenhouse gases (GHGs); greenhouse effect; geoengineering; catalytic converter; photochemical smog; acid rain; runaway global warming/greenhouse effect; albedo; internal combustion engine (ICE)

We are immersed in and breathing the ambient air around us. Unlike polluted water and toxic substances, which we can move away to avoid, we cannot easily escape polluted air. Any pollutants in the air can also get into our lungs and sometimes blood streams directly, so even low levels of air pollutants could cause significant health impacts. As a common resource shared by everyone, air quality can most easily suffer from the "Tragedy of the Commons" and its impact can be wide-ranging, sometimes globally because of the high mobility of air masses. If we consider **greenhouse gases (GHGs)** as air pollutants, the air quality issue becomes an issue that puts the wellbeing and sustainability of our planet Earth on the line.

4.1 The Earth's Atmosphere

4.1.1 Overview and History

As discussed in Chapter 1, the Earth's atmosphere is formed mostly from the degassing of the interior of the Earth. The favorable size of the Earth and its optimal distance from the sun allow it to keep the atmosphere without being "blown away" by solar winds, such as what happened to Mercury and Venus. The Earth is also blessed with water, and water vapor and carbon dioxide provided enough **greenhouse effect** to keep the young Earth at favorable temperature to eventually allow life to emerge and thrive. Early plant species changed the atmospheric composition over time, most notably during the Great Oxygenation Event (see Chapters 1 and 2) around 2 to 3 billion years ago that changed the atmosphere from a toxic mixture of hydrogen, methane, ammonia, and carbon dioxide to mostly nitrogen and oxygen.

4.1.2 Structure of the Earth's Atmosphere

The Earth's atmosphere has a layered structure as shown in **Figure 4.1**. The layer closest to the ground surface is called troposphere (12 km; 'tropo' means 'turning'), followed by stratosphere (12–49 km; 'strato' means 'layer'), mesosphere (49–85 km; 'meso' means 'middle'), thermosphere (85–700 km; 'thermo' means 'heat'), and exosphere (>700 km to about 10,000 km, 'exo' means 'outside'). Troposphere is the layer closest to the ground and is where most of the weather events (as well as air pollution) take place. Stratosphere has a layered structure and is much more stable than troposphere. Jet-powered aircrafts use this layer for traveling. Also notable is the ozone layer at the base of the stratosphere, which protects the Earth from harmful ultraviolet (UV) light from the sun. Mesosphere is the coldest layer of the atmosphere (−85 °C) and is where meteors burn up. The layer between mesosphere and thermosphere is Karman Line (~100 km height), which is generally considered to be the boundary of the Earth's atmosphere. The lower part of the thermosphere contains the ionosphere, which is a layer ionized by solar radiation and is important for our radio communications. Auroras take place at the base of the thermosphere. For reference, the International Space Station orbits the Earth in the thermosphere, between 350 to 420 km.

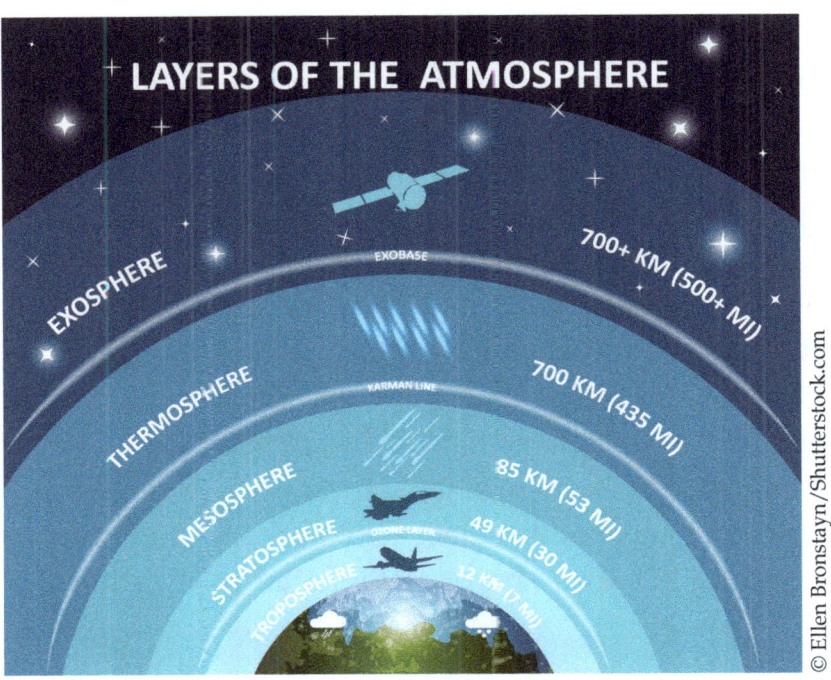

Figure 4.1 Layers of the atmosphere.

4.2 Air Pollution

4.2.1 Overview

The composition of the Earth's atmosphere is quite stable. For the average dry air, there is about 78% nitrogen and 21% oxygen. The rest is mostly argon, but there are many other gases with various amounts, such as carbon dioxide, ozone, radon, various air pollutants such as carbon monoxide, sulfur oxide, and so on. The ambient air could have various amounts of water vapor (up to 4–5%) as well.

Air pollution emerged as an environmental concern as early as at least 1300s in London when the English King Edward I banned coal burning in London due to extremely bad air quality. Air quality increasingly became an issue after the Industrial Revolution, when increasing amounts of fossil fuels are consumed to drive the world's economy and to support an ever-increasing global population. Most of the air pollutants, as will be discussed in more detail in the following sections of this chapter, are associated with combustion process either as part of industrial processes (furnace, turbines, etc.), or in vehicles/airplanes powered by **internal combustion engines** or jet engines.

4.2.2 Sources of Air Pollution

The combustion process could be quite complicated, involving a series of reactions with different radicals, so it will not be discussed here.[1] Due to impurities of nitrogen and sulfur in the fuel, as well as incomplete combustion, air pollutants such as nitrogen oxides, sulfur oxides, carbon monoxide, and fine particles can be produced. Nitrogen oxide could also be formed when nitrogen in the atmosphere, after being taken into the combustion chamber as part of the fuel-air mixture, is oxidized under high temperature and pressure. Carbon monoxide is formed due to incomplete combustion. Some hydrocarbons may be released due to incomplete combustion as well. Fine particles could be produced via the following mechanisms. For solid fuel such as coal, there will be ash. For internal combustion engines, turbines and jet engines, incomplete combustion could generate soot. Soot particles are agglomerates of small, roughly spherical particles mostly in the range of 0.01 to 0.05 micrometer (μm) of mostly carbon, but with some hydrogen as well. Soot is a serious air pollutant. Briefly, as shown in **Figure 4.2**, the following is a list of the most common air pollution sources:

- Fossil fuel power plants: as mentioned earlier, fossil fuel power plants burn fuels (oil, natural gas, coal) to produce electricity and are often the largest source of air pollutants, including SO_2, NO_2, particulate matter (PM), among others.

---

[1] A good reference could be Flagan and Seinfeld (1988).

Figure 4.2 Major sources of air pollution.

- Motor vehicles that use internal combustion engine, as mentioned earlier, can produce many air pollutants including PM (due to the soot formation), volatile organic compound (VOC, due to incomplete combustion), NO_2, CO, Pb (mainly when lead-based gasoline is used), and others. **Figure 4.3** shows the composition of the exhaust of a typical internal combustion engine.
- Factories and machinery could produce various air pollutants depending on their operations. Many gas-powered machinery produces similar air pollutants as motor vehicles. Some industries such as mining operations could produce PM that is particularly harmful.

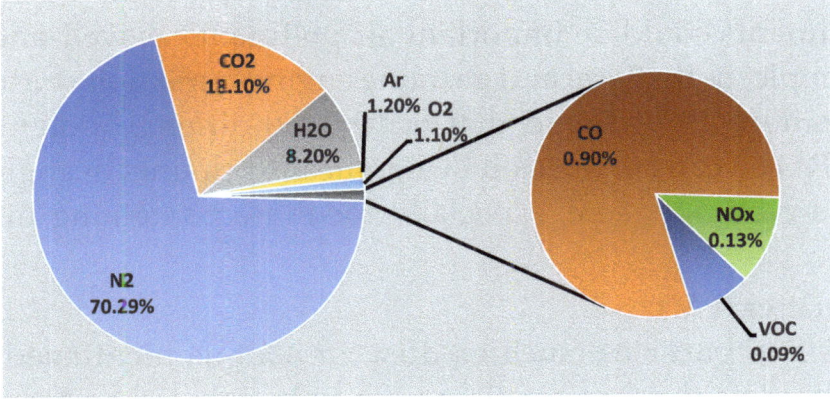

Figure 4.3 Mean composition of the exhaust gas from a typical internal combustion engine without catalytic converter (Abdel-Rahman, 1998).

Source: Created by Jian Peng

- Household or building heating could produce carbon monoxide and PM. For example, in northern China, coal-based household heating has been determined to be the chief sources of air pollution in winter.
- Wildfires can produce large amounts of PM and other air pollutants.
- Animals, especially domestic animals such as cattle, could produce large amounts of methane. Even though methane is not a criteria air pollutant, it is a major greenhouse gas. Cattle has been found to be a major contributor (37%) of human-sourced methane, which significantly contributes to global warming.

> **Knowledge Box**
>
> **Soot**—Soot is the small carbonaceous particle produced in the combustion of gaseous fuels and from the volatilized components of liquid or solid fuels. It is what causes the "black smoke" from cars, trucks. If you place a metal piece in the flame of a candle, you can collect some soot as well. Soot particles are agglomerates of small, spherical particles that has 5 to 10 layers of mostly carbon crystals that are similar to graphite. The chemical composition of soot particle is roughly C_8H. Soot is formed in flames following a complicated pyrosis of fuel molecules, with the content of hydrogen decreasing toward the end of combustion. Therefore, complete combustion will generate little or no soot. Due to its small size (0.01–0.05 μm), soot could become a significant component of PM and cause significant air quality impairments.

4.2.3 Main Air Pollutants

The most common air pollutants include PM, ground-level ozone (O_3), nitrogen dioxide (NO_2), sulfur dioxide (SO_2), carbon monoxide, lead, VOCs, and mercury. Some other pollutants could be important air pollutants as well under certain conditions. For example, both diesel and gasoline engines emission studies by the World Health Organization (1989) found that nearly 100 different polynuclear aromatic hydrocarbons (PAHs) can be detected in significant amounts in engine exhaust, and many of them are carcinogenic, especially for those containing nitrogen, such as nitroarenes.

- **Particulate matter (PM):**
  PM contains fine particles (such as dust, or soot as mentioned previously) or tiny water droplets that are small enough to be inhaled and absorbed to cause serious health problems. PM less than 10 μm (PM10) or 2.5 μm (**PM2.5**) in diameter pose the greatest problems, because they can get deep into the lungs, and

some may even get into the bloodstream. These particles come in many sizes and shapes and can be made up of many different chemicals. Some are emitted directly from a source, such as vehicles, smokestacks, power plants, construction sites, roads, or fires. Many particles and other air pollutants are secondary pollutants, i.e., they form in the atmosphere as a result of complex reactions of chemicals such as sulfur dioxide and nitrogen oxides, as will be discussed later in this section.

Knowledge Box

Smog—Smog is a portmanteau term based on "smoke" and "fog". It is a type of intense air pollution visible to the naked eye. Smog is composed of fine PM, nitrogen oxides, sulfur oxides, ozone, water droplets, and other materials. Fossil fuel burning from power plants, vehicles, factories, wild and man-made fires are the main sources of smog. Smog can often be formed by secondary reaction among the pollutants in the air through photochemical reactions (i.e., chemical reactions facilitated or triggered by sunlight), with ground level ozone being the main driver of the reactions. Secondary smog gained its notoriety in Los Angeles in the 1980s, when **photochemical smog** created an environmental crisis, see Figure 4.4. A contributing factor that makes conditions much worse for many major cities is a phenomenon called atmospheric inversion, where the air temperature increases and density decreases with elevation, making the air very stable and trapping the air pollutant over the cities.

Knowledge Box

PM2.5—PM2.5 are particulate matter 2.5 μm in diameter or smaller, and can only be seen under a microscope. For comparison, common human hair has a diameter of 50 to 70 μm, and a grain of fine beach sand has a diameter of about 90 μm. PM2.5 are produced from all types of combustion, including motor vehicles, power plants, residential wood burning, forest fires, agricultural burning, and some industrial processes. PM2.5 could also be generated by wind over dry land surface where there is no or little vegetation cover. In some cases, such dust storms or sand storms could cause health hazards. For example, Owens Valley (see Figure 6.13) and Salton Sea area in California, USA, have such issues. In both cases, active dust control measures have to be implemented. Beijing, China, has experienced many sand storms recently.

- **Ground level ozone:**
  Ozone in high altitude such as in the stratosphere is beneficial to us by shielding the harmful UV light (Section 4.1). However, at ground level, ozone is quite harmful to human health because it is a powerful oxidant that can irritate airways and cause inflammation and other damages to the respiratory system. Here ozone is not emitted directly into the air, but is created by chemical reactions between oxides of nitrogen and volatile organic compounds in the presence of sunlight. Breathing ozone can trigger a variety of health problems, particularly for children, the elderly, and people with lung diseases such as asthma. Ground level ozone can also have harmful effects on sensitive vegetation and ecosystems.

- **Lead:**
  Once abundant in the ambient air due to leaded gasoline, the concentration of lead in ambient air has dropped significantly, especially in the countries where leaded gasoline has been phased out. Once taken into the body through breathing, lead distributes throughout the body in the blood and is accumulated in the bones. Lead can adversely affect nervous system, kidney function, immune system, reproductive and developmental systems and cardiovascular system. Of these adverse effects, the impact on children and infants is of the greatest concern because it could cause developmental and behavioral problems such as learning deficits and low IQ. Lead can also affect the oxygen carrying capacity of the blood.

- **Carbon monoxide:**
  Carbon monoxide mostly comes from incomplete combustion. Breathing in carbon monoxide reduces the amount of oxygen that can be transported in the blood stream to critical organs like the heart and brain and can cause dizziness, confusion, unconsciousness and death. While very high levels of CO are only possible indoors, elevated ambient carbon monoxide could be a significant risk factor for people with some types of heart diseases that impair their ability to get oxygenated blood to their heart and brain.

- **Sulfur dioxide:**
  Sulfur dioxide comes mostly from burning of sulfur-containing fossil fuel, especially coals. Other sources of sulfur dioxide emissions include industrial processes, volcanoes, and engines that burn fuel with a high sulfur content. Short-term exposures to sulfur dioxide can harm the human respiratory system and make breathing difficult. Children, the elderly, and those who suffer from asthma are particularly sensitive to the effects of this air pollutant. Sulfur dioxide emissions could also lead to the formation of other sulfur oxides (SO_x). SO_x can react with other compounds in the atmosphere to form small particles and is an important cause of smog as discussed above and acid rain as discussed in the knowledge box.

> ### Knowledge Box
>
> **Acid Rain:** Acid rain is a phenomenon that the rain is exceedingly acidic (i.e., has very low pH value) due to air pollution. Natural rain is weakly acidic due to the dissolution of atmospheric carbon dioxide, which forms carbonic acid with water. However, in the area where the air is heavily polluted by sulfur oxides and nitrogen oxides from many sources, especially from power plants, strong acids such as sulfuric acid and nitric acid could be formed, rendering the rain more acidic. Acid rain will cause many environmental issues, such as corrosion of buildings and other exposed infrastructure; damage to plants and crops; and acidifying soils and natural waters.[2] In the United States, USEPA and other agencies established the National Atmospheric Deposition Program's National Trends Network for measurements of wet deposition. This program collects acid rain at more than 250 monitoring sites throughout the United States, Canada, Alaska, Hawaii and the U.S. Virgin Islands. The data shows that since 1990, significant emission reductions in sulfur oxides and nitrogen oxides have resulted in major decreases in acid rain nationwide.

- **Nitrogen dioxide:**
  Nitrogen dioxide is one of the nitrogen oxides (NO_x) produced during the combustion process when the nitrogen gas in the air reacts with oxygen under high temperature. Nitrogen dioxide can irritate airways in the human respiratory system and can cause or aggravate respiratory diseases, particularly asthma, leading to respiratory symptoms such as coughing, wheezing or difficulty breathing. Longer exposures to elevated concentrations of nitrogen dioxide may contribute to the development of chronic asthma and potentially increase susceptibility to respiratory infections. People with preexisting asthma, as well as children and the elderly are generally at greater risk. Recent studies showed that long-term exposure of nitrogen oxides may be associated with diabetes, poorer birth outcomes, premature mortality, and cancer. However, in most cases, nitrogen oxide as an air pollutant occurs with other air pollutants and it is uncertain whether the above health effects are independent from the effects of other traffic-related pollutants. NO_x also reacts with other chemicals in the air to form PM and ground level ozone, which are the main components of smog. It also interacts with water, oxygen, and other chemicals in the air to form nitric acid and causes/contributes to acid rain, as discussed in the Knowledge Box.

---

[2] As mentioned in Chapter 2, natural waters are mostly basic due to the buffering effect of carbonate ions. This buffering is quite beneficial because it ensures that the conditions in these waters are stable and protective of the aqueous ecosystem. Acid rain could break this balance, rendering these waters susceptible to small environmental changes and exposing the ecosystem to many potential harmful impacts.

> ### Knowledge Box
>
> **Nitrogen aerial deposition**—Atmospheric deposition of nitrogen could be an important source of nutrients to the surface waters. For example, research has found that atmospheric deposition accounts for as much as 30% in the Chesapeake Bay watershed in mid-Atlantic region of the United States, and nearly 40% of nitrogen input from the top eight tributaries of Chaohu Lake (China's fifth largest lake), Anhui Province, China. Due to the nonpoint source nature of atmospheric deposition of nitrogen, this poses a major cross-media management challenge to curb eutrophication issues in these two watersheds.

4.3 Air Quality Issues and Regulations

4.3.1 Air Pollution Regulations in the United States

Due to the environmental and health impacts of these air pollutants, United States and EU have both set ambient air limits for these pollutants. **Table 4.1** shows these limits, termed the National Ambient Air Quality Standards (NAAQS) for these pollutants set by USEPA. Primary standards provide public health protection, including protecting the health of "sensitive" populations such as asthmatics, children, and the elderly. Secondary Standards provide public welfare protection, including protection against decreased visibility and damage to animals, crops, vegetation, and buildings. They are usually not enforceable and more lenient. The averaging periods are also different, depending on the nature of each air pollutant and the period during which significant health effects may result.

Despite these limits, there are many 'nonattainment areas' in the United States, especially for Southern California counties (including Los Angeles), where levels of all six **criteria air pollutants** exceeded the limits. Surprisingly, this condition is actually a drastic improvement over historical conditions. Around the turn of the 20th century, the air quality in the Los Angeles metro area kept deteriorating, prompting the City to adopt increasingly stringent measures. However, it was not enough to curb the worsening air pollution as a result of booming population, robust industry, and increasing number of vehicles. In 1947, Los Angeles established the nation's first air pollution control program. Extensive research was conducted and ozone was found to be one of the key culprits. Regulations on garbage burning, factories, and oil refineries improved air quality. Starting from around 1975, all cars were required to install **catalytic converters** and resulted in significant improvement of air quality. In 1976, the South Coast Air Quality Management District (AQMD) was established to coordinate air quality management throughout Southern California.

Table 4.1 USEPA National Ambient Air Quality Standards (NAAQS) for six **Criteria Pollutants**. For detailed information, refer to USEPA (2016)

Pollutant		Primary/Secondary	Averaging Time	Level	Form
Carbon Monoxide (CO)		primary	8 hours	9 ppm	Not to be exceeded more than once per year
			1 hour	35 ppm	
Lead (Pb)		primary and secondary	Rolling 3 month average	0.15 µg/m³	Not to be exceeded
Nitrogen Dioxide (NO_2)		primary	1 hour	100 ppb	98th percentile of 1-hour daily maximum concentrations, averaged over 3 years
		primary and secondary	1 year	53 ppb	Annual Mean
Ozone (O_3)		primary and secondary	8 hours	0.070 ppm	Annual fourth-highest daily maximum 3-hour concentration, averaged over 3 years
Particle Pollution (PM)	PM2.5	primary	1 year	12.0 µg/m³	annual mean, averaged over 3 years
		secondary	1 year	15.0 µg/m³	annual mean, averaged over 3 years
		primary and secondary	24 hours	35 µg/m³	98th percentile, averaged over 3 years
	PM10	primary and secondary	24 hours	150 µg/m³	Not to be exceeded more than once per year on average over 3 years
Sulfur Dioxide (SO_2)		primary	1 hour	75 ppb	99th percentile of 1-hour daily maximum concentrations, averaged over 3 years
		secondary	3 hours	0.5 ppm	Not to be exceeded more than once per year

Note: There are numerous footnotes to this table. For more information, please refer to the website: https://www.epa.gov/criteria-air-pollutants/naaqs-table

The issue for Los Angeles is different from many other areas, especially after the sources other than automobiles have been controlled. Studies found that the main culprit for Los Angeles' air pollution issue is the photochemical smog, which is mostly the secondary pollution caused by ground level ozone, as well as hydrocarbons and nitrogen oxides. **Figure 4.4** shows one of such smoggy days in Los Angeles in the 1980s.

Knowledge Box

Photochemical Smog: photochemical smog is more common in sunny summer days when the sunlight triggers a series of reactions that produce ozone. There are both primary pollutants (i.e., those emitted directly from the sources, such as SO_x, NO_x, volatile organic compounds, carbon monoxide, PM, etc.) and secondary pollutants (i.e., those derived from primary pollutants, mostly ozone, as well as secondary PM). The chemical reactions include many complicated free radicals. This mechanism was discovered by Arie Haagen-Smit and Arnold Beckman.

Figure 4.4 A typical smoggy day in Los Angeles in the 1980s. Nowadays, the smog condition has improved significantly as a result of coordinated effort.

Los Angeles provides an excellent case study on the air quality management in the United States, perhaps in the world as well. There are still many other areas in the United States where one or more air pollutants exceeded the NAAQS, and these areas are called 'nonattainment' areas'. They are concentrated in the northeastern United States. To inform the public about air quality conditions, **air quality index** (AQI)[3] is often used. Nowadays, real-time AQI numbers are readily available and are the most important indicator of overall air quality. Table 4.2 shows how AQI can be interpreted.

4.3.2 Air Pollution Regulations in China

Since China opened its economy in 1978, its gross domestic product (GDP) has increased more than 250-fold. Along with this dramatic economic development came with many environmental quality issues, and deterioration of air quality, especially in industrialized areas such as northeast and southeast regions, has been significant and caused national and international attention (see Figure 4.5 as an example). In 1987, when air pollution first started to become an issue, China passed its first '**Clean Air Act**'. The law was amended in 1995 and 2000 before it was significantly amended again in 2018 to include improvement in monitoring, source control, enforcement, and integrated management. As of 2020, air quality remains a significant environmental challenge in China. However, the situation has seen stabilization and steady improvement since 2013.

---

[3] https://aqicn.org/map/world/#@g/2.0574/7.9102/2z

Table 4.2 Air Quality Index Table. The numbers are scaled against the corresponding NAAQS numbers, and a score of 100 indicates that the air quality meets the standard. In the United States, AQI is calculated using ozone, PM, carbon monoxide, and sulfur dioxide.

AQI	Air Pollution Level	Health Implications	Cautionary Statement (for PM2.5)
0 - 50	Good	Air quality is considered satisfactory, and air pollution poses little or no risk	None
51 -100	Moderate	Air quality is acceptable; however, for some pollutants there may be a moderate health concern for a very small number of people who are unusually sensitive to air pollution.	Active children and adults, and people with respiratory disease, such as asthma, should limit prolonged outdoor exertion.
101-150	Unhealthy for Sensitive Groups	Members of sensitive groups may experience health effects. The general public is not likely to be affected.	Active children and adults, and people with respiratory disease, such as asthma, should limit prolonged outdoor exertion.
151-200	Unhealthy	Everyone may begin to experience health effects; members of sensitive groups may experience more serious health effects	Active children and adults, and people with respiratory disease, such as asthma, should avoid prolonged outdoor exertion; everyone else, especially children, should limit prolonged outdoor exertion
201-300	Very Unhealthy	Health warnings of emergency conditions. The entire population is more likely to be affected.	Active children and adults, and people with respiratory disease, such as asthma, should avoid all outdoor exertion; everyone else, especially children, should limit outdoor exertion.
300+	Hazardous	Health alert: everyone may experience more serious health effects	Everyone should avoid all outdoor exertion

Figure 4.5 The smog-shrouded Forbidden City, Beijing, China.

> **Knowledge Box**
>
> **Air quality challenges in Beijing, China**—Beijing, the capital of China as well as of many historical Chinese dynasties, is known for its long history and ancient landmarks such as the Forbidden City and the Great Wall. In recent years, unfortunately, it became known for its poor air quality. The drastically increased number of motor vehicles, together with coal-fired factories and power plants were the main culprit. In 2008, when the City hosted the Olympic Games, the poor air quality forced some athletes to wear face masks and caused some controversies. The government forcefully shut down most of the heavy polluters during the Olympics to improve air quality. After the Olympics, more sustainable and effective measures, such as traffic control (only a subset of vehicles are allowed in different weekdays); source control; moving some factories out of the city; and stronger enforcement have resulted in steady improvement since 2013–2014, when the air quality in Beijing reached a low point (Figure 4.6).

Figure 4.6 shows a daily photo log of Beijing's mostly smoggy skyline during 2013–2014 (top) and the annual average PM2.5 values from 2008 to 2019 (bottom). Clearly, the air quality has been improving after the period of 2009–2014, when the city suffered the worst air quality after a brief period of improvement in 2008.[4] Currently, the average air quality is 'moderate', compared to 'unhealthy' from 2009 to 2017. One of the contributing factors could be that in 2017, Beijing shut down the last of its four coal-fired power plants and transitioned to natural gas.

4.3.3 Air Quality Issues and Regulations in Europe and Beyond

EU has similar limits for the key air pollutants discussed above. **Table 4.3** lists the air quality standards for EU as well as by the World Health Organization (WHO) as reference. These standards are mostly similar to those in the National Ambient Air Quality Standard of the United States (Table 4.1). The differences include the absence of limits for lead, carbon monoxide, and sulfur oxides.

---

[4] In 2008, the U.S. Embassy in Beijing installed an air quality monitoring device and broadcast the data to the world via Twitter. This prompted the city government to take air quality issues seriously, and start to share its own data, which previously was confidential information. One of the contributors to this book, Dr. Cindy Lin, was with USEPA at that time and participated in the effort in Beijing.

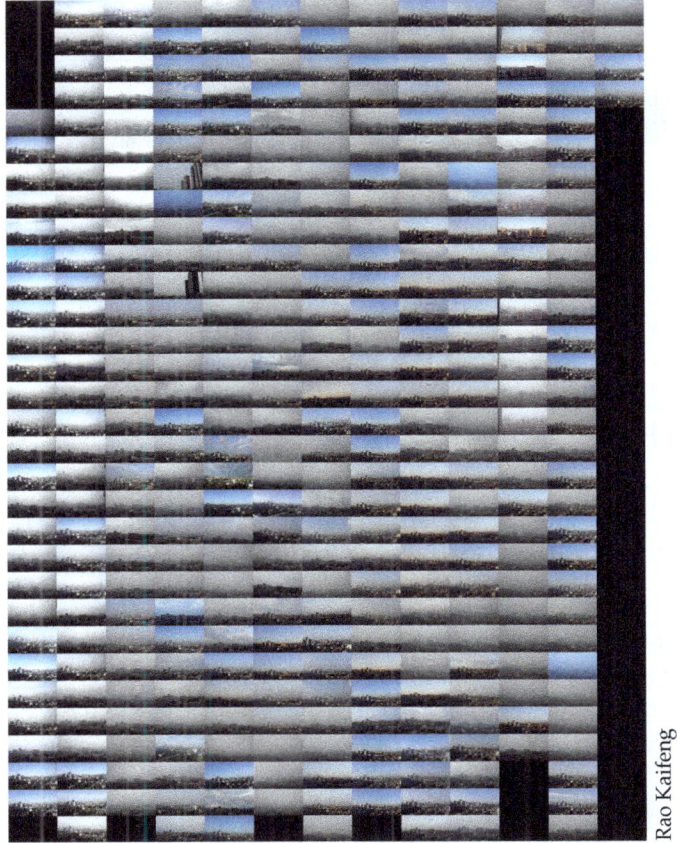

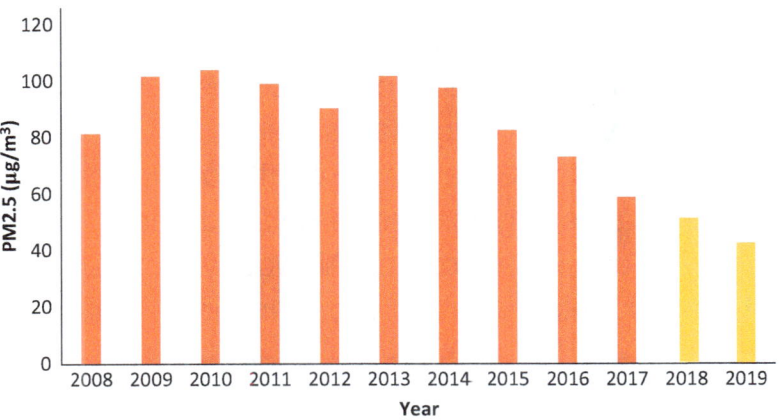

Figure 4.6 Top: daily photo log of Beijing's skyline from April 5, 2013 to April 4, 2014 (starting from upper left). During this period, Beijing's air was smoggy for more than 200 days. Bottom: Annual average PM2.5 (in μg/m³) in Beijing from 2008–2019, indicating a clear improvement since 2013–14. Red: unhealthy for sensitive groups. Orange: moderate.

Table 4.3 Air quality standards in the EU.

Pollutant	Averaging Period	EU Air Quality Directive		WHO Guidelines	
		Objective and legal nature and concentration	Comments	Concentration	Comments
$PM_{2.5}$	Hourly			25 µg/m³	99th percentile (3 days/year)
$PM_{2.5}$	Annual	Limit value, 25 µg/m³		10 µg/m³	
PM_{10}	Hourly	Limit value, 50 µg/m³	Not to be exceeded on more than 35 days per year	50 µg/m³	99th percentile (3 days/year)
PM_{10}	Annual	Limit value, 40 µg/m³		20 µg/m³	
O_3	Maximum daily 8-hour mean	Target value, 120 µg/m³	Not to be exceeded on more than 25 days per year, averaged over three years	100 µg/m³	
NO_2	Hourly	Limit value, 200 µg/m³	Not to be exceeded on more than 18 times a calendar year	200 µg/m³	
NO_2	Annual	Limit value, 40 µg/m³		40 µg/m³	

Source: EEA. Weblink: https://www.eea.europa.eu/data-and-maps/figures/air-quality-standards-under-the/table-1.eps/image_large

Note: Exceedances of upper and lower assessment thresholds shall be determined on the basis of concentrations during the previous five years where sufficient data are available. An assessment threshold shall be deemed to have been exceeded if it has been exceeded during at least three separate years out of those previous five years.

Knowledge Box

London air quality issues—The air quality issues in London, nicknamed 'The Capital of Fog', were well-known for centuries and were first caused by burning of soft coal for residential heating. Back in 1306, concerns over air pollution were sufficient for Edward I to (briefly) ban coal fires in London. Severe episodes of smog continued in the 19th and 20th centuries, mainly in the winter, and were nicknamed "pea-soupers". The Great Smog of 1952 darkened the streets of London and killed more than 12,000 people, prompting the establishment of UK's Clean Air Act in 1956. It was later found that the Great Smog death toll was caused by high level of sulfuric acid in the air caused by excessive sulfur dioxide, and the reaction was facilitated by high levels of nitrogen oxides. Nowadays, the air quality in London is much improved, but traffic-related air pollution remains significant.

Around the world, air quality challenges are not unique to major cities such as Los Angeles, Beijing, and London. **Figure 4.7** shows the air quality as measured by PM2.5 for the largest cities around the world. Of these cities, Dhaka, Lagos, Mumbai, and Beijing have the highest average PM2.5, while other major cities in developed countries such as Tokyo, New York, Los Angeles, and London boast good air quality.

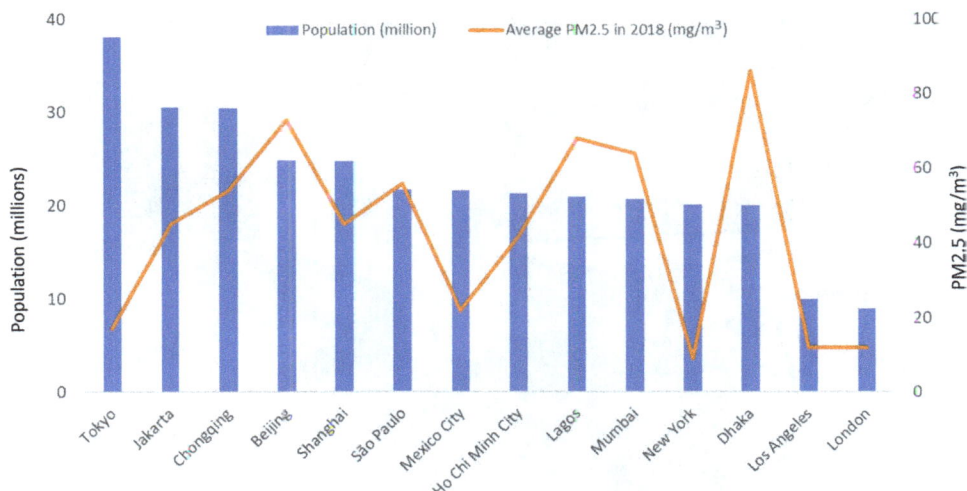

Figure 4.7 Megacities and air quality (PM2.5 data are 2018 annual average from the Ambient Air Quality Database, WHO, April 2018, retrieved 1/2/2020. Dhaka PM2.5 data were from the World Bank, 2019; Sao Paulo PM2.5 data were from the Berkeley Lab, http://berkeleyearth.lbl.gov/air-quality/local/Brazil/Sao_Paulo).

Source: Created by Jian Peng

4.4 Engineering Measures for Air Pollution Control

4.4.1 Overview

Many engineering solutions have been devised to tackle the above air pollutants. As shown in Figure 4.7, many of the world's largest cities are still faced with significant air quality challenges. At the same time, success stories in Tokyo, Los Angeles, New York, and London offered hope that with proper technology and management, it is possible to manage air quality in major cities. For the **criterion air pollutants**, USEPA published a guideline on control measures listing hundreds of different options for treating various air pollutants from a wide range of pollution sources (USEPA, 2013). Many such control measures are nonstructural and straightforward. For example, USEPA published new guidelines for many household utilities, including water heaters and furnaces that use new technology (including those described below) to curb the emission of air pollutants. For the general public, simply replacing old water heaters with new ones will reduce significant amount of nitrogen oxides. USEPA also banned open burning of garbage and other materials, including agricultural stubs, during the days when ozone levels exceed the threshold.

4.4.2 Vehicle Emission Control

USEPA requires all vehicles to conduct periodic "smog check" (**Figure 4.8**) to ensure that the vehicles meet the emission standard. Those that fail the smog test have to be repaired before they can be driven legally. Different states may have different standards,

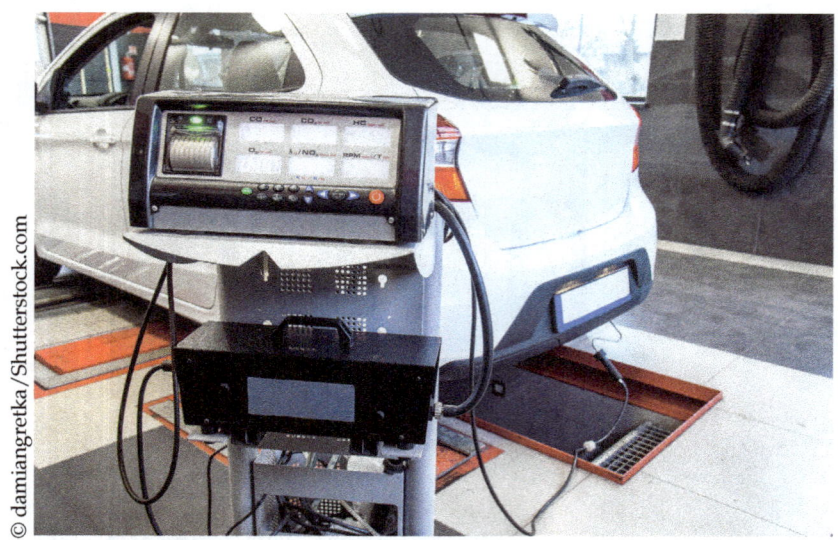

Figure 4.8 Smog check is being conducted at a station, where parameters such as hydrocarbons, carbon monoxide, carbon dioxide, oxygen, and nitrogen oxides are measured in the exhaust pipe.

but smog checks have been shown to be a very effective way of controlling vehicular pollution. China adopted similar practice since the 2000s, and currently is implementing fuel quality and emission standards similar to those in Europe and the United States (Wang et al., 2019).

Another significant progress toward lower emission of vehicle-related air pollutants is significant improvement in average fuel economy. As shown in **Figure 4.9**, from 1975 (when fuel economy started to be monitored systematically) to 2018, fuel economy for all vehicles in the United States improved from about 13 to 25.1 miles per gallon in 2018, a nearly 100% improvement. This trend is expected to continue into the future. One key driver for this encouraging trend is USEPA's Greenhouse Gas (GHG) program, which will be discussed in the next section (Section 4.5). This is quite impressive because the horsepower of the engines also increased significantly, and the average size of the passenger vehicles did not change.[5] The positive environmental impact of controlling fuel efficiency goes beyond greenhouse gas reduction because high fuel efficiency results in at least the same proportion in reduction of other air pollutants as well. In fact, the reductions of these other air pollutants have been much more significant due to improved engine technology, more stringent requirements on fuel quality, as well as postcombustion treatment technologies as described below. For example, the allowable level of sulfur in fuels has been reduced by 75% (from 120 to 30 ppm) from 2004 to 2006 alone, and by 2010, another 50% reduction to 15 ppm was required.

---

[5] Most of the passenger cars became smaller. However, due to American's love for large sport utility vehicles, which saw an increase in market share, the average vehicle size did not change between 1975 and 2019.

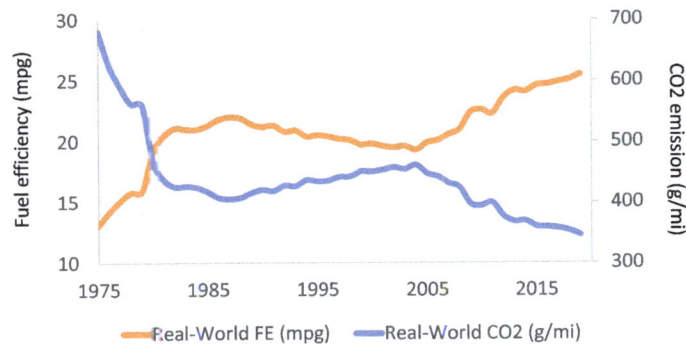

Figure 4.9 Fuel efficient trend in the United States from 1975 to 2019.
(Data source: USEPA; https://www.epa.gov/automotive-trends/download-automotive-trends-report).

4.4.3 Control of Other Air Pollutants

Carbon monoxide is a product of incomplete oxidation of fuel. To reduce carbon monoxide production, various oxygenated fuel additives have been invented and tested to improve the oxidation efficiency. Most notable additives include methanol, ethanol, and methyl-tert-butyl-ether (MTBE). However, MTBE was found to be an important pollutant to groundwater and drinking water sources as a result of leaking underground storage tanks (see Chapter 7). Even though MTBE's health effect is uncertain, it gives drinking water an unpleasant taste at very low levels. In 2000, USEPA drafted plans to phase out the use of MTBE nationwide by 2004, and many states banned MTBE before the deadline.

Lead came into air mostly because of the leaded gasoline (in the form of the antiknocking compound tetraethyllead or TEL) used until the 1980s, when it was banned due to the significant health impacts as well as its incompatibility with catalytic converters. Lead in paints that were manufactured before 1978 could also be a source of lead exposure. Thanks to these bans, overall lead levels in the environment have been decreasing steadily over the years. For example, the average level of lead in the blood of the children in the United States decreased from 12.8 to 2.8 microgram per liter between 1976 and 1991 (Pircle et al., 1994), owing mostly to the reduction of lead in ambient air originated from the combustion of leaded gasoline.

NO_2 emitted by vehicles is produced by the reaction of oxygen and nitrogen in higher temperature and is difficult to prevent, but recent technological advancement has enabled catalytic converters to be installed in each vehicle to remove NO_2 from the vehicle exhaust. In fact, catalytic converters are responsible for over 90% reduction of NO_2. For nitrogen oxides produced by other combustion processes, as in water heaters and furnaces, two common technologies are used. The first technology is low NO_x burners, usually by lowering the temperature of one combustion zone and reducing the amount of oxygen available in another. The second technology is selective noncatalytic reduction, which uses chemical reduction of nitrogen oxides to convert them into nitrogen gas and water vapor.

> ### Knowledge Box
>
> **Catalytic Converter**—Catalytic converter is a device to clean up the exhaust from an internal combustion engine and remove air pollutants. They can perform both reduction as well as oxidation reactions, i.e., by reducing NO_x to N_2, and to oxidize CO and hydrocarbons to CO_2, by using trace metals from the platinum group, such as platinum, palladium, and rhodium. Sometimes other metals are alused. They use high surface area materials such as aluminum oxide or titanium dioxide as the base material and spread the trace metals over the surface. As required by USEPA, all vehicles produced after 1975 need to be equipped with catalytic converters. This is one of the most important measures that helped improve the air quality in urban areas by removing much of the air pollutants from vehicle tailpipes.

PM, including large particles/ash/soot, or smaller particles such as PM10 and PM2.5, can be controlled in different manner depending on the sources and nature of the particles. Dust from construction sites and industrial activities can be controlled by enhanced management actions, such as water spray. Soot from vehicle exhaust is due to incomplete combustion and can be mitigated by using oxygenated fuel additives, which reduce production of carbon monoxide as well. For stationary sources using coal as fuel, such as boilers and gas turbines, there are many options to physically remove PM from the flue gas.

For control of sulfur oxides, the first step is to avoid using high sulfur fuel such as coals and crude oil before combustion as mentioned before. This is especially important for vehicle-related sulfur oxides because catalytic converters cannot remove sulfur oxides effectively. For other sources of sulfur oxides such as power plants and furnaces, flue gas can be treated using a range of technologies such as alkali scrubbing, gas-phase oxidation, ammodia treatment, and other newer technologies.

4.5 Greenhouse Effect and Management

4.5.1 Greenhouse Effect

Carbon dioxide (CO_2) used to be considered as harmless as water. However, in 2009, USEPA declared that it is a pollutant and poses a danger to human health and welfare, and it must be regulated in similar ways as other air pollutants. Compared with other

GHGs including water vapor, ozone, methane, nitrous oxide (N_2O), hydrofluorocarbons, perfluorocarbons, and sulfur hexafluoride, carbon dioxide is abundant and is the most prominent of these gases even though it is not the most potent GHG (see next section). Due to its countless sources all around the globe, carbon dioxide requires a coordinated, systematic, and global effort to be regulated effectively.

The Earth's energy comes from both internal and external sources. The internal source is from the internal heat from upper mantle and the crust in the form of plate tectonics, volcanic activities, and geothermal activities. The internal energy is mostly the remnant heat from the time when the Earth was first formed, as well as the heat from radionuclide decays. It accounts for less than 2% of the energy on the Earth's surface, which receives external solar energy of about 3.5 to 7 kWh/m^2, enough to power all of the surface processes such as atmospheric and ocean circulation, and virtually all of the biosphere activities, which are based on the primary producers that capture the solar energy and provide food for higher trophic levels.

Greenhouse effect used to be a natural phenomenon that is beneficial to the Earth, as discussed in Chapter 1. Under normal conditions, sunlight heats up the Earth by imparting solar energy to its surface. The Earth reflects much of the energy back into space during daytime, mostly by clouds and other surfaces with high reflectivity (or **albedo**[6]), and releases the rest of the heat during the night as long-wave radiation. The GHGs in the Earth's atmosphere effectively trap the long-wave radiation within the atmosphere that would have been lost to space (**Figure 4.10**). This greenhouse effect has been beneficial to life on Earth by maintaining a comfortable temperature on the planet (otherwise it would have been too cold, especially at night). Without greenhouse effect, the Earth would have an average temperature of $-18°C$ and too cold for most of the life. However, when the levels of GHGs are too high, excessive heat will be trapped on Earth and cause the Earth's average temperature to rise. The Earth will reach a new equilibrium at a higher temperature since higher temperature results in more longwave radiation that releases the heat back to space.

Knowledge Box

Greenhouse Effect—The term "greenhouse effect" is a misnomer because a real greenhouse allows sunlight to penetrate through glass or clear plastic. The glass or clear plastic then physically traps the heat and prevents it from escaping. This is different from the greenhouse effect we discussed above, where the trapping of long-wave radiation is the main cause.

---

[6] Albedo is the ability of a surface to reflect sunlight. The darker a surface is, the lower albedo.

Figure 4.10. Greenhouse effect.

The Earth's average temperature has increased about 1.5 °C in the past century. While seemingly insignificant, this has caused profound global impacts. Further increase will not only exacerbate the problem, it may also trigger a phenomenon called "runaway greenhouse effect" (Figure 4.11), where increased global temperature reduce the snow cover, raise the sea level (both will result more heat absorption due to reduced albedo), driving more carbon dioxide (due to lower solubility in warmer oceans) and water vapor (itself is a GHG) from the ocean to the atmosphere. This will form a positive feedback loop that may accelerate uncontrollably, posing existential threat to mankind and natural ecosystem. More discussions on sustainability issues will be presented in Chapter 12.

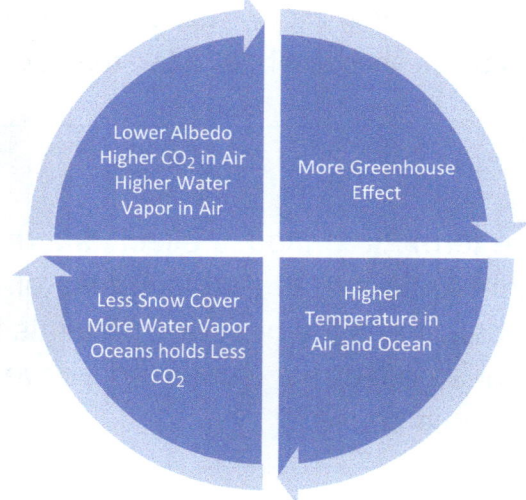

Figure 4.11 Runaway global warming feedback loop.
Source: Created by Jian Peng

> ### Knowledge Box
>
> **Runaway greenhouse effect on Venus**—Venus is much hotter than the Earth because it is closer to the sun and because of the so-called runaway greenhouse effect. It has an extremely dense atmosphere consisting of 96% carbon dioxide. Its surface temperature is 462°C. Venus may have had water oceans, but they would have boiled off under such a high temperature.

4.5.2 Other Greenhouse Gases

Of the GHGs mentioned in Section 4.5.1, not all GHGs contribute equally to the greenhouse effect, with water vapor contributing the most (60%), followed by carbon dioxide (26%), ozone (8%), methane, and other gases. However, water vapor level in the air is largely stable as a whole, but nearly all other GHGs have increased. Carbon dioxide and ozone have each increased by 40%; methane has increased more than 150%, and nitrous oxide by 20% due mostly to human activities.

> ### Knowledge Box
>
> **Methane and Cows:** It turns out that cattle are a significant source to the global methane production that may have contributed to the global warming significantly, so much so that scientists (Wallace et al., 2019) are looking for ways to deal with this issue with potentially a bioengineering method. In a 1000-cow study across four European countries, Wallace et al. (2019) identified a 39-species gastrointestinal microbial community that are linked to methane emissions. This finding could enable genetic manipulation to potentially reduce methane production without affecting other functions such as milk production.

4.5.3 Geoengineering Options for Atmospheric Carbon Dioxide

Unless and until humans can rely on energy sources other than fossil fuels, elevated and increasing level of carbon dioxide will stay in the atmosphere for a prolonged time, causing worsening problems in global warming and climate change. Removing carbon dioxide from the air is nontrivial because of its low concentration (~400 parts per million). However, there are some **geoengineering** options. For example, we can use more renewable biofuels such as ethanol made from corn, which has no "new carbon" (because corn extract carbon dioxide from the atmosphere). We could also remove carbon from carbon cycle by increasing our use of biochar, a type of charcoal that can be produced via pyrolysis process using common plant or plant-based materials. Since biochar is quite stable, it can be buried and used as an effective carbon sink. Enhanced weathering could be used to capture carbon dioxide from the air by exposing and facilitating in situ carbonation of silicates such as olivine, limestone, silicates, or calcium hydroxide (see Chapter 2). Carbon dioxide could also be removed from the air directly by carbon capture at places where it is generated in large amounts and high concentration, such as power plants. The carbon dioxide in the flue gas could be captured, concentrated, and disposed of using injection wells to store the gas deep in the geological formation. Using pure oxygen for combustion will increase the effectiveness of this method so carbon dioxide is the overwhelming product of combustion (instead of nitrogen). Lastly, we could use ocean fertilization to remove carbon dioxide from the surface ocean and atmosphere by adding nutrients (such as nitrogen, iron, etc.) to open ocean where the primary productivities are limited due to lack of these limiting nutrients. This oligotrophic condition accounts for over 90% of the global ocean. By fertilization, the primary productivities of the affected area will see a drastic increase, extracting carbon dioxide from the air and bicarbonates from the surface ocean. When the primary producers die off, they bring the captured carbon to the deep ocean and sediment, lowering the carbon dioxide from the atmosphere.

4.6 Other Air Quality Issues

4.6.1 Indoor Air Pollution

Recently, many research has indicated that indoor air, i.e. the air within homes and other buildings, can be more polluted than the outdoor air even in the area where the air pollution is the most severe. Since most people spend 90% of their time indoors, indoor air quality has great significance to human health.

There are many types of indoor air pollutants. One major type is biologic pollutants including bacteria, molds, viruses, dust mites, pollen, and so on. The recent COVID-19 global pandemic put indoor biologic pollution and human to human transmission of airborne virus in focus. Viruses for measles, chickenpox, and influenza can also easily transmitted in indoor air. Mold could be a significant indoor air pollutant because many molds produce mycotoxin and can cause allergic health effects or immune responses.

Indoor chemical air pollutants include carbon monoxide (gas stove), ozone (mostly from outside), tobacco smoke, VOCs (many sources including furnitures), radon (emanated from the ground), pesticides (indoor pest control), and so on. There could be many toxic materials in indoor air such as asbestos (from building materials), lead (from paints), arsenic (from wood treatment), among others.

Due to the health hazards of indoor air, building ventilation has become an important design consideration, and 'breathing building' is such a design principle that promotes natural ventilation and healthy indoor air.

4.6.2 Ozone Depletion

Ozone depletion refers to the depletion of the ozone in the stratosphere, which provides beneficial protection from the UV radiation from the sun. This term does not apply to ground-level ozone, which is both a significant pollutant and greenhouse gas. The main cause of ozone depletion and the ozone hole is halocarbon refrigerants, solvents, propellants and foam-blowing agents such as chlorofluorocarbons (CFCs) and other ozone-depleting substances. Once getting in the stratosphere through turbulent mixing, they release halogen atoms through photodissociation, which catalyze the breakdown of ozone (O_3) into oxygen (O_2). It was estimated that one CFC molecule can destroy up to 100,000 ozone molecules.

Ozone depletion and the ozone hole have generated worldwide concern over increased cancer risks[7] and other negative effects. The ozone layer prevents most harmful UV wavelengths of UV light from passing through the Earth's atmosphere. These wavelengths cause skin cancer, sunburn and cataracts, which were projected to increase dramatically as a result of thinning ozone, as well as harming plants and animals. These concerns led to the adoption of the Montreal Protocol in 1987, which bans ozone-depleting chemicals starting in 1989. Since then, ozone levels have stabilized and started to recover. It was estimated that the ozone layer will see full recovery by 2050. In 2019, NASA announced that the "ozone hole" was the smallest ever since it was first discovered in 1982. For this reason, the Montreal Protocol is considered the most successful international environmental agreement to date.

4.6.3 Global Deoxygenation

Global deoxygenation was discussed in a research paper by Martin et al. (2017). By looking at the oxygen level in the history of the Earth, including the Great Oxygenation Event itself, as well as more recent, accurate measurement of oxygen level on Earth since 1990, the authors discussed the causes of such a steady decline. Human activities such as deforestation, burning of fossil fuels (and associated rise in the global carbon dioxide levels and average temperature) may have caused or contributed to the global deoxygenation. It appears that the deoxygenation process will accelerate with time, potentially causing significant impact to human health.

The oxygen level in the atmosphere is controlled by several processes as part of its global biogeochemical cycle, such as production by primary producers; consumption by animals; consumption by fossil fuel burning; interaction with the ocean; and weathering of rocks. Since these processes are in delicate and dynamic equilibrium with the global biogeochemical cycling of other elements and materials (see Chapter 2), it is

---

[7] Skin cancer is one of the most common cancers in the world. One of the principal causes of skin cancer is ultraviolet radiation from the sun.

difficult to pinpoint the exact reason, or quantify the contribution from each possible process, how much they contribute to the decline of global oxygen level. Whatever the reason, the decrease in atmospheric oxygen level could potentially cause wide-ranging issues such as prohibition of photosynthesis and deoxygenation of the ocean. Without proper intervention, humans cannot survive beyond 3,600 years from the present time.

4.6.4 Long-Range and Global Transport of Air Pollutants

Global air circulation, as will be discussed in Chapter 11 (Figure 11.2), regulates the Earth's climate system. It also carries with it the air pollutants as discussed earlier, as well as many volatile pollutants such as polychlorinated biphenyls (PCBs), dichlorodibenzotrichloroethylene (DDT, both PCB and DDT are persistent organic pollutants, or POPs), mercury, etc. Due to the volatile nature of these pollutants, they can hitch a ride with air masses and join the global air circulation. With time, POPs from a heavily polluted area could become a regional issue, such as the situation about PCBs in the Great Lakes area in the USA–Canada border. Other examples include the large contribution of atmospheric mercury from China from coal combustion (Feng, 2005), as well as the transport of DDTs and PCBs from central America to the Great Lakes area. POPs have been detected in polar regions, including in the tissues of polar bears. As will be discussed in Chapter 5 and elsewhere, long-range transport of POPs often involved other environmental media such as soil and ocean as well (for example, see Figure 11.3 about the global ocean circulation). The 2011 Stockholm Convention on POPs called for elimination of 12 toxic compounds, many of which are pesticides. In addition, the treaty offers technical assistance to developing countries that need it to ratify the Convention.

Further Reading

Flagan, Richard C. and John H. Seinfeld. Fundamentals of Air Pollution Engineering. Englewood Cliffs, NJ: Prentice Hall, 1988.

Feng, Huiyun, Dongxia Wei, Xuan Li, Yuheng Zhao, and Zengliang Yu. Atmospheric nitrogen and phosphorus deposition in the Chaohu Lake Watershed, Anhui, China: Research and Considerations, in: *Proceedings of the Expert Consultation Summit on Integrated Environmental Management of Chaohu Lake*, February 4, 2018. Hefei, Anhui Province, China.

Abdel-Rahman, A.A. "On the Emission from Internal Combustion Engines: A Review." *International Journal of Energy Research*, 22 (1998), 483–513.

USEPA. Integrated Review Plan for the National Ambient Air Quality Standards for Particulate Matter, 2016.

Wallace, R.J. et al. "A Heritable Subset of the Core Rumen Microbiome Dictates Diary Cow Productivity an Demissions." *Science Advances*, 5 (2019): eaav8391

Linker et al, 2013, JWRA, on nitrogen atm deposition in Chesapeake Bay watershed.

USEPA. Integrated Science Assessment (ISA) For Oxides of Nitrogen—Health Criteria (Final Report, 2016). U.S. Environmental Protection Agency, Washington, DC, EPA/600/R-15/068, 2016.

Blasing T.J. Recent Greenhouse Gas Concentrations. Carbon Dioxide Information Analysis Center (CDIAC), Oak Ridge National Laboratory (ORNL). Oak Ridge, TN (United States), 2013, doi:10.3334/CDIAC/ATG.032.

Kiehl, J.T. and Kevin E. Trenberth. "Earth's Annual Global Mean Energy Budget." *Bulletin of the American Meteorological Society*, 78, no. 2 (February 1997): 197–208.

World Health Organization, 1989. Diesel and Gasoline Engine Exhausts and Some Nitroarenes, in: IARC Monographs on the Evaluation of Carcinogenic Risks to Humans, Vol. 46. IARC, Lyon, France. Weblink to the book: https://www.ncbi.nlm.nih.gov/books/NBK531303/pdf/Bookshelf_NBK531303.pdf

USEPA. 2013. Menu of Control Measures. Weblink: https://www.epa.gov/sites/production/files/2016-02/documents/menuofcontrolmeasures.pdf

USEPA, 2014, Air Quality Index—A guide to Air Quality and Your Health, February 2014, Research Triangle Park, North Carolina, USA. Weblink: https://www3.epa.gov/airnow/aqi_brochure_02_14.pdf.

European Environmental Agency (EEA), 2016. Air Quality Directive—Air Quality Standards. 2016. Weblink: https://www.eea.europa.eu/data-and-maps/figures/air-quality-standards-under-the/table-1.eps/image_large. Access date: April 26, 2020.

Pirkle, J.L. et al. 1994. "The Decline in Blood Lead Levels in the United States. The National Health and Nutrition Examination Surveys." *Journal of the American Medical Association*, 272, no. 4 (1994): 284–91.

Wang, Jin ey al. 2019. Vehicle Emission and Atmospheric Pollution in China: Problems, Progress, and Prospects. *PeerJ* 7 (2019): e6932, doi:10.7717/peerj.6932.

Martin, Daniel, Helen McKenna, and Valerie Livina. "The Human Physiological Impact of Global Deoxygenation." *Journal of Physiological Science*, 67, no. 1 (2017): 97–106.

USEPA, 2003, Long-Range Atmospheric Transport of Persistent Bioaccumulative Toxics from Central America. Technical Report 905-R-02-004. July 2003.

Jensen, Bjørn M. et al, 2015. "Anthropogenic Flank Attack on Polar Bears: Interacting Consequences of Climate Warming and Pollutant Exposure." *Frontiers in Ecology and Evolution*. Published: 24 February 2015, doi:10.3389/fevo.2015.00016.

Web Resources

The alarming decreasing trend observed by U.C. San Diego's Scripps O_2 Program: http://scrippso2.ucsd.edu/

The process of smog check in California: https://www.youtube.com/watch?v=F7eo308p_dQ

How California Air Resource Board (CARB) dealt with Los Angeles smog issues: https://www.youtube.com/watch?v=k2Ra8PRtXSU&t=147s

Drastic reduction of China's air pollution due to COVID-19: https://www.youtube.com/watch?v=cHSftdHskxs

Nonattainment Map of Counties in the United States: https://www3.epa.gov/airquality/greenbook/map/mapnpoll.pdf

USEPA Table of National Ambient Air Quality Standards: https://www.epa.gov/criteria-air-pollutants/naaqs-table

The positive impact of air quality in Los Angeles due to COVID-19: https://www.lamag.com/citythinkblog/air-quality-covid/

European Union's Air Quality: https://www.eea.europa.eu//publications/air-quality-in-europe-2019

New Design Trend-Buildings that 'breathe' like skin-https://www.usatoday.com/story/tech/2014/03/28/ozy-doris-sung-breathable-houses/7002915/

Questions and Exercises

1. Check your local air quality index by going to https://www.airnow.gov/aqi/aqi-basics/
2. By reading the material about the runaway global warming, can you think of the opposite process called "runaway global cooling"? How does that feedback loop work compare with that for the runaway global warming?
3. Elon Musk has an idea of building deep underground tunnels to allow trains to travel at very high speed in vacuum tubes. Could you think of the impact of this idea to reducing carbon dioxide emission?
4. The scenario of runaway global warming is rather alarming. What could be a few natural processes that could put a halt to such a positive feedback loop?
5. Albedo is an important controlling factor in the feedback loop for global warming. Can you think of several practicable ways to increase the albedo of the Earth as a way to control global warming?
6. Find a reference either on line or in the library about the trend in asthma incidences in a major city or a country of your choice. Use the corresponding AQI trend in the same period. Discuss whether there are causal relationship between the two.
7. Find a web resource and learn how internal combustion engines and catalytic converters work.

Chapter 5

Water Pollution

Learning Outcomes		104
Key Concepts		104
5.1	Water Quality Regulations	105
	5.1.1 The Clean Water Act	105
	5.1.2 The Water Framework Directive in the European Union	108
	5.1.3 Water Quality Regulations in China and India	109
5.2	Common Causes of Water Pollution	110
	5.2.1 Eutrophication	113
	5.2.2 Pathogen Contamination	115
	5.2.3 Toxic Contaminants	117
	5.2.4 Trash	124
	5.2.5 Sediment	126
5.3	Stormwater Issues	127
	5.3.1 Overview	127
	5.3.2 Urbanization and Stormwater Flow and Quality	128
	5.3.3 Stormwater Management and Low Impact Development	132
5.4	Water Treatment and Reuse	134
	5.4.1 Background and History of Water Treatment	134
	5.4.2 Overview of Water Treatment Process	135
	5.4.3 Water Recycling and Biosolid Management	137
5.5	Emerging Water Quality Issues	139
	5.5.1 Overview	139
	5.5.2 Microplastics	139
	5.5.3 Per- and Polyfluoroalkyl Substances (PFAS)	142
	5.5.4 Other Contaminants of Emerging Concern	144
	5.5.5 Long-Range Transport of Pollutants	145
5.6	Water Quality Monitoring	145
Further Reading		147
Web Resources		149
Questions and Exercises		150

Learning Outcomes

- Understanding of basic water chemistry in freshwater and marine systems
- Knowledge of common causes and sources of water pollution
- Knowledge about common water pollutants and their impacts, including nutrients, pathogens, toxicants, and other pollutants such as sediment, trash
- Knowledge of emerging pollutants and their sources
- Knowledge of management and treatment options for common and emerging pollutants
- Knowledge of basic water and wastewater treatment processes

Key Concepts

Clean Water Act; water quality standard; point source; nonpoint source; total maximum daily loads; stormwater; discharge permit; Water Framework Directives; eutrophication; harmful algal blooms; pathogen; toxicology; toxicity; trash; contaminants of emerging concern (CECs); microplastics; PFAS; endocrine disrupting compounds (EDCs); pharmaceuticals and personal care products (PPCPs)

In Chapter 1, we learned about the importance of water to the planet Earth and to all life on Earth. Without water, the Earth would be just like Mars or the moon: barren, lifeless, and desolate. Chapter 2 covered water cycles and other biogeochemical cycles in which water plays the important role as an effective erosional force, carrier, and solvent that distributes materials around the planet. With the advent of human dominance, however, water (especially freshwater) is increasingly used for human consumption for agriculture, industrial, and household uses. From a general thermodynamic standpoint, these human activities result in a high degree of organization and low entropy in human society. However, this condition came at the expense of waste heat and pollution (high entropy) in the environment.[1] In the process of human development, water gets contaminated by various pollutants such as fertilizers, pesticides, industrial chemicals, human pathogens, and so on. Much of the contamination is by design when water is used for cleaning or cooling, and proper treatment is needed before the water can be released back to the environment so it can still support

---

[1] The second rule of thermodynamics states that entropy (or randomness) will always increase. When we build highly organized things (crops, computer chips, buildings), we have to give out heat energy, waste materials, and sometimes, unfortunately, pollution of air and water, to balance out the entropy. To treat polluted water, more energy would be needed. So, it is always better (less costly and more environmentally friendly) to prevent the pollution before it happens.

beneficial uses. However, most of the pollutants we find in water are not meant to be there. Agricultural uses of fertilizers and pesticides should be controlled. Irrigation should consider water conservation and return flow be treated before releasing; **stormwater** in urban and suburban areas should be better managed through source control and green infrastructure. The fact that water pollution remains the most significant environmental issue tells us that not nearly enough has been done. Complicating the situation even further are the emerging contaminants and new chemicals whose environmental impact and behavior are unknown and treatment options are limited. In this chapter, we will look at these issues and explore the possible solutions.

5.1 Water Quality Regulations

5.1.1 The Clean Water Act

Chapter 3 mentioned the **Clean Water Act** (CWA) in the context of other environmental regulations in the United States and around the world. CWA is one of the most influential and successful environmental regulations in the history of the United States. Spurred largely by the environmental movements of the 1960s, the CWA was established in 1972 and was originally named Federal Water Pollution Control Act (FWPCA).[2] It was the first of a flurry of environmental regulations overseen by the United States Environmental Protection Agency (USEPA), which was created in 1970. For the CWA specifically, the triggers include the famous Cuyahoga River fire in Ohio (see Chapter 3), and Rachel Carson's book Silent Spring. The CWA was then significantly amended in 1977 and 1987, each time significantly improving its authority and effectiveness.

In a nutshell, the CWA can be viewed as a simple flow diagram as shown in **Figure 5.1**. As the first step, **water quality standards** are established (see the Knowledge Box) for a waterbody.[3] Periodically, the water quality is monitored and assessed against the standard. If the water quality meets the standard, no additional action is needed (the antidegradation policy still applies). If the water quality standard is not met, the water will be put on a list of impaired waters.[4] For these impaired waters, total maximum daily loads (TMDLs) will be developed to provide a pollutant diet, and a series of implementation measures will be taken to implement the TMDLs.

---

[2] The name "Clean Water Act" is actually named after the 1977 amendment of FWPCA.

[3] The CWA applies only to most surface waters (i.e., it does not apply to groundwater). The waters that are protected by CWA is called 'the Waters of the United States' or WOTUS, which itself is a complicated term. Without going into unnecessary details, the WOTUS could simply considered to include all important and major waters such as large rivers, lakes, wetlands, and coastal ocean.

[4] This list is called 303(d) list because it is specified by the section 303(d) of the CWA.

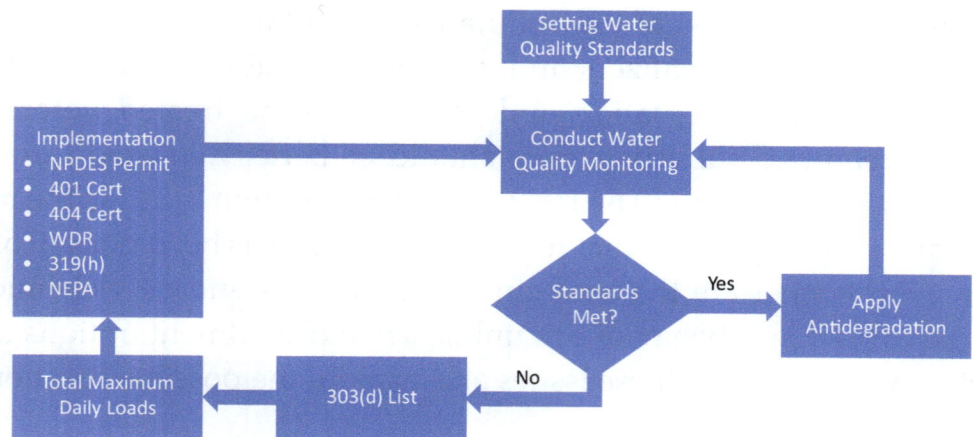

Figure 5.1 Clean Water Act flow chart.

Source: Created by Jian Peng

Note: WDR: Waste Discharge Requirements (a type of regulatory permit); NEPA: National Environmental Policy Act (a federal environmental review process; see Chapter 3).

These measures could include pollutant **discharge permits** (generally called National Pollutant Discharge Elimination System, or NPDES permits) for **point sources**; **nonpoint source** control measures; water quality certification for various projects; permits for dredge and fill of wetlands; and so on. Of these measures, NPDES permits are the most important and the most effective mechanism to control and protect water quality and are the core of the CWA itself.

> ## Knowledge Box
>
> **Water Quality Standard**—A Water Quality Standard for a certain pollutant under the CWA is more than just a number. It includes three components: designated uses, water quality criteria, and an antidegradation policy. The designated uses specify what beneficial uses such as swimming; habitat; drinking water; agriculture uses, etc., the water body supports. Water quality criterion is the number to protect such uses, and different uses require different levels of water quality (hence different criteria). For example, if the water is designated for drinking, there will be specific limits on toxic chemicals, fluorides, nitrate, hardness, iron, etc., that protect human health. If the water is for industrial use, pH, hardness and total dissolved solid will become important. If the water is used for farming, boron and alkalinity will become important. The antidegradation policy is a default requirement specifying that water quality for any waterbody should be protected against degradation even if it has met the applicable criteria. Water quality standard could also be narrative based on nuisance (e.g., odor or visual) or other descriptive measures (e.g., no toxic substance in toxic amounts).

Knowledge Box

Point source and nonpoint source—Under the CWA, a point source generally is a pollutant source that is more or less well defined, such as a pipe. A nonpoint source is usually more diffuse, such as discharges from a surface (for example, agricultural lands and open space) and atmospheric deposition. The Clean Water Act uses legally binding permits to regulate point sources. For nonpoint sources such as agricultural runoff and stormwater runoff, CWA uses voluntary or BMP-based measures rather than mandatory approaches. In 1987, the CWA was significantly amended to include urban stormwater runoff as point source discharges regulated by NPDES permits. Since then, urban stormwater management became one of the most active environmental fields.

Knowledge Box

BMP—Best management practice (BMP) can be measures, devices, guidelines, and activities for pollution control. Compared to the treatment devices and processes that are implemented to achieve specific pollutant load reduction or effluent water quality objectives, BMPs do not usually have strict, numeric performance goals. BMPs could include practices such as public education; runoff and erosion control at construction sites or in agricultural fields; housekeeping of an industrial site to minimize hazardous material spills or polluted stormwater runoff; stormwater low impact development practices such as bioswales and rain gardens, etc. BMP is an important concept for stormwater management and is the main tool for implementing stormwater regulations (see Section 5.3.3).

Knowledge Box

TMDL—Total maximum daily load (TMDL) is like a pollutant diet that specifies the upper limit of the amount of pollutant that a water body can receive and still meets water quality standards. Once such an upper limit is set, it will be allocated to point sources, nonpoint sources, and a margin of safety (usually 10% or 20% to ensure full protection under all conditions). A TMDL is useful because the point source allocations can then be implemented by NPDES permits.

The NPDES permit system is based on the principle that all point source discharges into the waters of the United States are prohibited unless they are allowed by an NPDES permit. The discharge limitations (i.e., the allowable concentrations or mass loadings of pollutants in the point source discharge) could be either based on technology or based on water quality conditions in the receiving water (i.e. the water body that receives the discharge). The technology-based effluent limitations (TBEL) are based on how much current treatment technology could remove pollutants from the effluent. USEPA has detailed regulations about these "demonstrated technologies" in its "NPDES Permit Writers' Manual." These types of permit usually have specific, numerical permit limitations for the effluent water quality associated with the treatment technology selected. They do not require that receiving water quality standards be met.

The other important type of effluent limitation is water quality-based effluent limitation (WQBEL), where the point source discharge is regulated based on the condition of the receiving water to which the effluent is discharged regardless of technical feasibility. When all point sources discharges have met their corresponding TBELs but the receiving water is still impaired, WQBELs may be imposed and they would be more stringent than the technology-based limitations. As shown in Figure 1, WQBELs will be based on TMDLs that are established for the receiving water(s).

The CWA also includes many other provisions that, taken together, protect the nation's waters to ensure their physical, chemical and biological integrity (or, in layman's terms, to make the waters fishable and swimmable). These provisions include funding programs for publicly owned treatment works (i.e., wastewater treatment plants), among others.

5.1.2 The Water Framework Directive in the European Union

EU's **Water Framework Directive**, established in 2015, is the main water regulation framework governing EU members on both freshwater (including groundwater) and marine waters. Compared to CWA, the Water Framework prescribes steps to reach common goals rather than adopting the more traditional limit value approach. Similar to CWA's goal of eliminating pollutant discharges to the nation's waters by 1985, the Directive aims for "good status" for all water bodies by 2027. However, currently 47% of more than 100,000 water bodies in the EU covered by the Directive are not achieving this goal.

In the Water Framework Directive, the assessment is fundamentally more broad-based than in the CWA. The Directive considers chemical, physical, and biological conditions in ways similar to the CWA, but its assessment scheme for biological conditions is more clearly defined and includes fish, benthic invertebrates, and aquatic flora. The hydromorphological conditions of a water body such as river bank structure, river continuity or substrate are also considered.

One distinct aspect of the Water Framework Directive is the introduction of River Basin Districts. The concept of River Basin is similar to watershed in the United States. In EU, large international rivers such as Danube and Rhine Rivers, whose watersheds span many countries, make such a policy a necessity. Representatives from different countries within the same watershed have to cooperate and work together for the management of the watershed in a holistic way. Under the Directives, these management schemes are called River Basin Management Plans, which are to be updated every 6 years.

5.1.3 Water Quality Regulations in China and India

In China, the Ministry of Ecology and Environment (MEE), formerly known as the Ministry of Environmental Protection, oversees environmental regulation and implementation for air, water, and soil/groundwater. The governing water pollution law is the Water Pollution Prevention and Control Act, which was first established in 1984 and revised in 1996 and 2008. During this time span, there was a general and severe degradation of water quality across the nation, as a result, the law was significantly revised and went into effect on January 1, 2018. The revised law mandates enhanced protection of drinking water sources, with severe penalties if the requirements are not met. The law also instructs the government to build sewage treatment and garbage disposal facilities in rural areas, and that standards be set on fertilizer and pesticide uses.

What is unique about this new law is that it brings the "river chief" system into being, with leading local officials assuming responsibility for addressing water pollution issues, including resource protection, shoreline management, pollution prevention and control, and ecological restoration. This system is necessary to mitigate the situation where many different agencies manage different aspects of water quality and water resources separately[5] and coordination and data-sharing have been lacking. River chiefs who achieve their goals will be rewarded, while those who fail in their responsibilities will be punished with fines and lose promotion opportunities. However, this linkage between water quality achievements and political career may have unintended, adverse impact on China's water management practices. In some cases, the schedule for assessment, permitting, planning, design, and implementation of major water projects have been cut to as short as two to three years (depending largely on when the official's term ends), making effective planning and design of optimal, long-term solutions difficult. Many short-term solutions such as diversions, in situ treatment, and dredging projects were hastily implemented with limited long-term benefits.

In India, the main water regulation is the Water (Prevention and Control of Pollution) Act of 1974, which provides for the prevention and control of water pollution to restore or maintain wholesomeness of water. In this law, the term "wholesomeness" means biological and physicochemical characteristics that harbors diverse aquatic flora and fauna, as well as various uses for humans. The implementation of the Water Act is carried out by the Central Pollution Control Board (CPCB), an agency under the Ministry of Environment, Forest and Climate Change. In addition to water, CPCB also manages air quality, solid waste, and other environmental issues. Relevant to CPCB's role is the Environment (Protection) Act of 1986 that added more functions and responsibilities to CPCB.

---

[5] In China this situation is termed "One Water Governed by Nine Dragons". For example, the same water body could be governed by agencies responsible for flood control; environmental quality; meteorology; land commission; construction and urban development; development and reformation; hydro power; agriculture; forestry; among others.

> ### Knowledge Box
>
> **The Ten Tenets for Water Protection of China:** On April 16, 2015, the Chinese Congress adopted the Action Plan for Prevention and Control of Water Pollution. Since there are ten major stipulations in the Action Plan, it is commonly referred to as "The Ten Tenets For Water." They include the following ten chapters: 1. Water pollution discharge control; 2. Pollution minimization and alternatives to manufacturing and energy industries; 3. Water conservation; 4. Scientific and technological support; 5. Market mechanism to improve efficiency; 6. Enforcement; 7. Water quality management; 8. Water security and safety; 9. Responsibilities and Accountability; 10. Public outreach and participation.

5.2 Common Causes of Water Pollution

There is no "pure" water in nature. As mentioned in Chapter 2, water is such an excellent solvent, it extracts chemicals from whatever it touches during its cycles through lithosphere, atmosphere, and biosphere. Water can also erode rocks and riverbeds and carry sediment and debris along to reshape the surface of the Earth. Hundreds of compounds and ions can be detected in every natural water sample, no matter how "pure" it is or appears. In a typical freshwater system, bicarbonate (HCO_3^-) is often the most abundant and the most important anion,[6] with carbonate, sulfate, chloride making up the balance of the rest. For cations, sodium, calcium, potassium are the most common (see the Knowledge Box). For marine water, which accounts for the bulk of the Earth's water, the composition is quite different due to much higher levels of ions with chloride and sodium dominating the ions (the Knowledge Box). In fact, water will not be stable in nature if it lacks ions that provide buffer against sudden changes in pH. This is important for the biota living in water.

> ### Knowledge Box
>
> **Natural Freshwater Chemistry:** Typical chemical composition of freshwater systems can be quite variable due to local geology, climate, hydrology, and many other factors, including environmental pollution. A typical river in North America would have the following major ions (ranked from high to low based on typical or average abundance): bicarbonate, calcium, sulfate, magnesium, sodium, chloride, silica, and potassium. Also refer to Section 2.3.2 on water cycle.

---

[6] An anion is an ion that carries negative charge(s). Similarly, a cation is an ion that carries positive charge(s).

> ## Knowledge Box
>
> **Natural Marine Water Chemistry**—The most abundant elements (other than water) in seawater, in the order of abundance, are chlorine, sodium, magnesium, sulfur, calcium, potassium, bromine, carbon, nitrogen, strontium, oxygen, boron, and silicon. These elements exist in marine water in various ions including chloride, carbonate/bicarbonate, and sulfate ions of metal elements. The abundance of these ions are ranked as follows: chloride, sodium, sulfate, magnesium, calcium, potassium, and bicarbonate. Note that seawater is not a simple concentration of freshwater over time. Due to precipitation of calcite and aragonite (the surface ocean is supersaturated with these two minerals) as well as other complex processes, the concentrations of these ions are not proportional to those in typical freshwater sources. Also see Section 2.3.2 on water cycle.

Unfortunately, water can often be contaminated by human activities. The pollution sources could be direct discharge of treated and untreated sewage and other waste water, or indirect discharges such as groundwater seepages and atmospheric fallout. Human activities also generate a wide range of pollutants on land, and nuisance flows and stormwater runoff could carry pollutants to receiving waters.

In addition to direct discharge of treated and untreated wastewater, which is historically important but increasingly less significant source of pollutants, there are many other, often more significant sources of pollutant discharges. The following is a list of common sources of pollution.

- Agricultural field irrigation water that could contain fertilizers, pesticides, and sediment
- Overirrigation by household that could contain fertilizers and pesticides
- Residential car wash or washing of driveways, sidewalks, roads, parking lots that could carry a wide range of pollutants such as dirt, oil and grease, surfactants, etc., to the receiving water
- Fire hydrant testing, swimming pool maintenance, air conditioning condensate, and other nuisance flow that could carry pollutants accumulated on roadways and curbs
- Accidental spill of sewage, fuels, chemicals or other liquids from pipes or containers
- Wind blowing land-based pollutants (such as trash and dust) into waterways
- Polluted air with nitrogen and sulfur compounds and other contaminants will result in both dry and wet deposition of these pollutants
- Illegal dumping of wastes

- Septic tanks that could overflow or seep into groundwater or surface water, carring pathogens and nutrients
- Chemicals and their degradation products such as salts, paints, preservatives used in the maintenance of roads, building, boats, and other infrastructure
- Stormwater washing off a wide range of pollutants (sediment, fertilizer, pesticide, and other toxic compounds, oil and grease, surfactants), and carry them to the receiving water in the stormwater runoff. Due to its significance and complexity, this topic will be discussed separately in Section 5.7.

As a result of these pollution sources, receiving waters often suffer from various impairments. The impairments could manifest in many different ways, such as **eutrophication; toxicity;** distressed biota or loss of biodiversity; odor; discoloration; turbidity; or other visual signs of degradation. In many cases, however, the impairments are determined by measured concentrations of various pollutants that exceed their corresponding water quality criteria. For example, USEPA sets national criteria on 126 toxic pollutants (termed "priority pollutants") and provides guidelines on other pollutants. States adopt these criteria and guidelines or develop their own (note: state water quality standards are subject to USEPA review and approval). If monitoring results for a waterbody indicate exceedances of the criterion for a pollutant, this waterbody will be deemed impaired and will be subject to a series of regulatory actions, including the development of **total maximum daily loads** (TMDLs) and subsequent requirements, as shown in Figure 5.1. **Figure 5.2** shows the most common causes of impairments in the United States. There are more than 60,000 TMDLs in the United States.

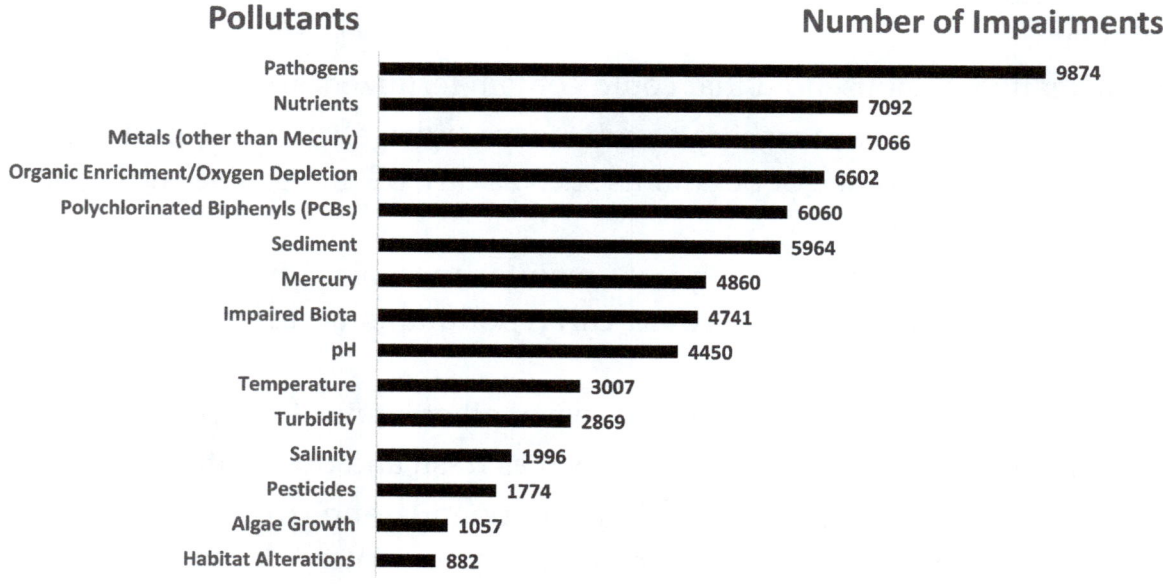

Figure 5.2 The leading causes of water quality impairments in the United States (2015).

5.2.1 Eutrophication

Eutrophication is a condition where oversupply of nutrients for aquatic primary producers spur excessive growth, mostly in the form of algal blooms. Such a condition brings about many undesirable consequences such as nuisance, depressed dissolved oxygen, **harmful algal blooms** (HABs), and even fish kill and other severe ecological impacts.

Nutrients (mostly nitrogen and phosphorus), the cause of eutrophication, could be from many different sources and could be natural or anthropogenic. The most common and important sources are human input from agricultural lands, where fertilizers, especially chemical fertilizers that do not bind to soils as well as organic fertilizers, are brought to creeks and other receiving waters as irrigation return flows (both surface and subsurface), or stormwater runoff. Runoff from other agriculture operations such as cattle farms and other animal husbandry operations and nurseries could be important sources of excess nutrients as well. Other sources of nutrients could come from the treated effluent of wastewater treatment plants; combined sewer overflows; atmospheric deposition (see Chapter 4); and urban storm runoff that carries nutrients from a variety of sources including fertilizer from landscaped areas, sediments, and other substances that contain nutrients. Stormwater itself could contain a significant amount of nutrients (see Section 5.3). In the United States, large water bodies such as the Great Lakes (especially Lake Erie), Chesapeake Bay and Gulf of Mexico all have various degree of eutrophication issues. In China, nearly all of its five largest lakes suffer severe eutrophication issues and algal blooms (**Figure 5.3**).

As discussed in Section 2.4.2 on nitrogen cycling, the global nitrogen balance has been perturbed profoundly by industrial nitrogen fixation for fertilizer production.

Figure 5.3 Left: algal bloom in Chaohu Lake, China's fifth largest lake, due to eutrophication caused by agriculture and urban development in its watershed of nearly 13,500 km². Right: Qiyu Creek, Fuyang, Anhui, China.

This artificially fixed nitrogen, together with production of phosphorus from mining, has greatly enhanced global agricultural productivity. At the same time however, these fertilizers often find their way into receiving waters and causing eutrophication. A small amount of input of nitrogen and phosphorus can cause eutrophication because under most natural conditions, the surface waters are oligotrophic, or the primary producers do not have sufficient nutrients to grow to its full potential even if sunlight, water, and atmospheric carbon dioxide are not limiting factors. Since primary producers, especially algae, have a general ratio of C:N:P of 106:16:1,[7] algae cannot grow if either nitrogen or phosphorus is insufficient. Depending on the relative abundance of nitrogen and phosphorus, a water body could be either nitrogen-limiting or phosphorus-limiting. As a general rule of thumb, flowing waters are more often nitrogen-limiting and nonflowing waters are more often phosphorus-limiting. Nitrogen fixation can be carried out by bacteria, alleviating nitrogen shortage for some water when it is severely limiting.

Knowledge Box

HNLC Waters—High-nutrient, low-chlorophyll (HNLC) regions are regions of the ocean where the abundance of phytoplankton is low and fairly constant despite the availability of macronutrients such as nitrogen and phosphorus. Micronutrients (e.g., iron, zinc, and cobalt) are generally available in lower quantities and include trace metals. HNLC regions are limited by low concentrations of metabolizable iron. Therefore, addition of a small amount of iron can drastically improve the primary productivities of these waters. In fact, ocean fertilization could potentially be used as a tool to reduce atmospheric carbon dioxide and control global warming (see Chapters 2 and 11).

To control nutrient input from agricultural operation, there are many best management practices (BMPs), such as stream bank stabilization; vegetated buffers; in-crop application (i.e. applying fertilizers directly to the root area of the crop); use of organic nutrient; irrigation control, etc. Many of these BMPs are listed on the AgBMP website hosted by the Ohio State University (https://agbmps.osu.edu/bmp). For urban area nutrient input, source control, use of organic nutrient, and irrigation control/water conservation may be the best way for source control. Other options include low impact development such as grass swales, rain gardens, and detention ponds/wetlands to infiltrate or treat the runoff, as will be discussed in Section 5.3 below.

---

[7] This ratio is called Redfield Ratio, see Section 2.4.2.

5.2.2 Pathogen Contamination

Pathogen contamination is also one of the most common water quality issues in the United States and around the world. It causes potentially acute and widespread health consequences, sometimes disease outbreak and even an epidemic. In fact, as shown in Figure 5.2, it is the leading cause of impairments in the United States. For COVID-19, research found that the virus could exist in surface water and wastewater, posing a potential threat to public health in addition to the more prevalent aerosol transmission. Historically, pathogen contamination of drinking water sources was the leading cause of human illness and death. After Luis Pasteur confirmed the ubiquity of microbes in the environment and Leewvenhoak invented microscope to view and study microbes, our understanding of microbiology and microbial water quality improved drastically. This prompted Alexander Cruickshank Houston to start chlorinating drinking water sources in the UK in the early 1900s to stop the spread of common waterborne diseases such as cholera, dysentery, and typhoid, saving millions of lives in the UK and around the world. In the US, Abel Wolman pioneered the chlorination of Baltimore's municipal water supply and contributed to the global effort on improving the safety of drinking water.

The most common human pathogens including many species of bacteria, virus, and protozoa. Even though domestic or wild animals could be sources of pathogenic contamination of waters, most of the issues are caused by human activities such as sewage spills, sewer line leaks, discharge of treated effluent, homeless encampments, domestic animals/pets, and other human-related sources that find their way to a water body. When the water body is used as a drinking water source or for swimming or other body contact recreation, the pathogens could cause gastrointestinal infection and other diseases.

Due to the difficulties in analyzing these varieties of pathogens, the practitioners usually use fecal indicator bacteria (FIB) that often co-occur with these pathogens and are easy to analyze. Therefore, the abundance of these indicator bacteria is a good indication of the occurrence and abundance of human waste. Even though healthy humans do not carry pathogens and their waste may not be harmful, the fact that human waste ends up in the water is a severe public health concern. This is especially true for large wastewater treatment plants where wastes from thousands of people are collected. Due to the virility (the ability for infection) of many pathogens, even a small amount of spilled wastewater could impact a large waterbody, rendering it unsafe for swimming or for use as drinking water source. So, it is important to find good microorganisms to indicate human waste in a receiving water. These criteria include the following:

- The organism should be present whenever enteric pathogens are present
- The organism should be useful for all types of water
- The organism should have a longer survival time than the hardiest enteric pathogen
- The organism should not grow in water
- The organism should be found in warm-blooded animals' intestines

Based on the above criteria, the following common FIBs are often used: total coliform, fecal coliform, *E. coli*, and Enterococci. There are a number of ways to analyze them, one of the common ways is to separate them through filtration and incubate the filtrate under favorable conditions overnight. Every cell will grow into a small colony, which can easily be counted visually (**Figure 5.4**). The water quality standards for drinking water and for ambient water that people swim in are very different, and for good reason. For drinking water, direct ingestion of pathogens poses grave health hazards, therefore the presence of indicators has to be eliminated. For recreational water where people swim in or conduct other indirect contact activities such as wading, kayaking, etc., the ingestion rate is much lower and the health risks are proportionately lower. For recreational water, based on a series of epidemiology studies that established the indicator bacteria levels corresponding to acceptable public health risks (usually measured by additional gastrointestinal illness per thousand swimmers), the USEPA set the recreational water quality criterion, as shown below (see **Table 5.1**). This criterion is largely in line with the World Health Organization's criterion.

Stormwater runoff could be a significant conveyance mechanism to carry indicator bacteria and pathogens to receiving waters during wet weather. When people swim in downstream waters, or use the water for drinking water source but the treatment is not sufficient, the pathogens will pose significant health risks since many pathogens cause diseases in low doses. However, sources of fecal indicator bacteria and pathogens are often different for stormwater, and high levels of fecal indicator bacteria, as often being found in stormwater and stormwater-impacted receiving waters, do not necessarily suggest human waste impact and associated health risks. Recent studies suggest that sources other than human wastes are often the dominant sources of fecal indicator bacteria, and not all nonhuman sources are free of pathogens. Effective management of these issues requires a robust source investigation and potentially costly mitigation and treatment. This has posed a considerable challenge for stormwater managers.

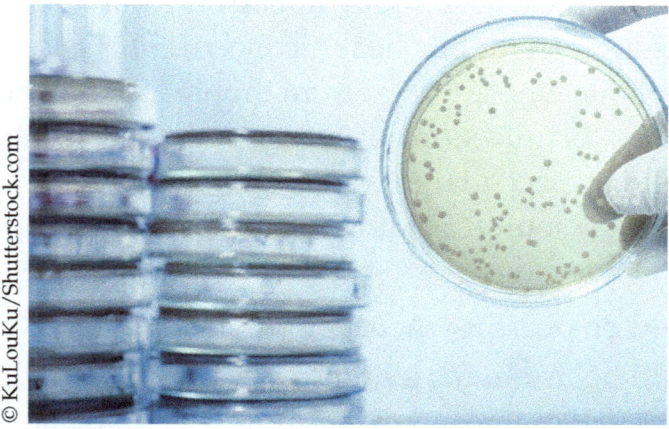

Figure 5.4 Measurement of fecal indicator bacteria is often done by incubation and counting of colony-forming units on a plate.

Table 5.1 USEPA Recreational Water Quality Criterion

Criteria Elements	Recommendation 1 Estimated illness rate 36/1000		Recommendation 2 Estimated illness rate 32/1000	
Indicator	Geometric Mean	Maximum	Geometric Mean	Maximum
Enterococci (Marine and Fresh Waters)	35	130	30	110
E. coli (Freshwater)	126	410	100	320

Note: the unit is in colony-forming unit (CFU) per 100 mL

5.2.3 Toxic Contaminants

Rachel Carson's monumental book "Silent Spring" brought everyone's attention to the issue of toxic contaminants especially pesticides, and the book was a reflection of increasing concern and alarm of the worsening situation in the nation's waters in the post-war United States. The book drastically raised public awareness of the ecological impact of a range of toxic pollutants, especially pesticides. To this day, people still often cite the book when similar issues are raised.

Toxicity is the degree to which a chemical or a substance can damage an organism or an ecosystem. An important concept of **toxicology** is that the effects of a toxicant are dose-dependent. In that sense, most, if not all substances are toxic, and the only difference is how much dose would produce a toxic effect. For example, even water can lead to water intoxication when one drinks too much of it, while a highly toxic substance such the venom from a poisonous snake could be harmless if the dose is low enough. Conventionally, only those substances that show toxicity in low levels (usually parts per million or lower) will be deemed toxins, and they will be the focus of this section.

To understand toxicity, and how water quality criteria are derived for toxic pollutants, we need to understand the dose-response curve and underlying science. **Figure 5.5** is a typical dose-response curve representing a series of toxicity tests, where groups of individual test organisms are subject to different levels of chemicals. After a given period of time, the response of the test organisms will be observed and recorded, and the results plotted on a dose-response curve. The horizontal axis represents the dose (usually in concentration) and vertical axis representing response (percent observed effect[8]). At a very low dose, no adverse effect will be observed.

---

[8] The effect could be tumor, lack of energy, or death, depending on study design and endpoint.

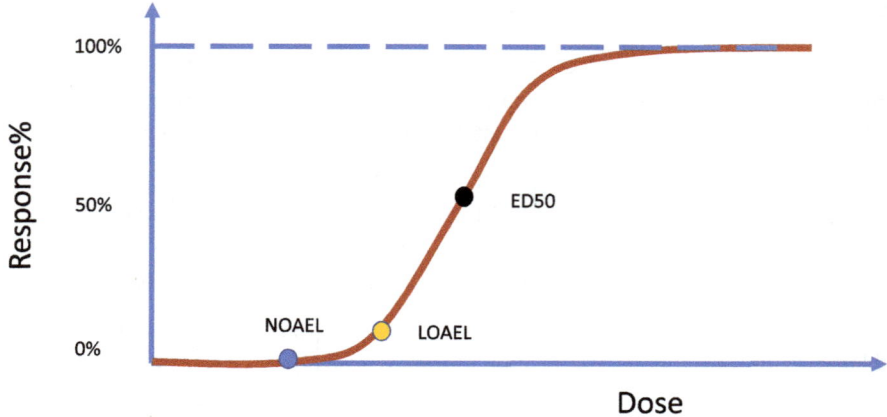

Figure 5.5 A typical dose–response curve for a toxic chemical.
Source: Created by Jian Peng

The highest such dose will be termed no observed adverse effect level, or NOAEL. With the dose increasing, more and more individual organisms will show the adverse effect. The low observed adverse effect level (usually set at 10%), or LOAEL, is usually used in setting water quality criteria. Common test organisms include rats, fish, algae, aqueous arthropoda (such as water flea *Ceriodaphnia dubia*, and the opossum shrimp *Americamysis bahia*), among others. Sometimes a pollutant-specific species will be used. For example, bluegill, which is one of the most sensitive species to selenium, is the species on which the selenium criterion was set. These species are commonly termed "sentinel species."

Knowledge Box

Toxaphene—Toxaphene is a mixture of over 600 different chemicals and is produced by reacting chlorine gas with camphene and the common chemical formula is $C_{10}H_{11}Cl_n$, where n could range from 5 to 11. Similar to other chlorinated pesticides, the addition of chlorine renders toxaphene its toxicity as well as it remarkable stability. It has been shown to be harmful to animals and is a possible carcinogen. Due to its hydrophobicity, it tends to bioaccumulate in fatty tissues, causing greater impacts to higher trophic levels. Over 34 million pounds of toxaphene were used annually from 1966 to 1976, mostly in the cotton field in the southern United States. It was banned in the United States in 1990 and globally in 2001 by the Stockholm Convention along with other chlorinated pesticides, including aldrin, chlordane, DDT, dieldrin, endrin, heptachlor, hexachlorobenzene, and mirex.

There are many different types of toxic pollutants. Of the 126 priority pollutants, there are 11 metals, 2 nonmetal (selenium and arsenic), one compound (cyanide), asbestos, and 111 organic pollutants. Many of these organic pollutants are groups of compounds, such as toxaphene and PCBs (see the Knowledge Boxes). They are either common industrial materials (such as naphthalenes, chlorobenzene, PCBs), pesticides, or their derivatives.

Knowledge Box

PCBs—Polychlorinated biphenyls or PCBs are a group of chlorinated industrial compounds with a chemical formula of $C_{12}H(10-x)Cl_x$ that are mostly used in transformers as dielectric fluids (i.e., as insulators) due to their excellent stability and other favorable properties. Because of their longevity, PCBs are still widely in use, even though their manufacture has declined drastically since the 1960s with the discovery of their environmental toxicity. Their production was banned by the United States in 1978, and by the Stockholm Convention on Persistent Organic Pollutants in 2001. It is a suspected carcinogen. The most common commercial products of PCBs in the United States are called Arochlors, but there are many other commercial names as well. Due to their stability and hydrophobicity, they are often found in polar regions. The Great Lakes region has elevated PCB levels due to historical industrial activities.

Knowledge Box

Selenium—Selenium is unique in that it is an essential micronutrient in low concentrations and selenium deficiency could cause stunted growth, heart diseases, and other problems. However, in slightly higher concentrations, such as over a few parts per billion in water, selenium could bioaccumulate and cause harm to many animals. This is because selenium is an essential component of a few enzymes that are important to the normal function of many animal species. However, in higher tissue concentrations it will substitute sulfur in some sulfur-bearing enzymes, impacting normal functions of these enzymes, causing many impacts including reproductive issues such as teratogenesis, or deformity of embryos. Most of the major causes of selenium contamination are power plants that burn selenium-bearing oil or coals. However, in places where natural background of selenium is high, it could become a difficult environmental issue to deal with.

- **Metals**

    The toxicity of metals is generally due to their binding with proteins and enzymes to alter or disable their normal functions. Metals can also cause neurotoxicity, generating free radicals which promote oxidative stress damaging lipids, proteins and DNA molecules. Lead is such an example, whereby it interferes with the activity of several essential enzymes. The brain is the most sensitive organ to lead poisoning because lead can behave like calcium and penetrate inside the brain. For young children, lead poisoning can impair the development of the brain and nervous system. Chapter 4 has more information about lead poisoning.

- **Pesticides**

    Of the toxic pollutants, few else have received so much attention as pesticides. According to Buchel (1983), the term "pesticides" could have different layers of meaning. The most commonly and widely used pesticides are actually 'crop protection products' to control arthropods, microorganisms, and weeds to protect agricultural plants.[9] A broader meaning covers those protective of humans as well, including the pesticides used against vectors of human diseases such as mosquitos, lice, fleas, and so on. Many such pests are also common nuisance that humans would like to control. Lastly, the term pesticide also covers chemicals for the control of rodents, slugs, snails, termites, and other household pests.

Pesticides could be either natural or man-made. There is a small group of naturally-occurring pesticides such as pyrethroids, alkaloids, and rotenoids, whose pest-fighting properties are the basis of many synthetic pesticides, such as synthetic pyrethroid pesticides that are still used widely today. The vast majority of pesticides are man-made, which can be classified based on chemical formula into organochlorine (or organochlorohydrocarbons), organophosphorus, carbamates, and inorganic pesticides. Based on specific target species, the pesticides can be classified into nematicides, acaricides (for the control of mites), molluscicides, rodenticides, fungicides, bactericides, herbicides, plant growth regulators, and so on. There are hundreds of different types of pesticides and thousands of different commercial products on the market. As of 2019, the global pesticide market is worth $84.5 billion and is expected to grow at a rate of more than 10% for the coming years. It is easy to see why pesticides are so important in modern society. It was estimated that the world's crop production would be cut in half due to pests without pesticides and other agrochemicals (excluding fertilizers).

Despite the benefits and necessity of pesticides for ensuring food safety and human health, there are numerous and potentially significant impacts to the environment, the

---

[9] For pesticides that control weeds, they are termed herbicides.

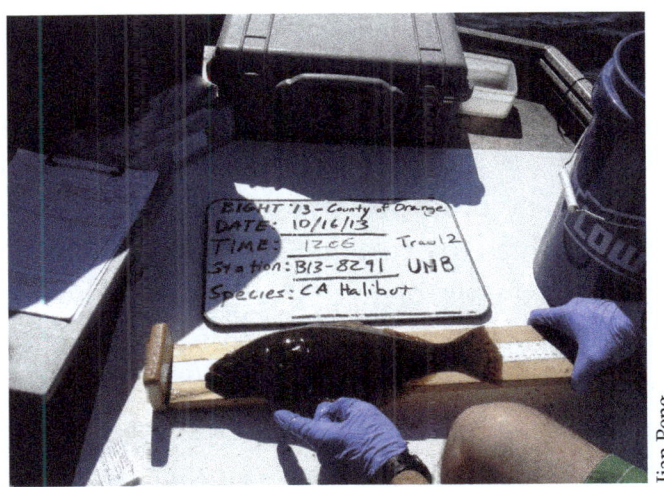

Figure 5.6 A California Halibut was collected for analysis for bioaccumulative compounds including DDT, PCB, and Chlordanes at Upper Newport Bay, California, USA. Halibut is an epibenthic fish living on the sediment surface and is an ideal sentinel species for sediment contamination.

ecosystem, and human health. The book "Silent Spring" is largely responsible for raising the awareness of the potential harm of pesticides such as DDT, chlordane, and other organochlorohydrocarbons. This book also resulted in the eventual ban of many pesticides and herbicides in the United States, such as DDTs in 1972 and chlordanes in 1988. Due to the stability and bioaccumulative nature of these persistent pesticides, some can still be detected in significant levels in the biota to this day (**Figure 5.6**).

USEPA and a few states[10] have a pesticide registration processes that involve scientific, legal, and regulatory review of the proposed use of any new pesticides and for their potential human and environmental risks. The risk assessment looks at human health (direct exposures and indirect exposures via food and drinking water), ecosystem effects, and risks to wildlife, amphibians, birds, pollinators, and aquatic organisms associated with use of the pesticide. This risk assessment, any identified mitigation measures, and an evaluation of the benefits of the product, will be the basis of approval or denial of the pesticide registration. For the approved new pesticides, USEPA must approve labeling with warning and use instruction. With the exception to United States Geological Survey (USGS) National Water Quality Assessment Project, no other nationwide pesticide surveillance program exists to monitor either legacy or current use pesticides, but a patchwork of Federal, state, University, regional, local government and nonprofit research institute monitoring programs do monitor heavily used pesticides. When USEPA conducts its periodic review of each pesticide (once every 15 years),

---

[10] Most states do not conduct scientific review of pesticide registration application. Only California has a robust review. Minnesota and New York take a cursory review.

these monitoring data could result in use restriction, modification, or even prohibition if adverse human health or environmental impacts are identified.

The above process is generally ineffective at preventing human and environmental impacts from pesticides, but has proven effective at identifying and mitigating pesticides risks in the long term. The main issue is that the cycle could take a considerable time (sometimes decades, such as the case for DDT). The length of time allowed pesticide manufacturers to replace the problem pesticides with new ones, which could turn out to be no better than the pesticides it replaces. This problematic cycle can be depicted by the "pesticide treadmill" shown in **Figure 5.7** (top). Such a process is subject to extensive and sometimes heated debates among different interest groups including the pesticide manufacturers; agricultural groups; regulators; and environmental advocates. The pesticide and agricultural groups are often frustrated about the long pesticide review and approval process, use restrictions, and prohibition of some pesticides that they believe should not be banned. Environmental advocates decry the proliferation of pesticides and believe that much more restrictions and prohibitions should be levied on current pesticides and new pesticides in the pipeline. The regulators often find it difficult to strike a balance in between.

With billions of dollars worth of pesticide and agriculture products on the line on one side, and on the other side potentially significant ecosystem and human health impact, the debate will continue for years to come. One possible solution is to break the treadmill by identifying whether the pesticide in question poses a threat (e.g., to water quality) using more robust, more specific predictive modeling to better link impacts to application locations and practices, and by using this information to restrict or disallow problematic uses before it becomes a problem.

Instead of traditional chemical-based pesticides, there have been some environmentally friendly alternatives to control agricultural and domestic pests. These alternatives include biological control;[11] pheromone traps;[12] use of organic and biodegradable compounds (such as use of orange oil to control termite); use of genetically engineered, pest-resistant crops;[13] use of physical methods (such as high-temperature steam for weeding, see **Figure 5.8**), and so on. As pests and weeds have coexisted with mankind throughout history,[14] the battle to control them will be a perpetual endeavor.

---

[11] There are quite a few success stories, such as use of parasitic wasps to control aphids; Australia's use of a beetle species to effectively control a cotton pest; and so on.

[12] A pheromone trap uses pheromones (chemicals that pests use to signal mating or aggregating) to lure insects. Pheromone traps are used for pest monitoring and control.

[13] Use of genetically modified organisms, or GMO, is a controversial topic. It is mentioned here as an alternative to pesticide.

[14] To be fair, these pests and weeds most likely existed long before humans became a species. In their perspectives (if they had any), humans have to be one of the most annoying pests.

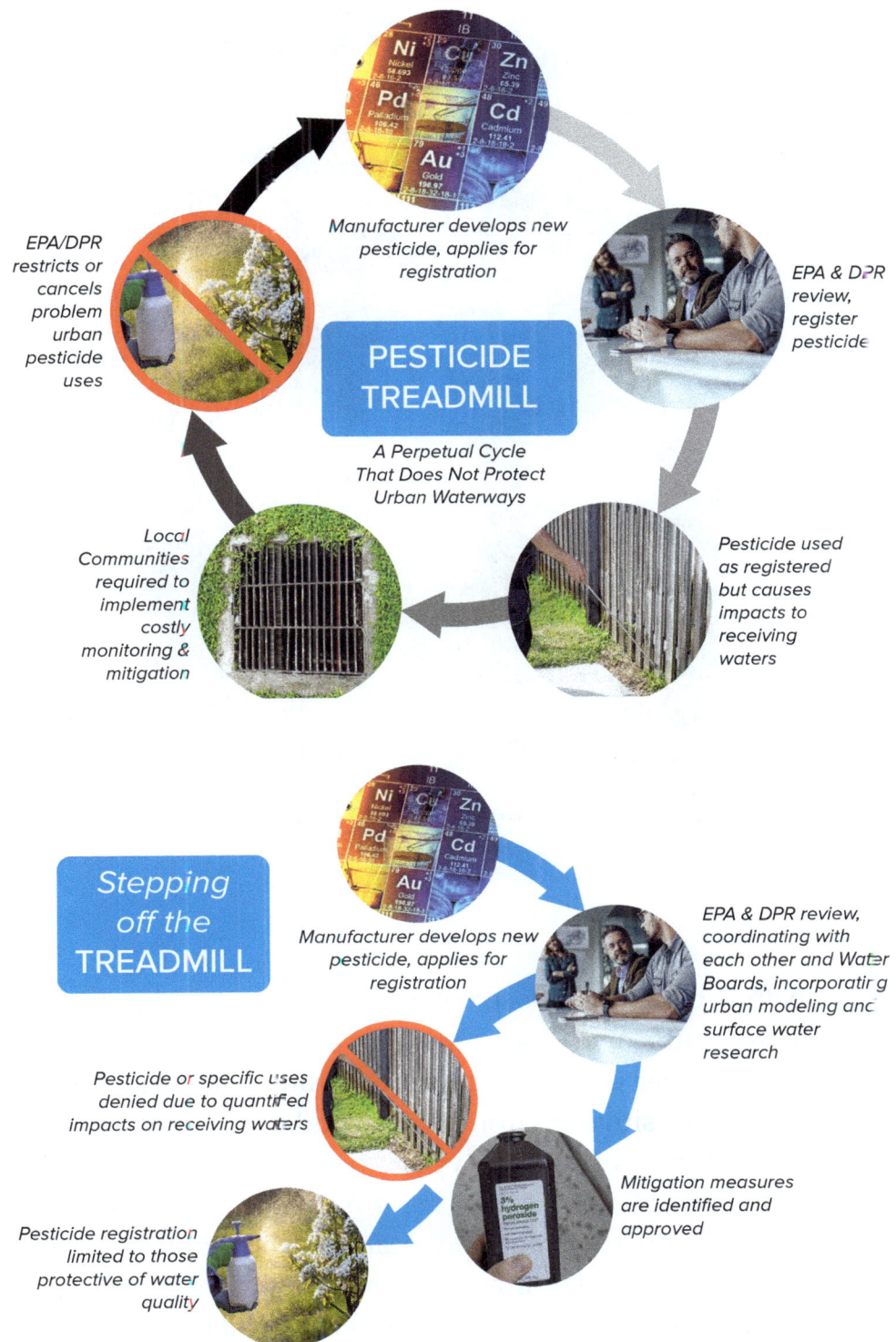

Figure 5.7 The Pesticide Treadmill (top) and a possible way to "step off the treadmill" (Figure 5.7, bottom) by identifying issues early to prevent problematic pesticides from entering the market before they could cause environmental and human health impacts Adapted from California's Department of Pesticide Regulation. images © shutterstock.com. ©Kendall Hunt Publishing.

Figure 5.8 A high temperature steam weeder is used instead of herbicide to control weeds in an office park in Orange, California, USA.

5.2.4 Trash

Trash and debris include garbage disposed of by humans such as plastic and other packing materials, cigarette butts, and other settleable and floatable man-made materials. Trash could be found in other environment settings, but since water could carry trash and debris to ecologically sensitive environments including the ocean, trash is an important water quality issue. Trash is a unique type of pollutants in that they do not always cause direct harm themselves. Rather they are a nuisance because they are unsightly (e.g., **Figure 5.9**). They also harbor other pollutants such as nutrients, pathogens, and toxic pollutants. Many plastic-containing trash can accumulate harmful organic pollutants including pesticides, so they could pose severe ecological threat to aqueous environment when freshwater and marine animals ingest these trash, causing physiological and toxicological harms. When plastic trash is ingested by animals by mistake, the animal could starve to death due to its inability to digest the plastics. Certain trash, such as fishing gears, could be especially harmful to some marine animals, causing injuries and deaths (see **Figure 5.10**). When plastic trash breaks down to smaller particles, their ability to concentrate hydrophobic pollutant will be enhanced and they will pose more

Figure 5.9 Left: A trash-impacted beach off the coast of Malaysia. Right: Tijuana River, on the United States–Mexico border, a trash boom intercepted a large amount of plastic trash.

Figure 5.10 Left: A dead sea turtle apparently tangled by fishing line off Texas Coast, Gulf of Mexico, USA. Right: A pelican in the process of swallowing a plastic water bottle. Trash pollution is more than unsightly. It has serious ecological consequences.

threat to ecosystem health. These plastic particles are termed "microplastics," which will be discussed in more detail in Section 5.5.2.

Under the USEPA, trash has not received much attention because it is neither a toxic pollutant nor a conventional pollutant due to its heterogeneity and diverse and diffuse sources. There is no water quality standard for trash, making it difficult to manage in similar fashion as other pollutants. However, USEPA does have an initiative called "Trash-Free Waters" to provide information on the impact, management, and research on trash pollution. In California, trash regulation started in around the early 2010s. The City of Los Angeles is subject to several trash total maximum daily loads, which set a goal of "zero trash" within 10 to 15 years. There is also a statewide trash policy (termed "Trash Amendments") that requires local governments to adopt a wide range of trash management measures to effectively control trash with a goal of zero trash by 2030.

Figure 5.11 Innercoastal cleanup day at Orange County, California, USA, hosted by the Surfrider Foundation and Orange County Environmental Resources in 2019. In the background is Santa Ana River, the region's largest river.

Since plastic materials are the most important type of trash, numerous initiatives have been underway to curb the generation of plastic trash from its source. For example, most of the nations around the world have enhanced their plastic bottle recycling effort. In the United States and Europe, more and more types of plastic materials are recyclable. Initiatives to ban single-use plastic bags, plastic straws, and polystyrene foams have gathered momentum in recent years. Public education efforts, such as "Innercoastal Cleanup Day" (Figure 5.11) in Southern California, have helped clean up the coastal environment and raise the awareness of the public on the environmental impact of trash. On the East Coast, the Baltimore Trash Wheel is famous in its effectiveness in removing trash from the Baltimore Harbor.

5.2.5 Sediment

Sediment is an important part of an aqueous environment and ecosystem. Erosion is also a natural phenomenon that has been shaping the surface of the Earth, as discussed in Section 2.4. Sediment is also one of the most common water pollutants because sediment-caused turbidity is one of the primary and easily observed water quality properties. Sediment in water could be from erosion of the natural environment or man-made landscape (Figure 5.12). At high levels, sediment could cause siltation or excessive sediment accumulation in rivers, lakes, and coastal ocean, resulting in habitat change, blockage of navigation, or direct impact of sensitive ecosystems such as submerged grass. Sediment could also carry a number of other pollutants such as bacteria, heavy metals, and organic pollutants. For certain

Figure 5.12 An exposed hillside (left) could be a significant source of sediment loading in rivers (right) and sediment-related impairment in receiving waters.
Photo credit: Jian Peng

environments, sediment is the main vehicle with which some pollutants migrate in the water environment. Therefore, proper management of sediment has many other benefits as well.

5.3 Stormwater Issues

5.3.1 Overview

Stormwater is water from precipitation such as rain or snow. As shown in Figure 2.6 on the global water cycle, precipitation is a key component of water cycle. In a typical water cycle, precipitation onto the land surface will infiltrate into the ground to replenish groundwater, or travel as shallow groundwater as "through flow" down-gradient (which will re-emerge into the streams as springs), or become surface runoff as rivers and streams. The rest of the water will go back to the atmosphere as water vapor, a process called evapotranspiration. In a natural landscape, about a third each will go to evapotranspiration, groundwater, and through flow, with only a small portion leaving as surface runoff. In an urban landscape, however, most of the surfaces are hardened impervious surface such as rooftops, roads, parking lots and buildings. The impervious surface allows little infiltration, and compacted soil allows little through flow. As a result, most of the precipitation (up to 70%) will turn into urban runoff immediately after the onset of the precipitation event. The lack of permeable surface and vegetation will significantly decrease evapotranspiration will worsen the "heat island" effect of the urban landscape. The lack of through flow will result in flashy runoff with very high volume, creating flooding hazards.

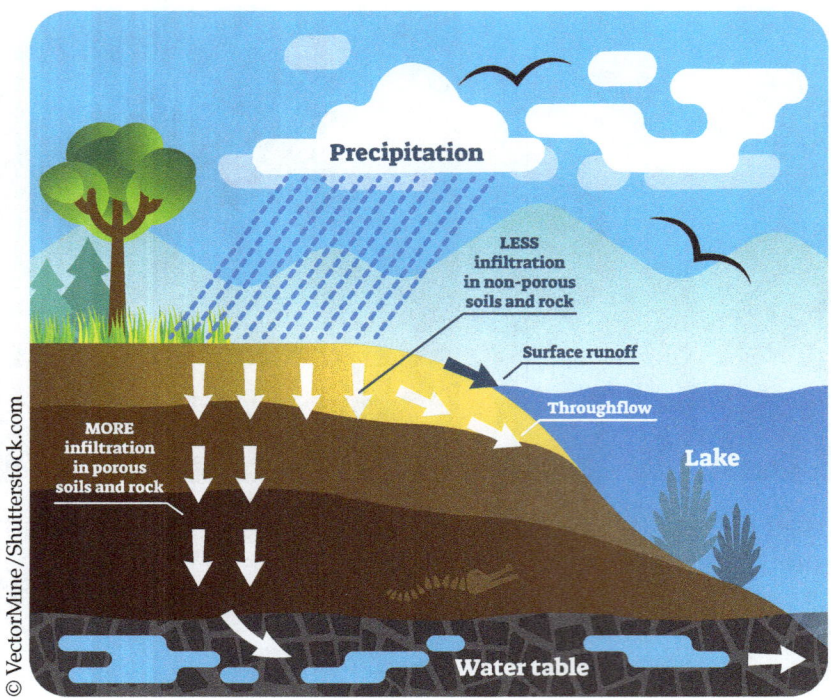

Figure 5.13 Water cycle related to stormwater.

For the same reason, the runoff will subside quickly after the storm, often leaving the creek dry in between storms. **Figure 5.13** is a simple representation of such a process.

5.3.2 Urbanization and Stormwater Flow and Quality

The contrast between the hydrographs of a typical creek for a typical storm event before and after urbanization is shown in **Figure 5.14**. For the creek before urbanization, the hydrograph will rise slowly and taper off slowly due to the dampening effects

> ### Knowledge Box
>
> **The Rational Method Formula:** For a small watershed, the volume of stormwater runoff can be calculated by a simple formula called the rational method $V = C \times i \times A$ where V is the runoff volume, A is the watershed area, C is the runoff coefficient (unitless), i is rainfall intensity. C is a constant for the watershed and is determined by perviousness of the land surface, soil moisture, vegetation and other factors. More urbanized watershed with higher imperviousness will have higher C values, resulting in higher runoff volume with the same rainfall intensity.

> ## Knowledge Box
>
> **Hydrograph**—A hydrograph (see Figure 5.14) is a useful tool to study the hydrology of a watershed. It is constructed by measuring flow of the creek draining a watershed, and plotting the flow with time. The peak flow and the timing of the peak flow can be calculated using a few empirical equations that takes into runoff generation process (such as the rational method above); runoff conveyance characteristics (i.e., how far it takes the runoff to reach the measurement spot), and so on. Some stormwater models such as USEPA's Stormwater Management Model (SWMM) can simulate these processes and generate hydrograph for any given precipitation events.

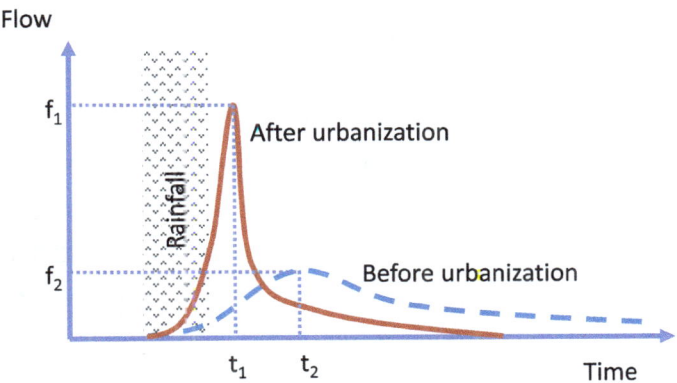

Figure 5.14 An example of hydrograph before and after urbanization. After urbanization, the stormwater runoff will peak earlier and have a much higher peak flow rate, posing a potential flooding risk.
Source: Created by Jian Peng

of infiltration and through flow on the surface runoff. In contrast, after urbanization, the hydrograph will undergo several changes. First, the surface flow will rise faster because there is little infiltration to absorb the stormwater. It will peak earlier for the same reason and because it takes less time for impervious surface to channel the stormflow to the creek. The peak flow will be higher as well. After peaking, the flow will decrease faster than its preurbanization condition due to lack of through flow.

The above changes in an urbanized watershed will cause many issues. Decreased infiltration will keep groundwater from being recharged, potentially causing groundwater overdraft and increasing land subsidence. A quickly decreased flow after a storm may leave too little water in the creek in between storms, making it harder for aquatic organisms to survive. Higher peak flow will present increased flooding risk.

> **Knowledge Box**
>
> **Flow Ecology**—Flow ecology is a field of environmental science that studies the relationship between stream ecology and flow. For urban streams, flow ecology could lend helpful insights on how to mitigate the impact of flashy storm flows and low flow or dry stream bed in between storms so that the ecosystem in urban streams can be maintained. Common values of ecological flow (i.e. the minimum flow that a healthy stream ecosystem can be sustained) include, for example, 10% of average annual flow; 90th percentile flow; lowest monthly average flow, etc.

To accommodate higher peak flows and avoid flooding, many urban streams are widened, deepened or armored with riprap or concrete to improve its flood capacity and avoid erosion (**Figure 5.15**, also refer to Figure 5.12). Compared with a natural meandering stream with a variety of habitats to support a healthy ecosystem, a highly modified urban stream often provides limited or no habitat.

Stormwater runoff could bring a range of water quality issues. In a natural watershed, stormwater will mostly infiltrate into the soil and recharge groundwater with little runoff. In an urbanized watershed, stormwater will wash off trash, dirt, oil, pesticides, fertilizers, and many other pollutants from impervious surfaces such as roads,

Figure 5.15 A typical and stoic urban drainage channel (Greenville-Banning Channel, Orange County, California, USA.

driveways, rooftop, parking lots, etc. In some cases, contaminants of emerging concern (see Section 5.5 for more details) could be found in stormwater discharge as well. Therefore, stormwater is not the source, but the conveyance mechanism of these pollutants. In urban environments, pollutants carried by stormwater from urban landscape is often the most significant source of pollutants in receiving waters. Therefore, stormwater management, as often required by stormwater discharge permits (see the Knowledge Box), is key to water quality improvement in nearly all urban watersheds. In fact, the United States amended the CWA in 1987 to re-designate urban stormwater from nonpoint source to point source.[15] The amendment allowed USEPA and state governments to regulate stormwater using discharge permits. In addition to general permits for urban areas, there are also permits for construction sites (mostly for sediment control), industrial facilities (mostly for industry-specific chemicals that could be exposed to stormwater), and sometimes transportation systems (for road-related pollutants such as oil/grease, metals, and sediment).

> ### Knowledge Box
>
> **First Flush**—First flush usually refers to the first storm event of the rainy season. In an urban environment, pollutants tend to accumulate on the land surface over time during the dry season. Therefore, the first storm events will see much higher pollutant loading in the stormwater runoff, as shown in **Figure 5.16** and **Figure 5.17**.

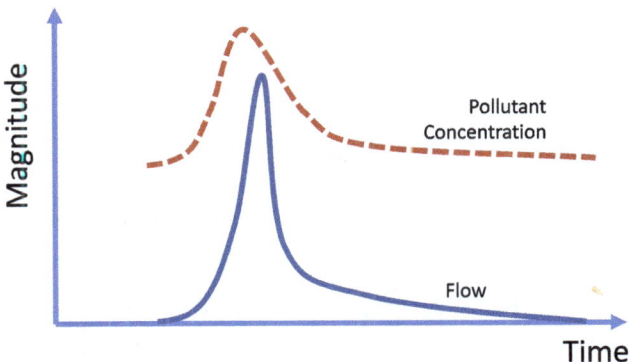

Figure 5.16 Schematic of a pollutograph for a first flush event.
Source: Created by Jian Peng

---

[15] Note that the Clean Water Act essentially manages point source discharges only. Before stormwater was designated as a point source, there was no effective regulatory tools to manage stormwater.

Figure 5.17 First Flush runoff in a river in Southern California, USA. Oil sheen, floating vegetation, debris, and trash are common.

> ### Knowledge Box
>
> **Stormwater Discharge Permit:** Under the CWA in the United States, a stormwater discharge permit is similar to other National Pollutant Discharge Elimination Program (NPDES) permits. There are significant differences between NPDES permits for regular point source dischargers (such as wastewater treatment plants) and stormwater permits. The most important distinction is that stormwater permits typically do not have numeric limitations. Instead, as long as the permittees implement best management practices to improve stormwater quality, and keep improving the practice in an iterative manner, they are deemed in compliance.

5.3.3 Stormwater Management and Low Impact Development

To deal with the above issues, nations around the world started to take actions since the early 2000s. In the United States, these practices are termed low impact development (LID) or green infrastructure (GI). In Australia and United Kingdom, they are called water sensitive urban design or WSUD (and drainage system built with this in mind is called sustainable urban drainage system or SUDS). In China, they are called sponge city (SC). In the EU, they are called integrated stormwater management (ISWM). For the most part, these practices focus both flood mitigation as well as water quality management for urban stormwater by several options, such as reducing imperviousness; increasing green space; promoting stormwater infiltration, storage, and reuse; and stormwater treatment. Figure 5.18 shows some examples of common stormwater LID best management practices.

Figure 5.18 An example of low impact development best management practice demonstration project for an office/warehouse complex in California, USA. Left: the project overview bulletin in front of one of the buildings. Right: examples of LID BMPs for the project. Clockwise from the top left: cistern; bioswale; treebox; pervious pavers. Not shown are porous concrete, permeable asphalt, underground cistern, flow-through planters, dry well, etc. The project retains 85% and treats 100% of the stormwater from a "design storm."

Photo taken by Jian Peng

Knowledge Box

Combined and Separate Systems: In stormwater management, it is important to understand how stormwater and wastewater are conveyed in relation to each other. Many cities, especially older ones, collects urban runoff and wastewater using the same system and treat the combined effluent in wastewater treatment plants. During storm events, especially large storms, the system could overflow, an event called combined sewer overflow (CSO). CSO is one of the major water quality issues in areas with combined sewers, such as some cities in the east coast of the United States; major cities in China; and many other cities around the world. Increasingly, the cities are separating stormwater conveyance systems from wastewater infrastructure. Such systems are often called municipal separate storm sewer system (MS4).

Currently, stormwater management is taking an increasingly integrated approach. In the United States, especially in California, stormwater management is now intrinsically linked to land development, urban watershed management, integrated water resource management, and water quality improvement projects. California Stormwater Quality Association (CASQA) is one of the most active and influential stormwater associations in the United States. Through stormwater-related activities, this organization is an important player in the state with influence well beyond traditional stormwater issues. In China, the sponge city effort has been touted as the key to many urban environmental syndromes. In Australia, the "Millenium Drought" during the early 2000s put stormwater and integrated water management issues into the spotlight. Around the globe, stormwater management is increasingly becoming a key component in integrated water management and is being treated as part of the whole water cycle that includes drinking water, wastewater, surface water, groundwater, as well as reclaimed water. This issue will be discussed in more detail in Chapter 6.

5.4 Water Treatment and Reuse

5.4.1 Background and History of Water Treatment

Water (including both drinking water and wastewater) treatment did not happen until about 150 years ago. Compared to wastewater treatment, drinking water treatment is relatively straightforward and will not be discussed in detail here. In some occasions, drinking water contamination could be linked to wastewater issues, when infectious diseases propagate due to contamination of drinking water sources by wastewater. By sourcing drinking water from protected streams and lakes or by digging wells, the communities can ensure the safety of their drinking water. For large water systems, this was more challenging until the 1910s, when chlorination was initiated in Europe that drastically reduced the health risk from drinking water contamination. However, safe drinking water remains a challenge in some developing countries.

Wastewater is generated after indoor water use such as drinking, washing (clothes, dishes, etc.), bathing/showering; flushing toilet, etc. In most cases, the wastewater is collected by sewer system and sent to a wastewater treatment plant to be treated and cleaned before discharging into a surface water. Historically (and currently in some developing countries), communities had to live by a river or lake to get drinking water they needed, and wastewater was not treated. There were no flushable toilets either. People use community or individual bathrooms or cesspools to collect feces and urine. After composting, these waste materials are used as fertilizers. The wash water would be put back into the river. As long as the community downstream is far enough, the river self-purifies the wastes generated by the community upstream through natural processes.

With population growth, communities grew larger, and closer together. The pastoral style of water use as described previously would become problematic. The river water may not be able to self-clean fast enough before the community downstream takes water for drinking and other uses, and contaminated water will cause health and environmental issues. Larger, more densely populated communities will not be able to handle large amounts of human waste for recycling, some may be disposed of back into the river or lake downstream, causing even more severe issues downstream. Even if the downstream community finds an alternative water source, the degradation of water quality by the upstream community may overwhelm the river or lake if wastewater continues to be discharged without treatment. In fact, such an issue happened in England, the oldest industrialized nation. Due to contaminated water supply, cholera outbreaks occurred in the 1830s to 1850s and killed thousands of people. In 1858, the "Great Stink" of River Thames prompted the construction of perhaps the first sewer system. However, the system simply piped wastewater to further downstream without treatment. It was not until the 1870s when Edward Frankland identified a porous gravel-based sewage treatment process in Croydon, England. Since then, sewage treatment technology has improved a great deal and the advancement of wastewater treatment is one of the key factors that lead to improved water quality and public health around the world.

5.4.2 Overview of Water Treatment Process

Since the 1970s when the CWA was enacted in the United States, a great number of wastewater treatment plans have been constructed and water treatment technology has improved. China saw a similar trend since the late 1970s when its economy took off, especially after 2000 when water quality problems started to worsen as the result of its rapid industrialization. In the United States, the amount of pollutants discharged from wastewater treatment plants has decreased more than 90%. For many pollutants such as heavy metals, the reduction is even higher. Europe has experienced a similar trend as the United States. However, China still has a long way to go, especially for smaller cities and towns, where there still is a significant portion of domestic sewage being discharged without treatment. In general, developed nations treat 70% of the generated wastewater. Lower income nations treat a smaller portion of their wastewater (upper-middle-income countries with 38%, lower-middle-income countries with 28%), and low-income countries[16] treat 8% of their wastewater. Overall around the world, 80% of all of the wastewater is discharged to the environment without treatment, posing serious environmental and health problems.

Depending on the wastewater amount, sources and strengths, and limitations on treatment plant space, wastewater treatment plants could use a range of different processes. However, the core processes are largely similar, as shown in **Figure 5.19**.

In a typical wastewater treatment plant, which is usually built downstream of the city near a waterway, raw sewage flows into the plant and goes through some physical screening and settling device such as bar screen (to remove large debris and trash to avoid clogging the treatment processes later. Grit chamber allows heavy solids such as

---

[16] The classifications of income levels for nations are based on the World Bank (2012) data.

Figure 5.19 A typical wastewater treatment plant. From the far end, there are equalization tanks, headworks (i.e. pump station), two circular primary clarifiers, several rectangular aeration tanks, and four secondary clarifiers. The sludge is collected from the aeration tanks and secondary clarifiers and sent to the digester, which is the spherical structure located at the far end near the equalization tank. Not shown in the photo are the bar screen and grit chamber, which are located near the headworks building.

coffee grinds, gravel, etc. to settle so they do not need to be treated. Solids intercepted through these two processes usually goes to landfill. The sewage then enters into a large chamber called primary clarifier, where the water moves so slow that most of the solids will settle out. The organic-rich solids, or sludge, collected from the bottom of the primary clarifier will be removed periodically. The effluent from the primary clarifier (called primary effluent), with about half of pollutants (most commonly measured by biochemical oxygen demand or BOD, see the Knowledge Box) removed, can then be disinfected and discharged back to the surface water. In many cases, however, primary effluent goes through more treatment called secondary treatment, or biological treatment because biological processes are employed to meet stringent effluent discharge limits or benchmarks. For most of the plants, the limits are based on concentrations of BOD and total suspended solid and so-called "30–30 rule" (i.e., both need to be within 30 mg/L) applies. In China, there are often limitations for many other pollutants such as total nitrogen, total phosphorus, and so on.

There are many different biological treatment processes such as activated sludge, integrated fixed-film activated sludge, trickling filter, membrane bioreactor, moving bed biofilm reactor, and so on. The basic principle behind the biological process is that microorganisms such as bacteria and protozoa take up suspended and dissolved organic matter in wastewater to grow and reproduce. These microorganisms could either be part of suspended sludge (such as the first two processes mentioned previously) or attached biofilm (the last three processes mentioned previously). They can then settle from the suspension or detach from the surface to be removed from the wastewater. In the meantime, most of the BOD could be removed from wastewater.

> ### Knowledge Box
>
> **BOD**—Biochemical oxygen demand, or BOD, is the key water quality parameter in the wastewater treatment industry to indicate the amount of easily degradable organic matter in the wastewater. The term comes from the way this parameter is measured—a given volume of wastewater is sealed for several days. The decrease in the oxygen content in the bottle at the end of incubation is its BOD, usually measured in unit of microgram of oxygen per liter of wastewater (mg/L). BOD could be measured for different days, but 5-day BOD is the most common and is termed BOD-5. Raw sewage has BOD of several hundred mg/L. Wastewater treatment process usually has a target of 30 mg/L BOD, but many plants can achieve better performance.

> ### Knowledge Box
>
> **The Pretreatment Program:** In the United States, industrial wastewater is mostly managed by the pretreatment program, where wastewater from each individual industrial dischargers such as petrochemical, textile, papermill, metal works, and so on, is treated to meet stringent industry-specific discharge limits before discharging into the municipal wastewater treatment system for further treatment with other municipal wastewater. For more information, refer to USEPA's national pretreatment program website at: https://www.epa.gov/npdes/national-pretreatment-program. A similar program for restaurants and other food-related facilities is the Fat, Oil and Grease (FOG) Program, where grease traps are installed for the wastewater from these facilities to trap FOG before they enter the sewer system to avoid clogging and odor issues.

5.4.3 Water Recycling and Biosolid Management

With improved wastewater treatment technology and increasing water shortage in many parts of the world (see next chapter), more and more attention is directed to water reuse where primary or secondary treated wastewater effluent is further treated to a degree that it could be used again. Depending on the quality of the treated water, it could be used for different purposes such as irrigation, industrial process water, or even potable water. While indirect and unintentional potable reuse is common, such as in a situation where a downstream city uses the water from a river that contains wastewater discharges from an upstream community, treated wastewater is increasingly being recycled in many places around the world. For example, Orange County Water District (OCWD) in Fountain Valley, California, USA operates the Groundwater Replenishment System, the world's largest of its kind. After treating the secondary wastewater effluent from a nearby wastewater treatment plant, OCWD treats the water using microfiltration, reverse osmosis, and advanced oxidation process to purify the water to a level that meets or exceeds drinking water standards. The purified water is then used to recharge the groundwater basin.[17] More detailed discussions on water reuse and recycling can be found in Chapter 6, Section 6.5.

Wastewater treatment plant generates a considerable amount of organic-rich sludge.[18] After dewatering the sludge to increase its solid content to 10% to 20%, the materia (now called biosolids) can be used as fertilizer or soil amendment after being treated through a digestion/composting process to deactivate the pathogens. The digestion process can recover the energy in the rich organic and produce biogas (mostly methane), which can be used to generate electricity. For large wastewater treatment plants such as those in the City of Los Angeles, California, the electricity generated from biosolids can be sufficient to power most of the wastewater treatment process.

Since wastewater treatment plants are all heavily regulated around the world, many plants have rigorous source control programs to ensure that only household wastewater

Knowledge Box

Biosolid—Biosolid is rich in organic material and nutrient and should be ideal for use as fertilizer. Right? Actually, it depends. Since biosolids are originated from human waste, there are a lot of bacteria and pathogens in them. If biosolids are digested in a very high temperature to kill essentially all pathogens, the treated biosolids are called Class A biosolids and can be directly reused as fertilizers or other useful materials. This treatment process is complicated and costly, so many treatment plants opt to use lower temperature for digestion, resulting in Class B biosolids that have more use restrictions.

---

[17] A portion of purified water goes to seawater barrier, but most of the barrier water flows back to the groundwater basin.

[18] In a sense, that is all a wastewater treatment plant does. Primary clarifier is to allow the solid to settle, and secondary/biological treatment is to use microbes to consume dissolved organic materials and turn them into biomass, which will then settle out in the secondary clarifier, leaving only the cleaner water behind. Biosolids are simply the solid material from the primary and secondary clarifiers.

and properly pretreated industrial wastewater are discharged into the sewer system. Improperly treated or illegally dumped industrial wastewater could cause exceedance of benchmarks in the discharged effluent or biosolids, adversely affecting the reuse or disposal of both. More severe contamination could even disrupt the biological treatment process. Solid materials such as dirt, wipes and rags could clog the sewage collection system. Improperly disposed household chemicals such as medicines, paints, and solvents are harmful to the wastewater treatment plant as well but are more difficult to control, and there are targeted public education campaigns to educate the public about these issues. If not properly handled, some of the chemicals, drug ingredients such as hormones, antibiotics, antidepressants, caffeine, and even antibiotic genes will end up in receiving water, causing potential health and ecological impacts, as will be discussed below.

5.5 Emerging Water Quality Issues

5.5.1 Overview

With thousands of new chemicals invented and produced every year, more and more new chemicals and pollutants end up in ambient water. Advancements in analytical methods have enabled us to detect and quantify an ever-increasing number of new pollutants that are unknown before, often at levels of parts per billion or lower. Such findings have been made in the U.S. coastal waters, Europe, and China's Yangtze River. Many such emerging contaminants, or **contaminants of emerging concern (CECs)** are not at toxic levels yet (refer to the dose-response relationship discussed earlier in this chapter), but some contaminants are important enough due to their persistence (such as PFAS, discussed below), toxicity (such as triclosan or N-nitrosodimethylamine, NDMA, a disinfection byproduct), or endocrine disruptive properties (such as estrogen), they are more and more routinely monitored in ambient waters. Sometimes, waste materials such as nanomaterials and microplastics (see below) could become emerging pollutants of concern due to special properties and risks associated with them. In places where treated wastewater is heavily reused, such as southern California and Singapore, the water agencies keep a close watch of these emerging contaminants.

5.5.2 Microplastics[19]

Global scale of plastics

Plastics are lightweight, durable, and cheap, which make them commonly used in a wide variety of domestic and industrial applications. The global production of plastic resins and fibers increased from 2 million metric tons in 1950 to more than 400 million metric tons in

---

[19] This section was written by Dr. Adeyemi Adeleye of UC Irvine.

2015, outgrowing most other man-made materials. The major use of plastics is packaging, which leads to them becoming waste in a relatively short time compared to other largely produced materials such as steel. Almost 80% of the plastic waste generated is accumulated in landfills or the natural environment, such as the oceans. In fact, it has been estimated that 4.8 to 12.7 million metric tons of plastic waste enter the oceans from land each year. The amount of plastics estimated to be floating on the surface of oceans is 0.006 to 0.25 million metric tons, which is a lot less than the total annual input. The rest of the plastics are either below the surface of the ocean or exist in extremely small sizes that are not easily accounted for. Large plastics are broken down by forces that are present in the environment, including heat and light energy from sunlight, mechanical energy from wave action, sand grinding, contact with animals, and degradation by microorganisms and other decomposers.

Microplastics are very small plastic pieces, generally, with sizes below 5 mm.[20] They are made up of common polymers such as polyethylene, polypropylene, polystyrene, nylon, polyurethane, polyethylene terephthalate, ethylene-vinyl acetate, polyvinyl chloride, acrylonitrile butadiene styrene and fluorocarbon polymer. Microplastics may be intentionally manufactured (primary microplastics) or result from the weathering of larger plastics (secondary microplastics). Microplastics are released from consumer products such as synthetic clothing, glitter, contact lens cleaners, jewelry, and vehicle tires. Industrial sources of microplastics include particles used in air-blasting, pre-production pellets (nurdles), Styrofoam used in shipping, wastes from textile industries, and dust particles from drilling or cutting plastics.

Four major classes of microplastics have been detected in the natural environment, including, fibers, fragments, microbeads, and nurdles. Some of them are shown in **Figure 5.20**.

- First, fibers or microfibers form the largest proportion of microplastics found in natural waters. They originate from intentional and unintentional sources such as fleece clothing, diapers, cigarette butts, etc. Laundry of clothing releases thousands of microfibers into raw wastewater, and a fraction of the fibers eventually end up with the final effluent. Washing a single garment can release 1,900 to 1,00,000 fibers, depending on textile properties (polymer, knit), washing conditions (temperature, friction, velocity, washing time), use and type of detergent and softener, and garment weathering. Microfibers from cotton and wool are degradable but those from fleece are not, which causes them to persist in the environment.
- Second, fragment microplastics originate from physical breakdown of large plastics and they tend to have irregular shape, depending on the cause of degradation. This class also include tiny pieces of Styrofoam from food containers, coffee cups, and packing materials.

---

[20] One of the reasons for this limit is that water quality regulations require that trash larger than 5 mm have to be controlled. Many trash control devices are designed accordingly, leaving microplastics unregulated.

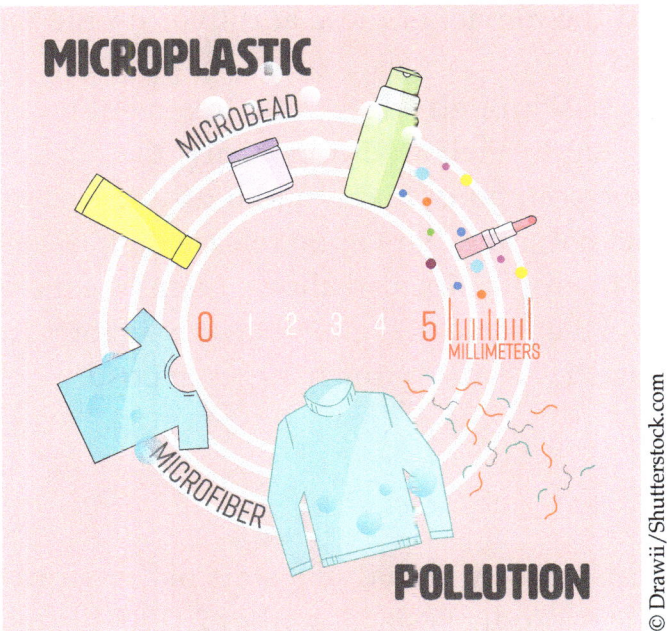

Figure 5.20 Microplastic pollution, which could result from personal care products such as lotion, shampoo, lipsticks, clothing, or breakdown of larger plastic items such as bottles.

- Thirdly, microbeads are manufactured circular microplastics measuring less than 1 mm. Their primary application is as exfoliants in health and beauty consumer products, such as facial scrubs, exfoliating soaps, and toothpaste, where their abrasive actions are desirable. Some cosmetic products may contain between 0.5% and 5% of microbeads, with an average size of 0.25 mm. One tube of toothpaste can contain up to 300,000 microbeads. Although the use of microbeads in certain consumer products has been banned in the United States, they are still present in natural waters due to their nonbiodegradability and ocean transport.
- Lastly, nurdles are small plastic pellets, which are melted and used by manufacturers to create larger plastic products. Due to their small size, nurdles sometimes spill out of transportation vehicles during delivery, which may then be mobilized by stormwater or other means.

Environmental impact of microplastics

The presence of large pieces of plastics in the natural environment is both an aesthetic and an ecological problem. The two main impacts of microplastics to the environment are toxicity to organisms and transport of pollutants. Microplastics are often mistaken for food by aquatic organisms. However, microplastics cannot be digested so their ingestion can clog the intestines of small organisms, which can lead to starvation and death. Microplastics may also bioaccumulate in the tissues of seafood and thereby enter the food chain via which higher organisms and humans can be exposed to them. In addition, the polymer additives in plastics may leach into the digestive tracts of aquatic organisms, and the impact of these additives on the well-being of organisms are not well understood.

Microplastics often pass through wastewater treatment process before reaching the natural environment. Due to their hydrophobicity, the surface of microplastics can attract persistent organic pollutants, toxins and heavy metals from wastewater, which may then be carried into natural waters upon release with treated wastewater effluent. The hydrophobicity of microplastics also allows them to concentrate some pollutants on their surface to a level that is much higher than the surrounding environment. Thus, uptake of such microplastics leads to much larger exposure dose of organisms to the concentrated pollutants.

Regulations

The European Chemicals Agency (Echa) will start phasing out "intentionally added" microplastics in consumer products starting from 2020. This regulation targets microplastics that are not necessary but have been added to products by manufacturers for convenience or profit. The California State Water Resources Control Board is working on adopting a definition of microplastics in drinking water by July 2020. In addition, by July 2021, the Board will adopt a standard methodology for testing drinking water for microplastics, adopt requirements for four years of testing and reporting of microplastics in drinking water, consider issuing quantitative guidelines, and accredit qualified laboratories in California to analyze microplastics in drinking water.

Knowledge Box

Microbeads and Microplastics: Many personal care products contains very large amount of microbeads and microplastics as filler or as abrasive. Many are manufactured nanoparticles. In a single use, an exfoliant wash can release 4,500–94,500 microbeads and a toothpaste can release around 4,000 microbeads.

5.5.3 Per- and Polyfluoroalkyl Substances (PFAS)[21]

Per- and polyfluoroalkyl substances (**PFAS**) are one of the most important contaminants of emerging concerns (CECs).[22] They are synthetic fluorinated organic compounds that have been widely produced and used since the 1940s in industrial, commercial,

---

[21] This section was written by Dr. Adeyemi Adeleye of UC Irvine.

[22] For interested readers, there is a 2019 movie "Dark Waters" about PFAS contamination, based on the 2016 *New York Times Magazine* article "The lawyer who became DuPont's worst nightmare" by Nathaniel Rich.

and consumer products due to their ability to repel oil, grease, and water. Their popular uses include oil and water repellent surface coatings for food packaging (such as pizza boxes), textiles (such as raincoat), furnishings, and cookware (such as nonstick pots), and firefighting foams used for extinguishing fires involving highly flammable liquids. The use of PFAS or PFAS-containing products can lead to their release into domestic and industrial wastewater, and different phases of the natural environment during different stages of the life cycle of the products. PFAS are hardly removed during conventional waste treatment, hence, their presence in wastewater poses a concern to the water utilities, challenging established practices such as water reuse and reclamation.

The main pathways by which PFAS are released to the environment are military and airports fire training (and fire response sites), industry that produce and/or use PFAS, landfill leachates, and wastewater treatment plants (particularly in biosolids that are often applied in agriculture). Due to their prolonged, wide use and persistence (as a result of carbon-fluorine bonds), PFAS are ubiquitous in the environment. They have been detected in house dust, air, natural waters, wastewater, sediment, and soil. PFAS have also been detected in environmental samples and wildlife in parts of the globe less occupied by humans, including the Arctic and the Antarctic due to the volatility of some PFAS, which enable them to be transported over a long distance. The detection of PFAS in water supplies is of significant concern for impacted water systems[23] due to potential adverse human health effects at low concentrations, including developmental effects to fetuses during pregnancy, kidney cancer, liver tissue damage, and thyroid effects. They have been found in the blood of nearly all people tested in several national surveys.

Although about 5,000 PFAS compounds have been produced globally, perfluorooctanoic acid (PFOA) and perfluorooctane sulfonate (PFOS) have been the most extensively produced PFAS in the United States. As a result, they are the two most detected PFAS in the environment and in drinking water. PFAS are generally made up of a hydrophobic backbone or "tail" (with 4–14 carbon atoms) and a hydrophilic functional group or "head", as depicted in the example of PFOA shown in **Figure 5.21**. The most common functional groups that make up the headgroups of PFAS are carboxylates, sulfonates, phosphates, and sulfonamides. Fluorine atoms are bonded to most, if not all, the carbon in the backbone of PFAS. The high content of fluorine (carbon-fluorine bonds) in PFAS causes them to persist in the environment and in the tissues of living organisms. Concerns about the persistence and health effects of PFOA and PFOS have led to phasing out their production in the United States. However, the compounds are still produced internationally and can be imported into the United States in consumer goods. Other PFAS compounds (mainly shorter-chains), and PFAS-containing goods and materials are still produced and used in the United States.

---

[23] Many water wells around the United States have been shut down due to PFAS concerns as of April 2020.

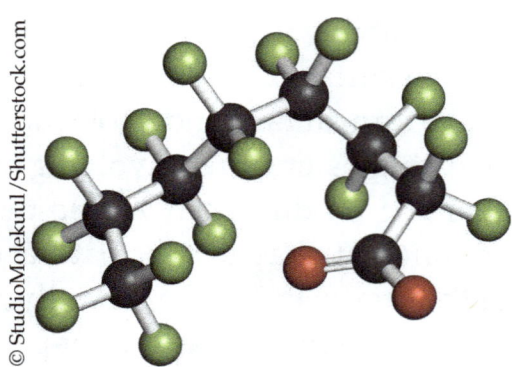

Figure 5.21 Chemical structure of perfluorooctanoic acid (PFOA). Green: fluorine. Black: carbon. Red: oxygen.[24]

In May 2016, the USEPA issued a lifetime health advisory of 70 parts per trillion (ppt) for PFOA and PFOS in drinking water. Based on the advisory, the EPA advised water municipalities to notify their customers when the levels of PFOS and PFOA in community water supplies exceed 70 ppt. Health advisories are not regulations and are therefore not enforceable. However, several states in the United States have established enforceable standards. California established a response level of 10 ppt for PFOA and 40 ppt for PFOS from January 2020. This implies that a water source should be taken out of service (many have been) or treated if the levels of PFOA and PFOS exceed the response level. In addition, California has a notification level of 6.5 ppt for PFOS and 5.1 ppt for PFOA. Water municipalities that detect PFOA and PFOS levels that exceed the Notification Levels are required to report these to the governing body for the areas where the water has been served, and are encouraged to inform their customers. California is still in the process of establishing regulatory requirements, or maximum contaminant levels, for PFOA, PFOS, and other PFAS commonly found in drinking water.

5.5.4 Other Contaminants of Emerging Concern

CECs are not necessarily new chemicals—some of them could have been in the environment for a long time but their presence and significance are only recently being evaluated, such as PFAS discussed in Section 5.5.3. CECs now often include **pharmaceuticals and personal care products** (PPCPs) such as prescribed or over-the-counter drugs like antidepressants, blood pressure prescriptions, ibuprofen; bactericides such as triclosan; and other personal care products such as sunscreens,[25] synthetic masks,

---

[24] There is a hydrogen atom connected to the oxygen atom to the right. It is not shown pursuant to organic chemistry convention.

[25] Recently it was found that some active ingredients in sunscreens could be harmful to coral reef. As a result, some places such as Hawaii have banned certain sunscreens.

and so on. Many of them are used by people in their daily life and are discharged with wastewater. Since wastewater treatment plants are not designed to treat these compounds, these compounds may be discharged to receiving water and impacting aquatic life and ecosystem health. For example, many CECs have been detected in receiving water and in biota in many places around the world.

Endocrine disrupting compounds (EDC) are compounds that could interfere with human endocrine systems. EDCs may include synthetic estrogens, naturally occurring estrogens, as well as many others, including some organochlorine pesticides discussed earlier. These compounds are structurally similar to some hormones and can act like one, hence disrupting normal functions. Many EDCs are CECs and/or PPCPs, so these compounds may overlap each other. Managing these pollutants poses unique challenges, including sampling, analysis, and mitigation. Due to lack of regulatory standards, most of the work remains on research level and is voluntary (i.e., not required by regulatory permits).

5.5.5 Long-Range Transport of Pollutants

As shown in Figures 11.3 and 11.13, global ocean circulation could move local or regional pollution to other places around the world. Migration of some animals could result in the propagation of some pollutants, especially the persistent organic pollutants (POPs, as discussed in Chapter 4) and other persistent and bioaccumulative pollutants. The example mentioned in Chapter 4 about POPs detected in polar bears should be a result of long-range transport of these POPs in both air and water. Another good example is the radionuclides that were spilled from Japan's Fukushima Nuclear Power Plant as the result of a tsunami on March 11, 2011.[26] The incident caused release of a number of radionuclides, most notably cesium-134, which amounted to about 26.5 kg. After two and half years, the radionuclides reached the west coast of the United States and Canada through Kuroshio Current and the California Current. Even though the radionuclides have been diluted to nonhazardous levels, this incident shows that long-range transport of contaminants is possible.

5.6 Water Quality Monitoring

Monitoring and assessment are foundational components in water quality management, as shown in Figure 5.1. Monitoring is to collect water, sediment, or other media samples for analysis of water quality and to make decisions based on the results.

---

[26] Nuclear energy and safety issues, including the Fukushima accident, will again be discussed in Chapter 10.

Assessment is to evaluate the analytical data to determine if the water quality has met the regulatory standard or other threshold/guidelines. If the water quality fails to meet the standards, regulatory and management actions will be needed. With technology advancement in analytical instruments, many pollutants can now be analyzed in parts per billion (ppb) or parts per trillion (ppt) levels. This allows for investigation of chronic and latent ecotoxicological or human health impact from low level contaminants (such as PFAS and other contaminants of emerging concerns). New analytical tools have emerged recently such as nontarget analysis,[27] DNA microarray,[28] and increasingly speedy and accurate toxicity test from *in vivo* (in the tissue) to *in vitro* (in the lab) to *in silico* (high throughput, predictive toxicology), that keep pushing the boundary of analytical sensitivity and versatility. On the other hand, new technologies such as in situ sensors, cloud-based computing and data processing, and machine-learning algorithms have made in situ and real-time monitoring and decision-making possible and allowed environmental managers to act quickly and effectively.

On the other end of the spectrum in water quality monitoring, more integrated monitoring effort has been initiated that takes into consideration biological conditions, such as toxicity, biological integrity, and a combination of other information to assess the overall health of aqueous environment. Biological components of water quality monitoring have some advantages over simple chemical analysis because they provide a holistic condition assessment of biological resources, which are often the ultimate target of protection. The organisms are also a better indicator of environmental health because their condition is a reflection of the integrated physical habitat and chemical exposure during their lifetime, not just the snapshot when a water sample is collected and analyzed (for example, see Figure 5.6). For example, sediment quality in California, USA and in Europe is now assessed using a combination of chemical, toxicological, and biological integrity parameters. Stream health is increasingly being assessed by the bioassessment tool in California (Figure 5.22).

---

[27] Non-target analysis (NTA) is a technique used for trace organics analysis that all potential pollutants, often amounting to thousands of both known and unknown compounds, are scanned. The rich information collected could be used for site characterization, pollution fingerprinting, and identification of new pollutants.

[28] DNA microarray is a collection of tiny DNA spots on a surface to be used to do a range of different tasks, including toxicity testing of some pollutants that is otherwise difficult or time-consuming.

Figure 5.22 Bioassessment in action in a mountain stream in California, USA. During the bioassessment process, where a total of 11 transects over a 150 m length of a stream will be monitored or sampled for physical habitat, water quality, benthic algae, benthic macroinvertebrates. A score for the California Stream Condition Index (CSCI) will be calculated using a complex algorithm that indicates the overall biological condition of the stream.

Photo credit: Jian Peng

Further Reading

Jambeck, Jenna R., et al. "Plastic Waste Inputs from Land into The Ocean." *Science*, 347, no. 6223 (2015): 768–71.

Geyer, R., J. R. Jambeck, and K. L. Law. "Production, Use, And Fate of All Plastics Ever Made." *Science Advances*, 3, no. 7 (2017): e1700782.

Prata, Joana Correia. "Microplastics in Wastewater: State of the Knowledge on Sources, Fate and Solutions." *Marine Pollution Bulletin*, 129, no. 1 (2018): 262–65.

Muir, Derek, et al. "Levels and Trends of Poly- and Perfluoroalkyl Substances in the Arctic Environment—An Update." *Emerging Contaminants*, 5 (2019): 240–71.

United States Geological Survey. Pesticide National Synthesis Project, website: https://water.usgs.gov/nawqa/pnsp/usage/maps/. Access date: February 15, 2020.

Carson, Rachel. Silent Spring, Boston, MA: Houghton Mifflin Company, 1962, ISBN 0-618-25305-x.

Drever, James I. The Geochemistry of Natural Waters: Surface and Groundwater Environments, 3rd Edition, Upper Saddle River, NJ: Prentice-Hall, Inc., 1997, ISBN: 0-13-272790-0.

Garrels, R.M. and F.T. McKenzie. Evolution of Sedimentary Rocks. New York: W.W. Norton, Chapter 6, 1971.

Buchel, K.H. Chemistry of Pesticides, USA: John Wiley and Sons, 1983, ISBN 0-471-05682-0.

Fletcher, T. et al. "SUDS, LID, BMPs, WSUD and more—The Evolution and Application of Terminology Surrounding Urban Drainage." *Urban Water Journal*, 12, no. 7 (2015): 525–542, http://dx.doi.org/10.1080/1573062X.2014.916314.

USEPA, 2010. NPDES Permit Writers Manual. EPA-833-K-10-001 September 2010. Washington DC, USA.

Cao, Y., et al. 2017. A Human Fecal Contamination Score for Ranking Recreational Sites Using the HF183/BacR287 Quantitative Real-Time PCR Method. September 2017.

Urban Water Resources Research Council. Pathogens in Urban Stormwater Systems, August 2014.

US Environmental Protection Agency (USEPA). Recreational Water Quality Criteria, November 2012.

Dufour A. et al. (eds.). Animal Waste, Water Quality and Human Health. USA: World Health Organization, USEPA, and International Water Association Publishing, 2012.

Wang, Ruo-Nan, et al. "Occurrence of Super Antibiotic Resistance Genes in the Downstream of the Yangtze River in China: Prevalence and Antibiotic Resistance Profiles." *Science of the Total Environment* 651, Pt 2 (2019): 1946–57.

Harb, Moustapha, et al. "Background Antibiotic Resistance and Microbial Communities Dominate Effects of Advanced Purified Water Recharge to an Urban Aquifer." *Environmental Science and Technology*, 6, no. 10 (2019): 578–84.

Mazor, R. et al. Bioassessment of perennial streams in Southern California: A report on the first five years of the stormwater monitoring coalition's regional stream survey. Southern California Coastal Water Research Project. Technical Report 844. Costa Mesa, California, USA, 2015.

World Bank, 2012. The World Bank data; GNI per capita, Atlas method (current US$). http://data.worldbank.org/indicator/NY.GDP.PCAP.CD.

Sato, T. et al. "Global, Regional, and Country Level Need for Data on Wastewater Generation, Treatment, and Use." *Agricultural Water Management*, 130 (2013): 1–13.

Brooks, B.W. et al., 2020. Toxicology Advances for 21st Century Chemical Pollution. One Earth, 2, April 24, 2020.

Stone, W.W., R.J. Gilliom, and J.D. Martin. An overview comparing results from two decades of monitoring for pesticides in the Nation's streams and rivers, 1992–2001 and 2002–2011: U.S. Geological Survey Scientific Investigations Report 2014–5154, 23 p., 2014, http://dx.doi.org/10.3133/sir20145154.

Web Resources

Fukushima Daiichi nuclear disaster: https://en.wikipedia.org/wiki/Fukushima_Daiichi_nuclear_disaster

Think before you flush: https://thinkbeforeyouflush.org/what-to-flush/

USEPA National Pretreatment Program: https://www.epa.gov/npdes/national-pretreatment-program

USEPA SWMM model: https://www.epa.gov/water-research/storm-water-management-model-swmm

https://finance.yahoo.com/news/global-pesticides-market-opportunities-strategies-120900830.html. Access date: February 15, 2020

California Stormwater Quality Association (CASQA), http://www.casqa.org

Antibiotic resistance genes in EU: https://www.ecdc.europa.eu/en/antimicrobial-resistance

USEPA's pretreatment program: https://www.epa.gov/npdes/national-pretreatment-program.

A good list of best management practices to control agriculture nutrient by the Ohio State University: https://agbmps.osu.edu/bmp

Methods for biological pest control: https://en.wikipedia.org/wiki/Biological_pest_control

Information about pheromone trap for biological pest control: https://en.wikipedia.org/wiki/Pheromone_trap

We are living in the Plastic Age: https://www.smithsonianmag.com/smart-news/are-we-living-plastic-age-180957817/

United States Geological Survey (USGS) National Water Quality Assessment Project, Pesticide National Synthesis Project data: https://water.usgs.gov/nawqa/pnsp/usage/maps/compound_listing.php

Internet of Water—a water data infrastructure initiative to meet the 21st Century Challenges: https://internetofwater.org/

USEPA National Aquatic Resources Surveys: https://www.epa.gov/national-aquatic-resource-surveys

The National Water Quality Portal—Managed by USEPA, it is a national scale portal for sharing water quality monitoring data with hundreds of millions of data entries: https://www.waterqualitydata.us/

The Johns Hopkins University professor Abel Wolman and his contribution to chlorination of drinking water: https://en.wikipedia.org/wiki/Abel_Wolman

The Baltomore Trash Wheel: https://en.wikipedia.org/wiki/Mr._Trash_Wheel

Questions and Exercises

1. Total maximum daily load is a tool that prescribe how much pollutants can be discharged via point sources and nonpoint sources without exceeding the water quality standards. The formula is: TMDL = (all point source loads) + (all nonpoint source loads) + (margin of safety). Can you think of any reasons why a margin of safety is needed?
2. List the advantages and disadvantages of biological monitoring for water quality assessment.
3. List several reasons why stormwater is challenging to manage.
4. List several low impact development types and explain how they can improve stormwater quality and/or reduce flooding risks.
5. Why does stormwater runoff usually have a steep rising trend before the peak of the flow but a more gradual falling trend after the peak flow (see Figure 5.14)?
6. Use online resources to identify the main pollutants in the drinking water and surface waters (rivers, lakes, and coastal ocean) in the area where you live. Identify their sources and explore possible management options.
7. Identify (maybe with the aid of a map) where stormwater generated from where you live will end up, be it a river, lake, or ocean. Identify potential pollution sources and what you could do to minimize or eliminate the pollution.
8. Identify the drinking water supplier and wastewater treatment facility that serve the area where you live. Read their water quality report and identify any pollutants of concern.

Chapter 6

Water Resources

Learning Outcomes	152
Key Concepts	152
6.1 Overview	152
6.2 Water Quality and Water Resources	155
6.2.1 Nexus Between Water Quality and Water Resources	155
6.2.2 Novel Potable Water Production Technologies	156
6.3 Water Shortage and Drought	158
6.3.1 Global Water Shortages	158
6.3.2 Water Security and Resilience	160
6.3.3 Water Resource and Sustainability	161
6.4 Water Supply Infrastructure and Its Environmental Impacts	161
6.4.1 History of Water Supply Infrastructure	161
6.4.2 Dams around the World	162
6.4.3 Ecological and other Impacts of Dams	163
6.5 Water Reuse and Recycling	168
6.5.1 Overview of Water Reuse	168
6.5.2 Types of Potable Water Reuse: Direct and Indirect	168
6.6 Desalination	169
6.6.1 Overview of Desalination	169
6.6.2 Desalination Process and Potential Impacts	170
6.7 Water Balance and Integrated Water Management	171
6.7.1 Overview of Water Balance	171
6.7.2 Case Study: Water Balance for North Orange County, California, USA	172
6.8 Water Footprint	174
6.8.1 The Concept of Water Footprint	174
6.8.2 Water Footprint of Common Items	174
6.8.3 Water Footprint and Sustainability	175
Further Reading	177
Web Resources	178
Questions and Exercises	178

Learning Outcomes

- Knowledge of the connections between water quality and water resources
- Appreciation of the fact that only a very small portion of the water on Earth is available for human consumption
- Knowledge and understanding of the global water resources issues, especially water shortage
- Knowledge of basic components of a water supply infrastructure
- Knowledge and understanding of the principle of water balance
- Knowledge of how to calculate water footprint

Key Concepts

Water resource; water resilience; water security; (California) State Water Project; Roman aqueduct; Bay-Delta; California Water Fix; water footprint; tiered water rate; desalination; Water Reuse and Recycling; water balance

6.1 Overview

As discussed earlier, water is of critical importance for all living organisms on Earth. In fact, without water, there likely will be no life on Earth. During our search of extraterrestrial life or other habitable planets, one of the most critical criteria is the existence of water, especially in liquid form. As shown in Figure 2.3, there is quite a sizable amount of water on Earth, and 71% the Earth's surface is covered by water. However, if we examine how much of it is actually available for human consumption, the result is shockingly depressing. As shown in **Figure 6.1**, of all of the water on Earth, only 2.5% is freshwater. Of this small amount of freshwater, nearly 69% exists in glaciers and ice caps in polar regions and is unavailable to us. The rest of 30% is groundwater.[1] Only about 1.2% is surface freshwater and only 21% of this 1.2% is in rivers and lakes that we can readily use. Therefore, the prose 'water, water everywhere, nor any drop to drink'[2] that lamented by a mariner who sees water everywhere in the ocean but cannot drink it, it is almost as fitting to describe the **water resources** situation on Earth as a whole.

In many places, including in some developed countries, water shortage is a worsening crisis. For example, in California, there is an ongoing threat of water shortage in southern California despite major efforts to move water from northern California and from the Colorado River. For the Colorado River basin, the water resources have been allocated to several states (including California, Arizona, Nevada, among others) that the river (one of

---

[1] Even though there is groundwater overdraft at some places, overall, we are tapping into a very small fraction of the global groundwater resources.

[2] From the poem titled The Rime of the Ancient Mariner, by Samuel Taylor Coleridge.

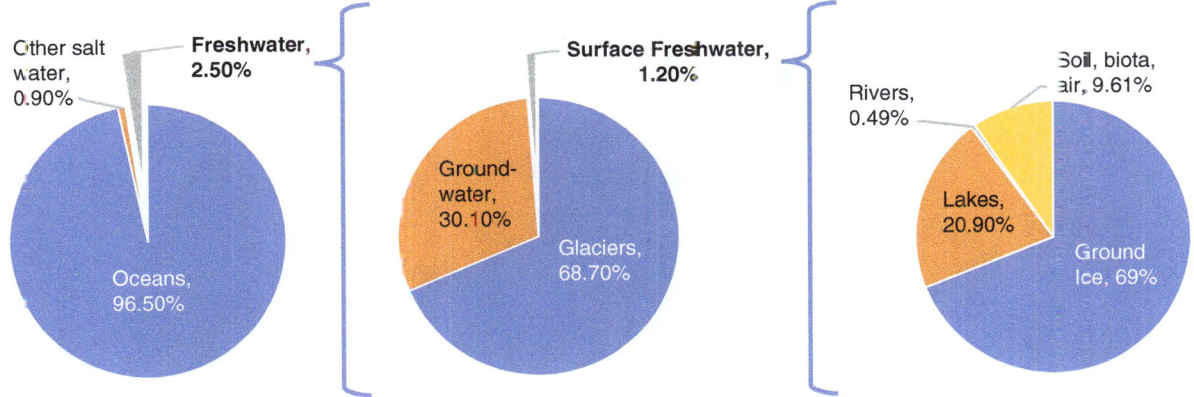

Figure 6.1 Global water resources (recreated from U.S. Geological Survey, Water Science School. https://www.usgs.gov/special-topic/water-science-school). Only 0.0064% of the global water exists as freshwater in rivers and lakes.

the largest in the United States) could dry up before reaching the Gulf of California, its natural final destination. In the United States as a whole, the total water withdrawal increased from 680 gigaliters per day in 1950 to 1220 gigaliters per day in 2015, a nearly 80% increase.[3] In China, water from Yangtze River is moved thousands of kilometers north to supplement increasing water needs in the dryer regions in northern China. In the arid Middle East, water issues could be the cause of regional conflicts, such as those in the basins of Jordan River; Tigris-Euphrates River; and Nile River. In many other developing countries, the situation can be quite challenging too. Roughly 1 billion people lack access to safe drinking water, and another 2 billion lack adequate sanitation.

> ### Knowledge Box
>
> **The Los Angeles Aqueduct**—The Los Angeles Aqueduct (Figure 6.3) was built in 1913 by the Los Angeles Department of Water and Power under the supervision of chief engineer William Mulholland. The system delivers water from Owens River in the Eastern Sierra Nevada Mountains to Los Angeles, California. This 675 km aqueduct uses gravity alone throughout its course and is still in operation more than 100 years later. The aqueduct has spurred rapid growth of Los Angeles and surrounding areas. By 1928, the area of Los Angeles grew more than sevenfold to 440 square miles. However, the City quickly outgrew the water supply from the aqueduct and more aqueducts, such as the California Aqueduct and Colorado River Aqueduct (Figure 6.2) were subsequently built. It is safe to say that without the Los Angeles Aqueduct and other water projects, there would be no Los Angeles as an international metropolis, nor would there be Hollywood, the global motion picture capital, and movie blockbusters such as "The Water World."

---

[3] Data source: https://waterdata.usgs.gov/nwis/water_use/

Figure 6.2 The California **State Water Project** water transfer network in California, USA. Recreated from California Department of Water Resources (2020).

Figure 6.3 A section of the Los Angeles Aqueduct, part of a 675 km water conveyance infrastructure that brings water from Sierra Nevada to Los Angeles.

Water resource issues are rarely isolated issues. Prehistoric humans established settlements near the water sources. Nomads chase water resources throughout the year. In the modern society, where a vast population has mandated civil engineers to explore the available water resources to the extent possible by drilling for groundwater and by moving water from one place to another.

6.2 Water Quality and Water Resources

6.2.1 Nexus Between Water Quality and Water Resources

In many cases, water quality issues may be the cause of water resources issues because human consumption requires high quality water. Freshwater with pollutants or other water quality issues, if not removed effectively and efficiently with a reasonable cost, will cause water resources issues. For example, the megacity Shanghai in southeast China sits next to Yangtze River (China's largest river that accounts for a third of all available water resources in China) and some other river networks (such as Huangpu River) that provide more than enough freshwater for its residents of more than 20 million people. However, the City has faced water resources challenges over the years and was forced to move its municipal water intake many times due to water quality issues. Currently it takes water from four different locations, with two newly constructed reservoirs built from a sandbar-formed island in the middle of Yangtze River (**Figure 6.4**). Even so, these reservoirs are faced with challenges in upstream water quality issues and seawater intrusion during high tides. The same is true for many other regions within the Yangtze River watershed where water quality issues often cause strain on water availability.

Knowledge Box

Water Quality Challenge of Yangtze River—There is significant water quality challenge faced by Shanghai when they take their drinking water source from the Yangtze River. Being the largest river in China and 6,300 kilometers in length, Yangtze river is in the receiving end of more than 2 billion tons of treated wastewater from over 6,000 wastewater discharge points along the river. Since the 1980s, water quality of the river deteriorated steadily with economic growth throughout its vast watershed (1.8 million square kilometers). In the mid 2010s, the situation became severe enough to raise attention from China's central government. In 2020, China's president proclaimed that there will be a national strategy to protect the Yangtze River watershed. Since then, water quality of the river has improved but many challenges remain.

In the United States, the Safe Drinking Water Act and the Clean Water Act regulate the safety of water supplies. For water resources that cannot meet regulatory requirements, further treatment or alternative water sources have to be explored. The recent Flint,

Figure 6.4 Drinking water sources for Shanghai, China. The City used to draw water entirely from Huangpu River. With population increase and pollution concerns, the water intake moved upstream several times, and added three more reservoirs.

Michigan water crisis in 2014 is a prime example. City of Toledo, Ohio, USA experienced a contamination of its drinking water supply by microcystin, a toxin caused by harmful algal blooms, in 2014.

In both cases, public health crises took place, causing severe water resource issues. For a good overview of national water quality and water resources data, please refer to the National Water Information System (NWIS) website managed by the United States Geological Survey (https://waterdata.usgs.gov/nwis). In its report published in 2021 on the monitoring results of 1,200 wells surveyed in 46 states (representing 70% of the volume of the groundwater supply in the US), USGS found that 41% of these wells have at least one of the 109 pesticides that were analyzed. Several wells have pesticides at levels that could threaten human health. If these wells are decommissioned, there will be water supply issues because limited options are available at these locations.

6.2.2 Novel Potable Water Production Technologies

The important linkage between water quality and water resources could also be shown by several new water supply techniques. In a desert or at remote military bases where there is no water supply, water could be harvested from the air by cooling and condensation of water vapors either naturally or by powered devices that create a cold surface to facilitate condentation. Many "survivor guides" point out ways to treat unsanitary water in the wild, which could contain unhealthy minerals or pathogens. These treatments could be simple alum, filtration devices, chlorination pills, and so on. In space shuttle or space station, astronauts' wastes such as shower, condensate, and urine (not fecal waste!) are completely recycled to produce 3.6 gallons of water each day. This water recycling scheme conserves water and minimizes payload. It can easily be imagined that if humans were to colonize exoplanets such as Mars or other places, the first and foremost concern would be reliable and sustainable ways to produce clean and safe water.

> ### Knowledge Box
>
> **Water Recycling System at the ISS:** The International Space Station (ISS, Figure 6.5) has a complex and closed water system to recycle every drop of water available, for a very good reason: it costs more than $1,800 to send a pound of payload to space. So, the astronauts drink a filtered mixture of recycled shower water, wash water from the sink, sweat, and urine. The station also keeps some water in reserve in case of an emergency. The ISS is split into two sections, one run by the United States that recycles all of the above sources, the other by Russia, which does not recycle urine and produces less water. Occasionally, the American astronauts will go over to the Russian side to fetch more 'liquid bodily waste' for recycling. The two sides of the ISS disinfect their water two different ways. Americans use iodine, and Russians use silver.

Figure 6.5 Yucky cheers: ISS-19 Crew members drink water from the Water Recovery System on the International Space Station: Gennady Padalka (center) and Flight Engineers Mike Barratt (right) and Koichi Wakata (left).

> **Knowledge Box**
>
> **Water purification in the wilderness**—Untreated natural water may appear to be clean but it could contain dangerous pathogens such as cryptosporidiosis, protozoa, parasitic worms, cholera, typhoid, and other viruses (see Chapter 5 for more details on microbial water quality). If the water is not devoid of living organisms, which is a good indicator of toxicity, the safe way of treating water in the wilderness is filtration or disinfection, or both. There are many commercial products that one could use to purify water of unknown quality. There are also other ways to collect clean water if you have tools and materials. These include solar still, morning dew, or from plant transpiration. One can never fully appreciate the value of clean water unless there is a short supply or worse, a life and death situation.

6.3 Water Shortage and Drought

6.3.1 Global Water Shortages

Figure 6.6 showed a dramatic difference in the annual average per capita water uses in different countries around the world. The United States, with over 1,200 cubic meter water uses per capita per year, is nearly 100 times more than that for Uganda. Water use pattern is determined by many factors, such as national wealth, water use habits, and available water resources. Figure 6.7 depicts the evolution of water use data for three groups of nations from 1901 to 2010. For the traditional developed industrial

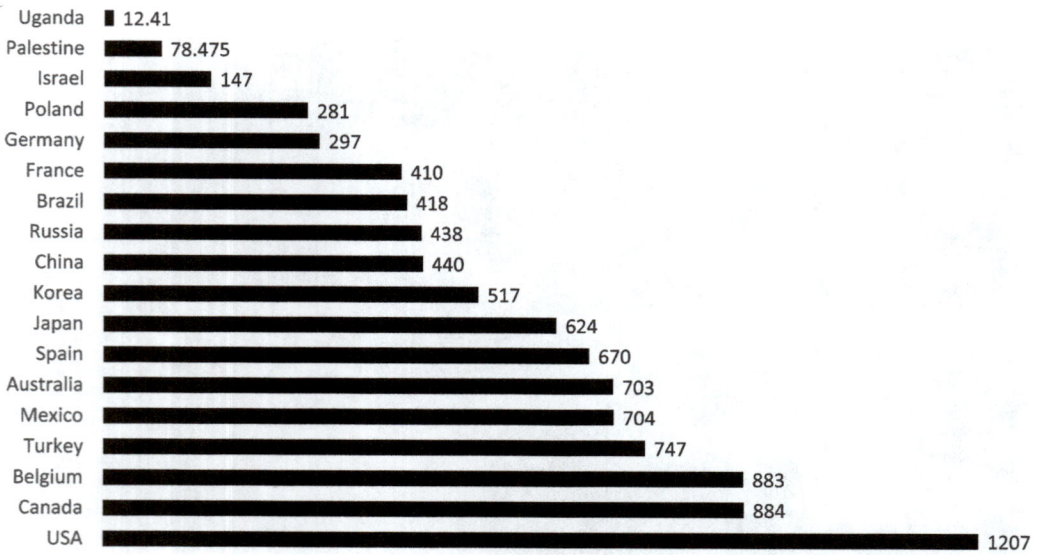

Figure 6.6 Per capita annual water consumption (cubic meter) by nation. (data source: Statista and Worldonmeters.com). Created by Jian Peng

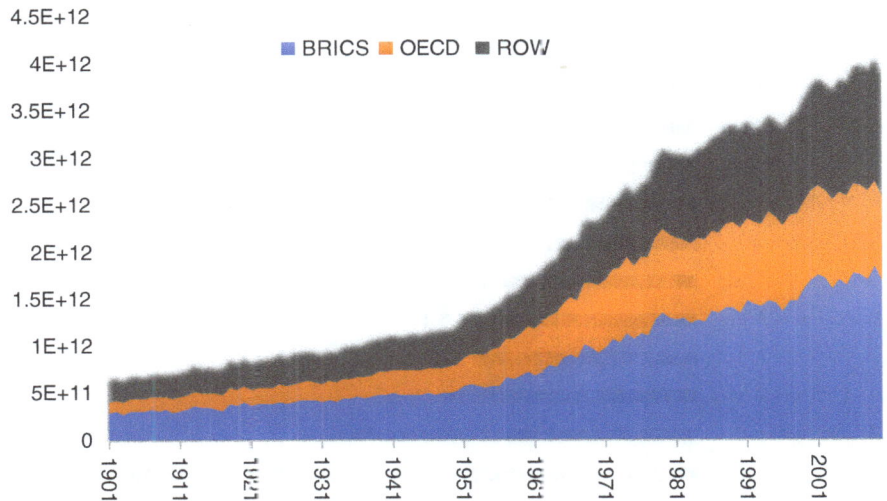

Figure 6.7 Water consumption rates trends from 1901 to 2010 for BRICS (Brazil, Russia, India, China, and South Africa); OECD (Organization of Economic Co-operation); and ROW (rest of the world).

Source: ourworldindata.org/water-use-stress. Created by Jian Peng

countries such as the United States, western Europe (the Organization of Economic Co-operation, OECD), their collective water consumption has remained largely even after the 1980s. For the developing countries that see quick economic developments in recent years, such as Brazil, Russia, India, China, and South Africa (so-called "BRICS" countries), a steady increase can be seen. The rest of the world (ROW) has a similar trend as the OECD countries. All three groups of the countries see an increase in water consumption. The overall global water consumption has increased nearly sevenfold during the last century, putting a significant strain in global water resources.

Also shown in Figure 6.6, where many nations with very low per capita water consumption are not shown, there is great disparity in water consumption for different nations. **Figure 6.8** shows the availability of renewable freshwater resources by region. The disparity among different regions around the world is even greater. For the driest areas such as Northern Africa and Central Asia, the available water resources are a tiny fraction of those for South America and Asia. In these dry and economically challenging areas, water scarcity is a significant challenge. Overall, nearly half of the people (3.4 billion in 2010) on Earth live in regions where their freshwater supply is insecure. A key consideration in water resources and water supply is **water security** or **water resilience**. Most of the communities can meet their water demand one way or another, but many do not have enough capacity or resilience in the face of drought, natural disaster, or other incidents that could impact their main water supply. For example, one major issue in the Los Angeles area is the threat of major earthquakes or other factors that could cause water infrastructure failures. Such failures could disrupt its water

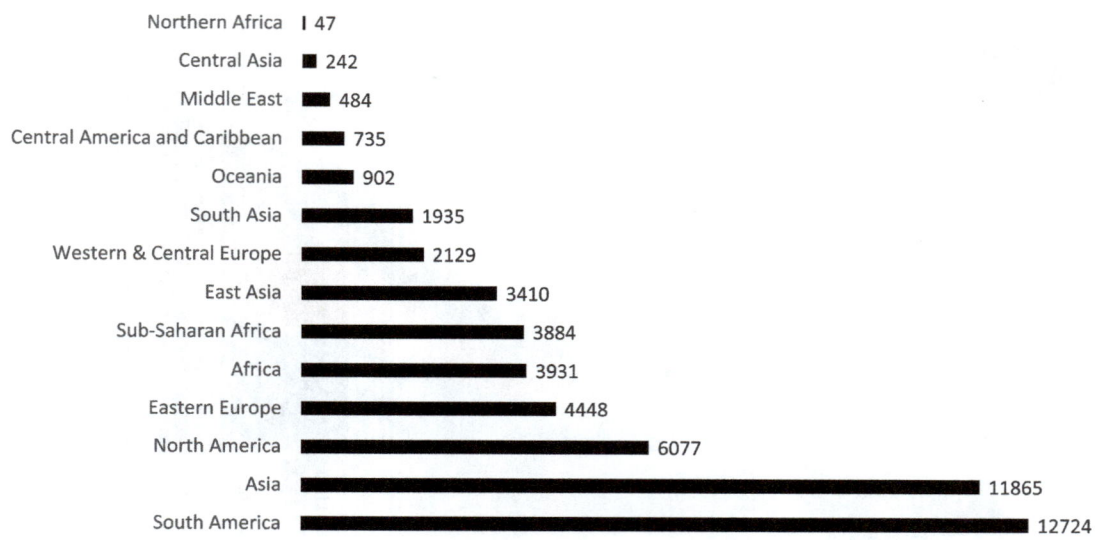

Figure 6.8 Renewable internal freshwater resources (cubic kilometer) by region in 2014.
Data source: The World Bank. Created by Jian Peng

supply from aqueducts and storage reservoirs. With little water sources locally, this threat is fairly significant. In fact, such an incident happened in 1928 on the St. Francis Dam that was one of the key water supply infrastructures for the city. The city was forced to change its water supply strategy since then.

6.3.2 Water Security and Resilience

Another important consideration in water security and resilience is the preparedness against drought and climate change. This is critical to every city, especially for those which are already facing the threat of water shortage. As will be discussed in Chapter 11 of the book, human-induced or natural climate change will definitely result in more climate anomalies. For some regions, this could mean extreme drought period that some cities simply cannot deal with. For example, Australia suffered the Millennium Drought during the 2000s with 2006 being the driest year on record. The drought placed extreme pressure on agricultural production and urban water supply in much of southern Australia. It has led to the construction of six major seawater **desalination** plants to provide water to major cities, and to changes in the management of water. To help cope with the challenge, the citizens of Melbourne responded by saving 47% of drinking water use. In 2019, the South Africa's capital city Cape Town was three months away from completely running out of water. Already being a nation with one of the lowest per capita water usage, South Africans responded by even more drastic water conservation, groundwater exploration, source area protection, invasive plant species clearing, and catchment management, among others, that helped avert the crisis.

> ### Knowledge Box
>
> **All Bets are Off:** In the book "The Day New York Went Dry" by Charles Einstein (not related to Albert), the protagonist Marlowe was pondering several options for New York City, which was facing a water shortage. There were four options. First, it needed to rain a lot to alleviate the water shortage. Second, the City needed to implement water rationing by shutting down the water system for hours every day. Third, the City could start taking water from the Hudson River and building a new treatment plant to treat the river water for use. Fourth, water conservation, especially for residents. In the end, however, when asked about the options if there is a drought, Marlowe quipped, "(T)hen all bets are off."

6.3.3 Water Resource and Sustainability

Ultimately, to deal with water shortage in the long run, mankind will have to consider some fundamental issues such as overpopulation, climate change, technological innovation, and water conservation measures. With limited resources at hand, water agencies and local governments are often unwilling to commit sufficient funding to improve water security unless pressured by real water shortage. Maybe they simply need situations such as Australia and South Africa to fundamentally improve their water resilience. Otherwise, the damage to the society could be catastrophic.

6.4 Water Supply Infrastructure and Its Environmental Impacts

6.4.1 History of Water Supply Infrastructure

Because water is so important, water supply infrastructure is often the first thing to be built before a community can be established. In the history of mankind, these infrastructures have been simple canals, pipes, and weirs to divert water from creeks to agricultural fields or communities for use. Increasingly, greater numbers and larger scale projects are built to cope with increasing numbers of population and associated agriculture and industries. Among the historic water infrastructure, **Roman aqueduct** (Figure 6.9) are one of the best examples despite the fact that they were built more than two thousand years ago. The Romans constructed them to bring water from outside to their cities for public baths, restrooms, fountains, and other everyday uses. They also supported many other uses such as farming, mills and mining. All Roman aqueducts are gravity driven and some even go underground. These aqueducts were the reason why Rome was able to support more than a million residents.

Figure 6.9 Roman aqueducts.

6.4.2 Dams around the World

Dams are another important type of water supply infrastructure to elevate water level in a river or lake for diversion. The top five dam-building nations around the world are China, USA, India, Japan, and Brazil. Currently, about 15% of the world's rivers flows are controlled by dams. Below is a list of number of dams in different countries/regions:

- In the United states, there are more than 84,000 dams, with more than 8,100 major ones that are taller than 50 feet, or those with a normal storage capacity of 5,000 acre-feet or more, or those with a maximum storage capacity of 25,000 acre-feet or more (e.g. **Figure 6.10**). However, dam-building activities have decreased markedly in recent years.

Figure 6.10 Left: Hoover Dam in Nevada (left) is the largest dam measured by trapped water volume (32 billion cubic meter). Oroville Dam in California (right) is the tallest dam in the United States (235 m).

Figure 6.11 The Three Gorges Dam.

- In China, there are more than 87,000 dams with one of the largest dams in the world, the Three Gorges Dam (**Figure 6.11**). With the Yangtze River watershed alone, there are more than 50,000 dams (see the Knowledge Box).
- In Europe, there are no statistics for all of the dams of different sizes, but there are 1,172 large dams. Similar to the United States, dam-building activities have decreased markedly in recent years in Europe.
- Worldwide there are 57,000 large dams (15 meters or higher) and 300 "giant dams" that are 150 meters or taller. Based on the same criteria, China has 23,000 large dams, and United States has 9,200 large dams. The slight discrepancy is due to unit conversion on height and volume.

6.4.3 Ecological and other Impacts of Dams

Dams are important infrastructure that generate clean hydropower (see Chapter 9) and provide crucial water resources to billions of people around the world. However, building large dams also displace millions of people (estimated at 40–80 millions, mostly in China and India). Dams are also damaging to river ecosystems, resulting in fragmentation and loss of habitat connectivity and fish migration routes. Since 1970, there has been 83% decline in freshwater species. Pollution and fisheries may have contributed to the decline, but Dams may have caused the majorities of the decline because many fish species cannot spawn after their migration routes are cut off.

Dams cut off water flow as well as sediment transport, a critical natural process. The accumulated sediment by dams hamper their capacities and is costly to remove. For example, the removal of Glines Canyon Dam in Washington, USA, unleashed 200 million tons of sediment accumulated behind it. While the sediment was supposed to be transported to downstream unobstructed, a sudden release of such a large amount of sediment caused short-term, localized impact as well. The retention of sediment by dams will cause erosion and other hydromodification in the downstream portions of river sections. This is because that in a river system, suspended and bedded sediment reach a state of dynamic equilibrium under a given flow regime (i.e. flow rate and discharge rate).

When there is too much suspended sediment in the water, sediment will accumulate. Conversely when there is too little sediment, as in the case behind a dam, bedded sediment will be scoured by the river flow, causing erosion.

The Conowingo Dam on Susquehanna River that flows into Chesapeake Bay is an excellent example of how dams can have significant benefits as well as potential issues. This dam has helped reduce sediment and nutrient input to Chesapeake Bay for the protection of the Bay's ecosystem and aquaculture. However, it has trapped so much sediment that its utility has diminished. Similar to other large dams, it has significant local ecological impacts. Currently, only one-third of the world's longest rivers remain free-flowing, and the rest are hampered to different degrees by dams and other infrastructure.

Knowledge Box

The Three Gorges Dam: The Yangtze River basin covers 20% of the geographical territory of China and sustains 400 million people, or 43% of the country's population. The Yangtze River region makes up more than a third of China's freshwater reserves, contributing 42% of China's GDP and 73% of the country's hydropower. Intensive development, however, has taken its toll on China's mother river. It also has over 50,000 dams of different sized cross its watershed, with the famous Three Gorges Dam being the largest on the River and in the world, with 22,500 megawatts of installed capacity and a volume of 39 km^3 of water. The dam also drastically reduced the flooding risk for millions of people downstream and facilitated river navigation. However, its ecological impacts are also significant. More than 1.3 million people were forced to relocate. Many species, including the iconic Baiji, a freshwater dolphin (*Lipotes vexillifer*), is now considered extinct (see Chapter 8).

Knowledge Box

Dam Removal: Due to their significant ecological impact, dams are increasingly facing pressure to be removed. Especially in the developed countries, more dams are being removed than built. For example, over 3,500 dams or other barriers have been removed across Europe, including the biggest dam removal in Spain last year. An ongoing historical river restoration project in Estonia will remove 8 to 10 dams and open up 3,300 km of river basin. In the United States, the largest dam removal took place in 2014 when the Glines Canyon Dam was removed (**Figure 6.12**), restoring 70 miles of river ecosystem for salmon migration. This record will soon be broken in 2022 when four dams along the Klamath River in northern California will be removed at a cost of $440 million and will restore more than 400 miles river system for salmon.

Figure 6.12 Glines Canyon Dam removal. Glines Canyon Dam was a 210-foot (64 m) high concrete arch dam built in 1927 on the Elwha River within Olympic National Park, Clallam County, Washington, USA. It was built to generate electricity for loggers and military bases. There was another, smaller Elwha River Dam downstream. The two dams cut off the migration route of wild salmon, which once numbered at 400,000. The dam removal project completed in August 2014 and was the largest dam removal project in the United States.

In addition to ecological impacts, there are also safety concerns. St. Francis dam disaster, as mentioned before, was an example. Natural disasters could cause dams to fail, causing severe damages. In fact, this is one of the main arguments against the construction of the Three Gorges Dam. Sometimes the dams may be near geologically unstable locations (many rivers in fact trace fault lines). The Three Gorges Dam actually sits on two large fault lines. In some cases, large dams create such a large body of water behind it, causing a significant shift in weight distribution that may cause earthquakes. There are also subtle changes, such as changes in microclimate, water quality, and so on.

Knowledge Box

The Three Gorges Dam and the Earth's Rotation: Can Three Gorges Dam slow down the Earth's rotation? The answer is yes—if you are a math and physics nerd. By raising the water behind the dam, the Dam and the water behind it will slow down the Earth's rotation and increase the length of our days by 0.06 mill seconds. This is because the Earth's angular momentum has to be conserved. When a large amount of water is raised higher and rotate with the earth, it will acquire more angular momentum because it becomes farther from the center of the Earth. Next time when you had a really long day, you have something to blame.

Figure 6.13 Owens Valley. A mostly dry Owens Valley with Sierra Nevada mountain range in the background. The Los Angeles Aqueduct, built in 1913, removed most of the water from Owens River and soon dried up the lake. It is now one of the largest sources of dust in the United States.

In addition to dams, other water resources infrastructure could have environmental impacts as well. In California, the State Water Project alone consumes about 19% of the energy resource for the entire state. As will be discussed in Chapter 9, this large energy expenditure will cause many environmental and sustainability issues. A major future project that is intricately linked to the State Water Project is the so-called "Bay Delta Conservation Plan" that has been in limbo for years. One of the key concerns of this project is its profound impact on the ecology of the **Bay-Delta** (see the Knowledge Box). Another water project, the Los Angeles Aqueduct built in 1913, essentially dried up Owens Lake and Owens River, drastically changed the ecosystem and hydrology in Owens Valley in Sierra Nevada (**Figure 6.13**). Besides environmental issues, water resources infrastructure could bring significant water right issues. Due to its complexity, this book will not cover this topic. But the reader is encouraged to explore it outside of this course.

Knowledge Box

California Bay-Delta Issue—The Bay-Delta issue, in other words, "The California Water War," is difficult to describe clearly without writing another book. California's Bay-Delta is a 1,150 square mile floodplain fed by the Sacramento and San Joaquin rivers. The Bay-Delta is the hub of California's water system, supporting a $3 trillion economy and a population that is breaking the 40 million mark. Water conveyance projects started in the early 1900s moved more and more water out of the Bay Delta to the vast Central Valley (the most productive agricultural land in the United States) and Southern California, where two thirds of Californians live but with only a third of California's water resources. With increasing amounts of water being pumped out of the Bay Delta, more environmental issues such as seawater intrusion, threat to local species such as delta smelt and steelhead salmon, started to emerge. In the early 1980s emerged the idea of a Peripheral Canal to move the intake of the state water project further north and bypass the heart of the Bay-Delta to tackle these issues. However, local farmers and environmentalists objected to the project and it was not approved. Since then, the situation keeps getting worse but the project keeps morphing from its original 400-foot wide, 30-foot deep canal to a pair of giant (40-foot diameter) underground (150-foot below ground) tunnels, to the current single underground (190-feet below ground) tunnel configuration. The project underwent numerous environmental impact assessments, changes in lead agencies, several name changes (Peripheral Canal—Bay Delta Conservation Plan—**California Water Fix** and Eco Restore—Delta Conveyance), and cost increases from the original 2 to 4 billion to the current $15 billion. It is unclear if the current project, which is a painful compromise that few fully support, could survive voter approval and environmental review. Figure 6.14 shows some representative perspectives of the Bay-Delta issue. There is no easy fix to California Water Fix.

Figure 6.14 California's Bay-Delta Issue in pictures. Clockwise from the top left: 1. California Aqueduct, which conveys water from Bay-Delta to central and southern California. 2. Edmonston Pumping Plant, which is the most powerful water lifting system in the world, lifts water 600 m over the Tehachapi Mountains. 3. San Luis Reservoir, which is the intermediary in the path of water transfer. 4. A sign posted by Central Valley farmers, who support the State Water Project.

6.5 Water Reuse and Recycling

6.5.1 Overview of Water Reuse

Water reuse and recycling start after the wastewater treatment process, as shown in Figure 5.18. The treated effluent could be discharged back to surface water, or be recycled for use as irrigation or industrial process water. For regions with water shortages, the treated effluent could be treated further with different processes for more uses, including specialized industrial uses or even potable uses. These treatment processes (commonly termed as "tertiary treatment" because it takes place after the secondary wastewater treatment) usually include enhanced filtration (such as microfiltration, ultrafiltration, nanofiltration), reverse osmosis, advanced oxidation (usually by oxidation agents such as ozone and hydrogen peroxide), and final disinfection. In some cases, the wastewater is treated to near or even better than drinking water quality before being reused, such as the case at the Orange County Water District, located in Fountain Valley, California, USA (**Figure 6.15**).

6.5.2 Types of Potable Water Reuse: Direct and Indirect

Water reuse could be indirect or direct. There are many examples of indirect potable reuse (IDPR). For example, a community takes water from a river for use and discharges treated wastewater back to the river downstream of the intake. When another community downstream takes in the mixture of treated effluent and river water for use, IDPR is taking place. In fact, IDPR is fairly common throughout the world in large

Figure 6.15 Water purification process at Orange County Water District's Groundwater Replenishment System. The schematics of reverse osmosis process, the key step in the cleanup process, is shown in Figure 6.17.

river watersheds such as Mississippi River in the United States, Yangtze River in China, and Nile in Africa. Sometimes, IDPR could be less obvious. For example, in the Santa Ana River watershed in Southern California, the river is often 100% treated wastewater in the dry season. In arid Southern California, where water resources are precious, even this effluent is not unused. In fact, the water agency downstream cleans up the water with a series of constructed wetlands before releasing the water to a series of spreading basins. The water infiltrates into the underground aquifer and is then pumped by water retailers downgradient. Indirect potable reuse could involve mixing and stabilization. In San Diego, California, USA, recycled water is mixed in a lake, which is a drinking water source. The process prolonged the residence time of the recycled water to 6 month or longer, rendering it safe to drink after routine treatment.

6.6 Desalination

6.6.1 Overview of Desalination

Desalination is a process that (as the term implies) removes salt from water to make potable water. As the Earth is covered with water on 71% of its surface and sea water accounts for 96% of all the water there is on the planet, most desalination targets ocean water as the virtually unlimited source of water.

Desalination is used where freshwater cannot be obtained otherwise, such as in arid regions (Australia, middle east, etc. see **Figure 6.16**), ocean-going ships, etc. For example, Kuwait gets nearly all of its potable water from desalination. Currently about 1% of the world's population depends on desalination as their main water sources.

Figure 6.16 One of the eight desalination plants in Abu Dhabi, Dubai. Desalination accounts for 24% of all water resources for this Mideastern country. The rest is from groundwater and recycled water.

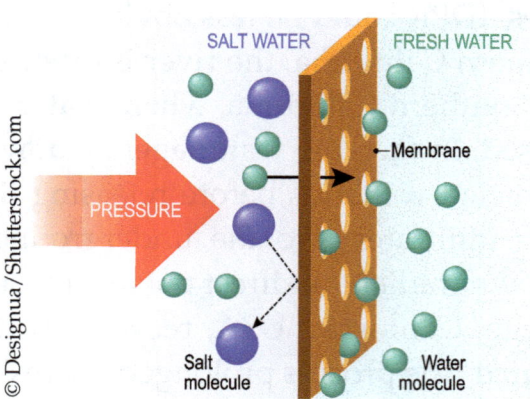

Figure 6.17 Schematics of the reverse osmosis process. Osmosis is a natural process where freshwater can penetrate a membrane to a saline solution. Reverse osmosis is an opposite process, where pressure is added on the saline solution and freshwater is pushed out, leaving concentrated saline solution behind.

Knowledge Box

The Desalination Process: Desalination can be done through many processes. These processes can generally be divided into two types, membrane-based (mostly reverse-osmosis, which is most common) and thermo-based. Membrane-based process uses reverse osmosis and is the predominant process used by most of the desalination plants (see Figure 6.17 for a diagram of reverse osmosis process) due to its lower energy consumption and overall cost. Thermal processes include solar desalination, flash distillation, vacuum distillation, and so on. They have seen decreasing use recently.

6.6.2 Desalination Process and Potential Impacts

The main drawback of desalination is its energy consumption. Currently most of the desalination processes consume about 3 to 15 kilowatts hour per cubic meter of water produced, with reverse osmosis process near the low end of energy consumption (and hence lower cost). Because of this, the costs of desalination could often be many times more expensive than other sources. For example, in Southern California, the cost for desalination water is about three times that for water imported from Northern California, and more than five times that of the water produced locally.

Environmental impact can often trump high costs as the deciding factor for the viability of desalination project. Desalination-associated environmental impact include greenhouse gas generation (due to its high energy use); harms to marine organisms through impingement and entrapment at the intake structures; and the by-product of desalination, which is concentrated salt water or brine. Depending on the process efficiency, the brine could have different levels of salt. The concentration will also affect the density of brine and its ability to mix with ambient sea water. If the mixing is not complete and salty, heavier brine sinks to the bottom of the ocean, there could be significant ecological consequences.

> ### Knowledge Box
>
> **The World's Largest Desalination Plant:** A large desalination plant will be built at the Taweelah Power and Water Complex, Abu Dhabi, United Arabic Emirates. With the design capacity of 902,000 cubic meter per day (240 million gallons per day), it will be the world's largest desalination plant with a cost of $900 million. Based on the reverse osmosis process, the project will set a world record by utilizing the lowest amount of energy per gallon of desalinated water produced.

6.7 Water Balance and Integrated Water Management

6.7.1 Overview of Water Balance

Water balance is the mass balance for water for a given system, be it a home, a community, a factory, a watershed, or a region. For example, a water balance schematic (sometimes also called a water budget) could be established for an average household in the United States (**Figure 6.18**). Due to the fact that urban water resources could involve both natural and human aspects of water, and these two aspects interact profoundly with each other, it is beneficial to consider both when a water balance is established. As shown in Figure 2.6 on water cycles, a water balance is a water cycle within a given system. For example, for the household water use in the United States as shown in Figure 6.18, it could be taken as a water balance where potable water flows into a household, gets used by different processes, and most of the water will exit the household as wastewater (with some losses such as evaporation). If irrigation of gardens is considered, the pattern could be quite different because irrigation in the United States could often be more than that for indoors use. By establishing such a water balance, one could see that, given a finite water supply, what could be saved if there were a water shortage. For example, during the Millennium Drought, citizens at Melbourne was able to cut back 47% of potable water use by shortening shower time, optimizing irrigation, and water reuse, etc.

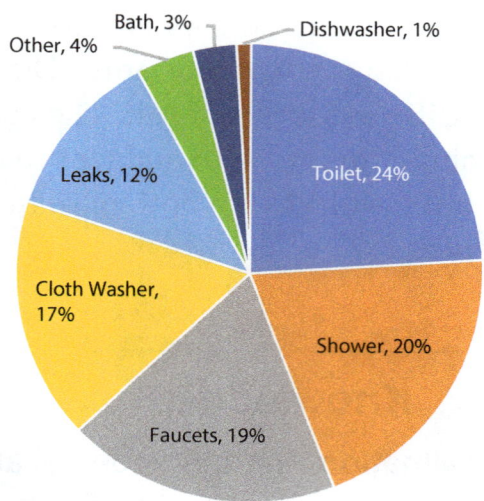

Figure 6.18 Household water uses (total usage is 520 liters per day per household).
Source: De Oreo et al. (2016). Created by Jian Peng

6.7.2 Case Study: Water Balance for North Orange County, California, USA

For a larger region, water balance could involve many different components and could get quite complicated. One such example is the water balance for the north Orange County, California, USA where water resources have a quite complex pattern (**Figure 6.19**). In this system, which includes about 2.5 million people in 26 cities, there are several sources of water. The natural source includes local precipitation, which generates surface runoff, evapotranspiration, shallow groundwater, and incidental recharge. The man-made sources of water include upstream (Santa Ana River; the largest river in the area) stormwater capture and release; dry weather baseflow in Santa Ana River; State Water Project from Bay/Delta in northern California (see Figures 6.2 and 6.14, and Section 6.4.4) and Colorado River; as well as wastewater residuals generated by many wastewater treatment plants along the upstream portion of the Santa Ana River. A 500-m-deep groundwater basin underlying the area provides an excellent storage capacity for many such sources during wet weather. During dry weather, the groundwater can then be extracted for use. Therefore, the groundwater basin in the area is one of the most critical components of the water balance and provides critical "equalization" for various sources, which could be transient and variable. After drinking water is used, it turns into wastewater and is treated in the local wastewater treatment plant (Orange County Sanitation District, OCSD). Most of the wastewater, after secondary treatment, flows to Orange County Water District (OCWD)'s Groundwater Replenishment System (GWRS) for treatment (see Section 6.5.1 for details). The treated water is then recharged back to the groundwater basin for reuse forming a nearly closed loop.

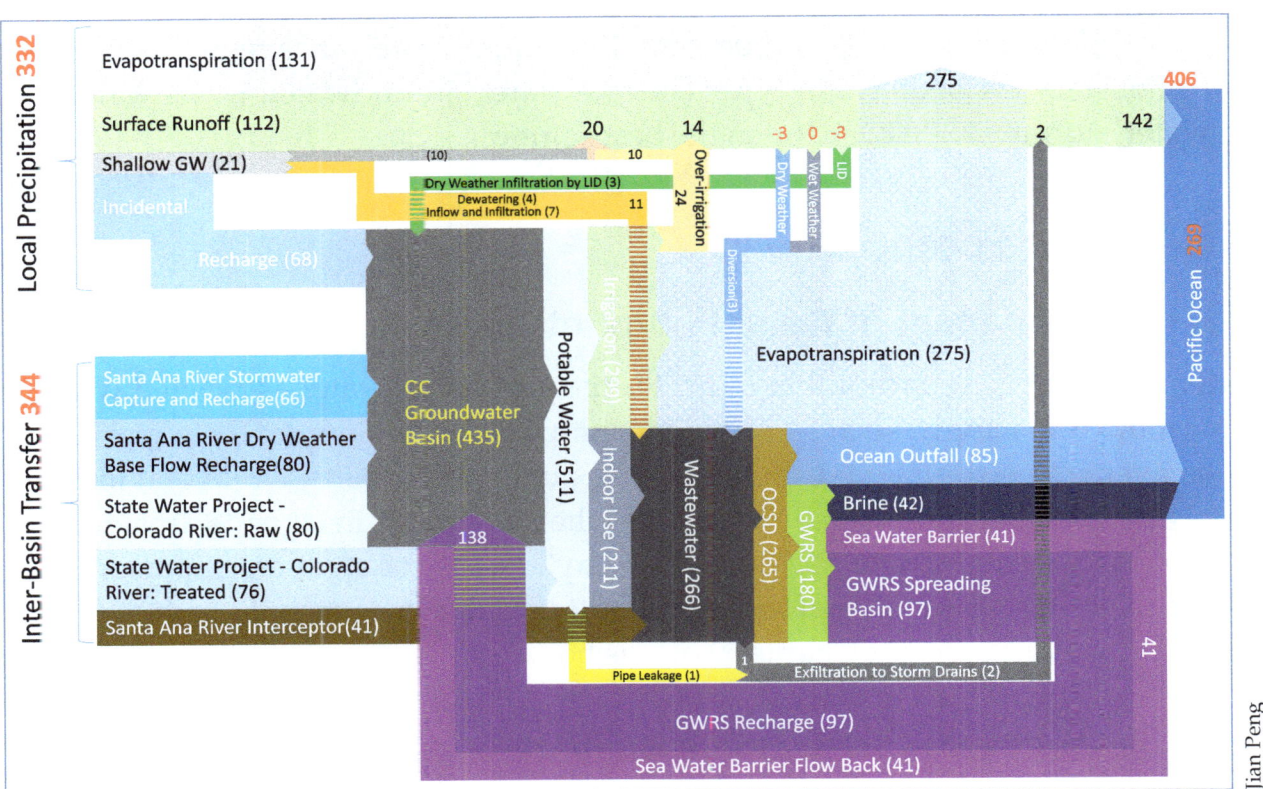

Figure 6.19 North Orange County Water Balance. Unit: gigalitre. Area: 1250 km²; population: 2.6 million. Interbasin Transfer include sources outside of the groundwater basin.

It can be seen that for Orange County, a significant portion of the water is recycled. GWRS, being the largest groundwater recharging operation in the world, is undergoing a final phase of expansion to increase the capacity by 30% and will recycle 100% of the wastewater generated in the region. Compared to other cities in the area that rely overwhelmingly on imported water, north Orange County has essentially maximized the utilization of available water resources. Still, meeting local water demand is challenging due to a significant mismatch between the population and local precipitation. More stormwater capture, water conservation, and even desalination may be needed to achieve local water self-sufficiency.

Another interesting observation is that evapotranspiration is the largest sink of water in North Orange County, with a total of 406 gigaliters of water losing to the atmosphere every year. The majority source of evapotranspiration is residential irrigation (Orange County has very little agriculture). Since Orange County has a semi-arid climate, maintaining a lush lawn or a garden may not be the best use of precious water resources. In addition, as discussed in Chapter 5, irrigation nuisance flow carries many pollutants to receiving waters. However, this habit has proven to be difficult to change. Currently, this is one of the key issues of environmental public education in Orange County and in southern California. Therefore, to best tackle the water quality and water resources issues in Southern California and beyond, integrated water management schemes that take a holistic view of all water types (surface water, groundwater, waste water, drinking water, stormwater, etc) are often the best solution.

6.8 Water Footprint

6.8.1 The Concept of Water Footprint

Water footprint is a concept that indicates the amount of water associated with a certain product or activity, such as water consumption. For example, the water footprint for an average person in the United States is about 500 gallons, while that in Norway is 300 gallons, and China 80 gallons, and less than 10 gallons in Uganda. In other words, these are the average water footprint for a day's living for an average person in these countries. In this sense, water footprint is quite similar in concept to carbon footprint, which will be discussed in Chapter 9 of this book.

6.8.2 Water Footprint of Common Items

Water footprint could also be for products, especially about food. **Figure 6.20** is a good example. While it is easier to understand water footprint for food items based on water consumption for agricultural operations, water footprint for things like a T-shirt would need several conversions to be calculated. For example, for a T-shirt, one would need to calculate the water footprint to produce the cotton as an agriculture product, the water footprint for textile processing, tailoring, printing, and even selling. For example, if a machine produces fabric that is enough for 1,000 T-shirts, the water this machine

Figure 6.20 Water footprint of several common items.

consumes will be part of the water footprint. The machinery itself will have a water footprint as well and it should be added. Other steps of T-shirt production, such as printing, washing, tailoring, and so on, can all be calculated as such and added. Therefore, the final number of 2,700 liters may seem like an unrealistically large number, it is reasonable because it accounts for all of the processes associated with each step of production. Some calculation schemes of water footprint also include the amount of water polluted during the production process.

6.8.3 Water Footprint and Sustainability

Water footprint has many implications on water conservation, sustainability, and lifestyles.[4] On a global scale, most of the water (72%) is used for agricultural production. Imported products represented 22% of water footprint, and other domestic products and industrial products account for the rest (**Figure 6.21**). For example, from the statistics about the drastic differences in water consumption in different countries (see Figure 6.6), one may find that there are certainly water resource shortages in countries such as Uganda. On the other hand, water usage in the United States may be wasteful. In time of drought or water supply issues, a review of water footprint, i.e., water consumption portfolio, will help conserve water. These activities include shortened shower time, turn off faucet while brushing teeth, use water-efficient dishwasher or laundry machine, etc. Water footprint information may also be useful in making lifestyle choices, such as the selection between

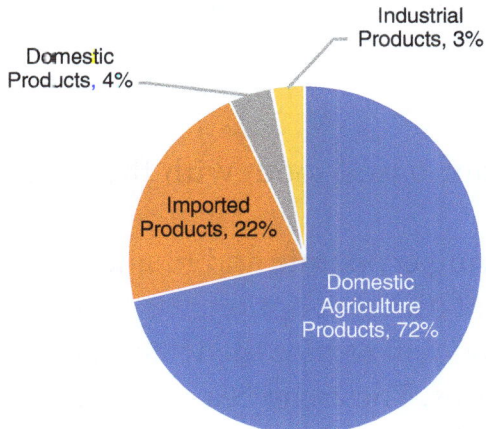

Figure 6.21 Global water footprint by sectors (Mekonnen and Hoekstra, 2010a). Also refer to Table 6.1 for more information.
Source: Created by Jian Peng

---

[4] Sometimes water footprint could mean volume of water polluted per unit of time. For simplicity, this accounting is not used in this book.

Table 6.1 Water Footprint of common products.

Product	Global average water footprint, L/kg
almonds, shelled	16,194
apple	822
banana	790
beef	15,415
bread, wheat	1,608
butter	5,553
cabbage	237
cheese	3,178
chicken	4,325
chocolate	17,196
cotton lint	9,114
cucumber	353
dates	2277
eggs	3300
groundnuts, shell	2782
leather (bovine)	17093
lettuce	238
maize	1,222
mango/guava	1,800
milk	1,021
olive oil	14,430
orange	560
pasta (dry)	1,849
peach/nectarine	910
pork	5,988
potato	287
pumpkin	353
rice	2,497
tomatoes, fresh	214
tomatoes, dried	4,275
vanilla beans	126,505

Source: Mekonnen and Hoekstra (2010a, b).

animal-based food and plant-based food, with the latter having much lower water footprint than the former.

Water footprint could bring economic and financial implications. For an industrial process, a conscious choice of different options, or decisions on a process improvement to reduce water consumption will reduce the water footprint of the process and all the products this process produces. These choices are often due to financial incentives provided by water providers, which often use **tiered water rates** to encourage water conservation. During severe drought period, there could be government regulations mandating tiered water rates. Table 6.2 is an example of tiered water rate for a water district in Southern California, where the rates for wasteful water use could be nearly 10 times higher than that for conservative uses for exactly the same source of water. The base volume, where the tiers are based on, is determined by a number of factors such as types of homes, outdoor landscape area, number of residents, weather, etc.

Table 6.2 Tiered water rate for residents in the Irvine Ranch Water District, Orange County, California, USA.

Tier	% of Base Volume Single Family Home	% of Base Volume Multi-Family Home	Rate (Per 1000 L)
Tier 1 Low Volume	0–40%	0–50%	$0.52
Tier 2 Base Rate	41–100%	51–100%	$0.71
Tier 3 Inefficient	101–140%	101–120%	$1.72
Tier 4 Wasteful	141+	121+	$4.81

Further Reading

Yang, S.L, Milliman, J.D., Lia P., and Xu K., "50,000 Dams Later: Erosion of the Yangtze River and its Delta." *Global and Planetary Change*, 5, no. 1–2 (January 2011): 14–20.

California Department of Water Resources, 2020. California Waterways Map. Available at: https://water.ca.gov/What-We-Do/Education/Education-Materials. Accessed May 1, 2020.

Charles Einstein. The Day New York Went Dry, New York, NY: Fawcett Publications, Inc, 1964.

Karl Weber. Last Call at the Oasis—The Global Water Crisis and Where We Go From Here. New York, NY: Public Affairs, 2012.

Zhao, Xu, Junguo Liu, Hong Yang, Rosa Duarte, Martin R. Tillotson, and Hubacek Klaus. "Burden Shifting of Water Quantity and Quality Stress from Megacity Shanghai." *Water Resources Research* 52, no. 9 (2016): 6916–27.

C.J. Vorosmarty et al. "Global Threats to Human Water Security and River Biodiversity." *Nature*, 468 no. 7321 (2010): 334.

DeWolf, Wendy. Engineering Clean Water, *Yale Scientific Magazine*, April 3, 2011

Mekonnen, M. M. and A. Y. Hoekstra. The green, blue and grey water footprint of farm animals and animal products. Volume 1: Main report. UNESCO-IHE., Institute for Water Education, 2010a, 50 pp.

Mekonnen, M. M. and A. Y. Hoekstra. The green, blue and grey water footprint of crops and derived crop products. Volume 2. Appendices main report. Value of Water Research Report Series No. 47. UNESCO-IHE Institute for Water Education, 2010b, 1196 pp.

DeOreo, William B., Peter Mayer, Benedykt Dziegielewski, and Jack Kiefer. Residential End Uses of Water, Version 2. Denver, Colorado: Water Research Foundation, 2016.

Bexfield, L. et al, 2021. Pesticides and pesticide degradates in groundwater used for public supply across the United States: occurrence and human health context. Environmental Science and Technology, 2021, 55(1), pp. 362–372.

Web Resources

USGS National Water Information System: Web Interface for surface water: https://waterdata.usgs.gov/nwis/sw

Flint (Michigan, USA) Water Crisis: https://en.wikipedia.org/wiki/Flint_water_crisis

Water resources issues—what you need to know: https://www.youtube.com/watch?v=-dUp8VQoFsc

Orange County Water District's Groundwater Replenishment System: https://www.youtube.com/watch?v=M4r3u9MXd-g&t=83s

Water-related violence: https://www.theguardian.com/global-development/2019/dec/31/water-related-violence-rises-globally-in-past-decade?from=groupmessage&isappinstalled=0

Water Recycling in Space Station: https://www.mentalfloss.com/article/67854/how-do-astronauts-get-drinking-water-iss

Infrastructure Report Card about Dams, ASCE< 2017: https://www.infrastructurereportcard.org/cat-item/dams/

Gline Canyon Dam Removal time-lapse video: https://www.youtube.com/watch?v=bUZE7kgXKJc

Discussions on the planned dam removal at Klamath River: https://www.earthisland.org/journal/index.php/articles/entry/largest-dam-removal-project-klamath-river-science/

The history of California's Bay-Delta: https://www.kcet.org/redefine/a-brief-history-of-californias-bay-delta

A good collection of water resources and water uses around the globe: https://ourworldindata.org/water-use-stress

National water use statistics: https://www.worldometers.info/water/

Questions and Exercises

1. The County of San Diego, California, USA, relies nearly entirely on imported water to support its 10 million people. In 2015, the Poseidon Desalination Project was completed to supply 10% of the County's drinking water, but at a cost 2 to 3 times more than the imported water (the plant cost $1 billion to build). What do you think could be the reason behind this major investment?

2. The Three Gorges Dam is one of the largest dams in the world. It has significant benefits for hydropower and flood mitigation. List at least five risks that it may have.

3. Observe Figure 2.6 about global water cycle. With the astounding number of dams and reservoirs, what do you think are the ways these dams interfere with the global water cycle?
4. After reading the information about the complex Bay-Delta issue in California, what do you think is the best solution?
5. Carefully examine the water balance for Orange County (Figure 6.19) to see how the mass balance of water from different sources are balanced. Do you think Orange County could eventually achieve water self-sufficiency by recycling all its wastewater?
6. Use Table 6.1 and calculate the water footprint of your last breakfast or lunch.
7. Check the numbers in Figure 6.19 and see if the water from different sources are indeed all balanced. After checking the numbers and getting familiar with the water balance, answer the following questions:
    a. If there is a severe drought, which of the water sources will be the most heavily impacted?
    b. What will be the impact of reduced amount of imported water?
    c. In the face of a water shortage or drought, what are possible ways to 're-balance' the water supply?
    d. How could the following practices help the above situation: more water recycling; xenoscape; stormwater capture and reuse?
8. Given the situation shown in Figure 6.19, would you be supportive of desalination to supply part of the water needs in Southern California? Why or why not?

Chapter 7

Groundwater and Soil Contamination

Learning Outcomes		182
Key Concepts		182
7.1	Groundwater Hydrology	182
	7.1.1 Groundwater and Aquifer	182
	7.1.2 Groundwater Resources	184
7.2	Groundwater and Soil Contamination	187
	7.2.1 Groundwater and Drinking Water	187
	7.2.2 Soil and Groundwater Contaminants	187
	7.2.3 Human Health Risk Assessment	192
7.3	Groundwater and Soil Remediation	193
	7.3.1 Underground Storage Tanks	195
	7.3.2 Soil Investigation	197
	7.3.3 Groundwater Investigation and Conceptual Site Model	197
	7.3.4 Soil and Groundwater Remediation	201
Further Reading		204
Web Resources		205
Questions and Exercises		205

This chapter benefited from the review from Scarlett Zhai, PhD, PE, California Department of Toxic Substances Control.

182 Contemporary and Emerging Global Environmental Challenges

Learning Outcomes

- Knowledge of the basic groundwater hydrology
- Understanding of groundwater hydrology as part of the water cycle
- Knowledge about major sources of groundwater contamination
- Knowledge of the basic behavior of different pollutants in soil and groundwater
- Knowledge about key remediation processes for groundwater and soil contamination

Key Concepts

Groundwater hydrology; groundwater contamination; aquifer; water table; unconfined aquifer; confined aquifer; piezometric level; vadose zone; Superfund; (USEPA) National Priorities List; aquitard; aquiclude; maximum contamination level (MCL); monitored natural attenuation (MNA); Underground Storage Tank (UST); VOC; BTEX; MTBE; TCE; LNAPL; DNAPL

Compared to air and surface water contamination, which we can see, breathe, and take in, soil and groundwater contamination can remain "invisible" for a long time before the issue manifests itself. When that happens, such as the case in the Love Canal (see Figure 7.5 and related discussions) and other places, the situation is often fairly dire and cleanup is difficult and expensive, as is the case for all Superfund sites. Therefore, it is important to understand the soil and groundwater contamination and control the contamination at its source and before it takes place, if at all possible. For a site with contaminated soil and groundwater, a good understanding of groundwater hydrology and cleanup techniques is needed to restore the site and protect human health.

7.1 Groundwater Hydrology

7.1.1 Groundwater and Aquifer

Groundwater, as its name suggests, is the water below ground surface. As shown in Figure 6.1, groundwater represents the vast majority of freshwater resources on earth, second only to glaciers that humans cannot use under normal conditions. Groundwater exists in soil pore spaces or in the fractures of rock formations. The soil or rock formation is called an **aquifer** if they are water-bearing. Similar to surface water, the groundwater also has a surface below which the groundwater is saturated. This surface is called a **water table**. If one drills a well into the aquifer, the water level in the well will be the same as the water table. The unsaturated zone above the water table is called **vadose zone**. An aquifer with such a water table is called an **unconfined aquifer**. A **confined aquifer** is one that is overlain with one or more impermeable or poorly permeable layers (such as clay or silt). These layers are called **aquitard** (poorly permeable) or **aquiclude**/aquifuge (impermeable).

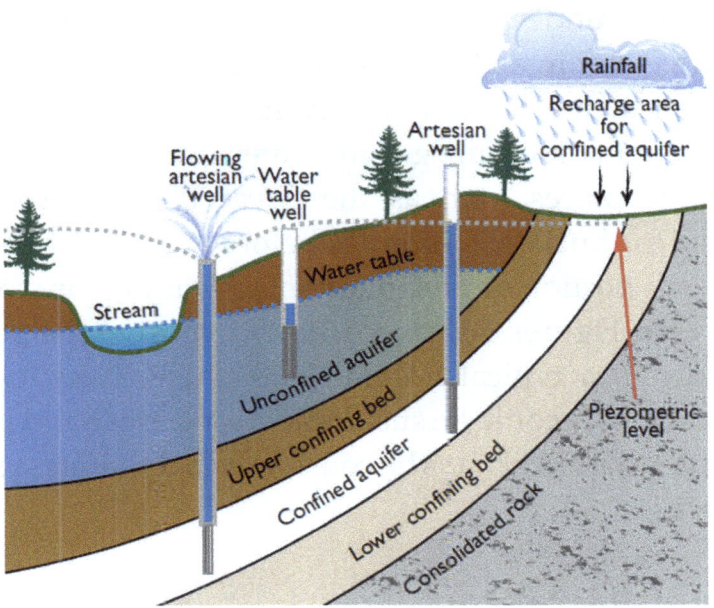

Figure 7.1 Diagram of a groundwater system (USGS, https://www.usgs.gov/media/images/artesian-wells-can-bring-water-land-surface-naturally, Public Domain).

For confined aquifers that may often be pressurized, the **piezometric level** may be above the actual water level. If a well is drilled into a confined aquifer, the water level in the well will rise to the piezometric level. If the piezometric level is above the ground level, the water well will flow out of the well without pumping. A schematics of a typical groundwater system is shown in **Figure 7.1**. Note that the low permeability layers are called "confining bed" instead of aquitard or aquifuge/aquiclude.

Knowledge Box

Darcy's Law: Groundwater exists in porous media underneath the ground surface and will flow downgradient similar to surface water. Unlike surface water, the groundwater table could have significant slope with the slope being determined by permeability and the level of source water, and groundwater will 'flow' from higher elevation to lower elevation. The flow of groundwater is impeded by the porous media, which could have very different permeabilities (termed 'conductivity' in groundwater hydrology) due to differences in the proportion of void spaces as well as how connected these void spaces are. Groundwater also has a tendency to flow in the direction of the least resistance according to Darcy's law:

$$Q = -KA \, dh/dl$$

where:
Q = rate of water flow (volume per time)
K = hydraulic conductivity
A = column cross sectional area
dh/dl = hydraulic gradient, that is, the change in head (h) over the length of interest (l).

In many cases, especially under dry weather conditions when precipitation-induced runoff is minimal, the water in a river is 'surfacing groundwater' if the water level in the river is below the water table. Conversely, if the water level in the river is above the water table of the surrounding groundwater, the river will be recharging the groundwater.[1] In some places such as the Great Plains,[2] groundwater travels in the sandstone aquifer for hundreds of miles before reaching the surface. In a natural watershed, only a small portion of precipitation will end up as storm runoff. Recall in Chapter 5 (especially see Figures 5.13 and 5.14) that most of the precipitation will end up being throughflow in a typical natural watershed. Even for a flowing river, a significant portion of the flow could be subsurface in the loose and permeable riverbed.[3] Therefore, groundwater should always be considered a dynamic system and an integral part of the water cycle.

7.1.2 Groundwater Resources

Under natural conditions, groundwater is a sizeable reservoir in the global water resource portfolio (Figure 6.1) and important water resources. In many places, groundwater is the chief source of drinking water. For some urban water systems, such as the one shown in Figure 6.19, groundwater basin can be both an important source of water and an underground reservoir that makes sustainable water resources management possible. In the United States, 71% of the more than 51,000 community water systems in the United States rely on groundwater.[4] **Figure 7.2** is a map of the major groundwater basins in the United States.

Groundwater is often extracted from an aquifer by a pump. As a result of the pumping, there will be a cone-shaped depression on the water table near the well (see Figure 7.1 and refer to the Knowledge Box). If there are many pumping wells close to each other, the general groundwater level will drop significantly.[5] This drop in

---

[1] In this case, the section of the river will be called a losing reach because the flow downstream will be lower than that upstream. Conversely, the section will be called a "gaining reach" if the flow downstream will be higher than upstream due to the input of groundwater.

[2] The Great Plains is a long (3,200 km) stretch of a broad expanse of flat land in the forms of prairies, steppes, and grasslands in the United States and Canada to the east of Rocky Mountains. In Canada, it is called the Canadian Prairies. The total area of the Great Plains is about 1.3 million square kilometers.

[3] This portion of the river flow is called "'underflow," which is difficult to measure.

[4] Note, however, that most of these water systems are small ones. For very large systems, they tend to use surface water. Therefore, in terms of the amount of water used, only 23% of all water resources are from groundwater, and the rest is from surface water.

[5] This can be visualized by overlaying these "imaginary" three-dimensional cones together, the net results would be a nearly uniformly lowered groundwater table for the area.

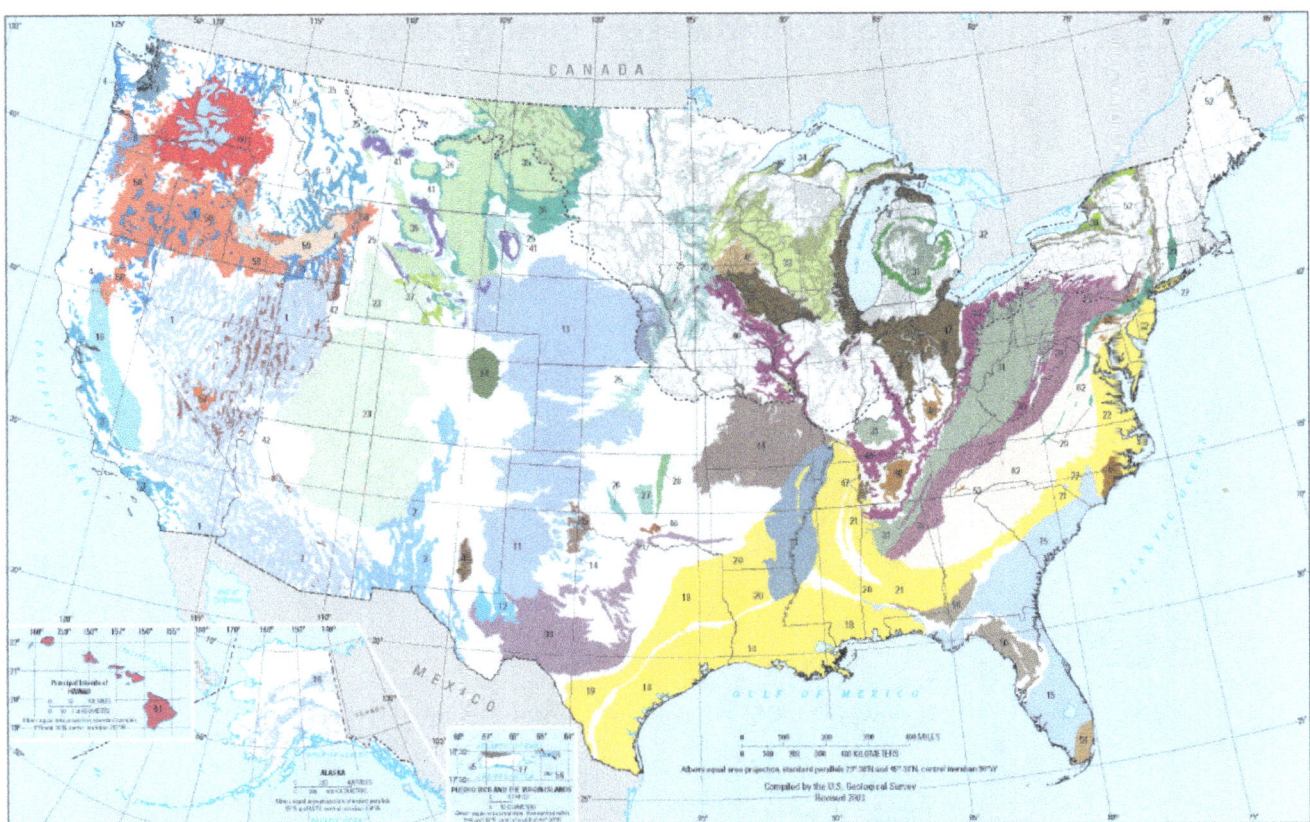

Figure 7.2 Map of the principal aquifers of the United States.

groundwater level will sometimes cause ground surface subsidence because aquifer, after losing water, will "shrink." due to loss of pore spaces Ground subsidence is a common problem for many major cities where water is overdrafted. This will in turn cause many other issues such as sea water intrusion, water quality degradation, and so on. Overdraft could also be caused by regional drought because ultimately all groundwater basins are recharged by precipitation (see Figures 2.6 and 5.13). To monitor groundwater levels, the United States Geological Survey (USGS) has a national network of more than 18,000 groundwater wells (**Figure 7.3**) that are monitored continuously, daily, or periodically on water level and water quality. These wells are part of a wider collection of 850,000 wells that have been compiled during the course of **groundwater hydrology** studies over the past 100 years. Information from these wells is served via the Internet through NWISWeb, the National Water Information System Web Interface (https://waterdata.usgs.gov/nwis?nwisweb_overview=).

These "groundwater watch" web pages group related wells and data from these active well networks, and provide basic statistics about the water-level data collected by USGS water science centers for Cooperative Programs, for Federal Programs, and from data supplied to USGS by its customers through cooperative agreements.

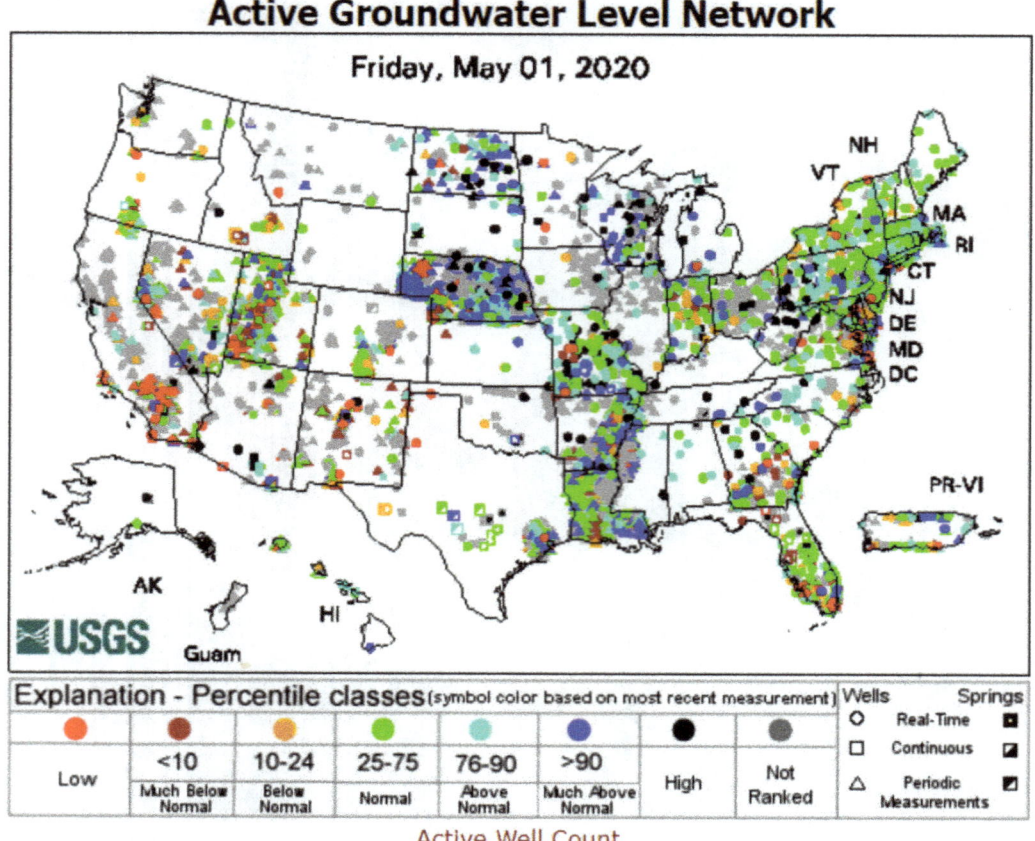

Figure 7.3 USGS National Active Groundwater Level Network. In the United States, the USGS manages this national database for wells that are locally managed. Weblink: https://groundwaterwatch.usgs.gov/usgsgwnetworks.asp#.

> ### Knowledge Box
>
> **Well Drawdown:** On Figure 7.1, the depression near the "flowing artesian well" as indicated by the dashed line on top is actually the piezometric level, not the water table itself. The actual water table is the dashed line below. If the "water table well" next to it extracts the water for use, there will be a depression (called "drawdown") on the water table as well. This depression is formed for the same reason as the one that would have formed in a water table of an unconfined aquifer. The shape of the cone will be dependent on the pumping rate. The faster the pumping rate is, the steeper the center of the cone will be. This makes sense if you consider Darcy's Law. Everything else being the same, the rate of flow will be dependent on the gradient of the water table or piezometric level.

7.2 Groundwater and Soil Contamination

7.2.1 Groundwater and Drinking Water

Both surface water and groundwater are commonly used as drinking water sources. As discussed in Chapter 5, surface water could be subject to many natural and manmade contamination. The treatment of surface water for potable uses, as a result, could involve complex processes and generally involves coagulation, sedimentation, filtration, and disinfection. Sometimes other processes may be involved, such as membrane filtration or advanced oxidation process in order to ensure that the product water meets the water quality standards.[6] In comparison, groundwater usually has more stable water quality and is generally free of pathogens and suspended materials. This is due to the natural filtration of these impurities by sediment particles when groundwater moves through the aquifer. Therefore, most of the potable water system treats groundwater in simpler processes such as filtration and disinfection. Sometime, such as the City of Downey in Southern California, USA, the groundwater has been pumped and sent directly to the end user without any treatment.

However, groundwater could contain many organic and inorganic chemicals and other naturally-occurring impurities such as fluoride, arsenic, boron, salts, and so on, that could impact or limit its beneficial uses. Anthropogenic pollution can also be a significant threat to groundwater by contaminating groundwater or overlying soil, or both. These contaminants include organics, inorganics, and some other contaminants of emerging concern (CEC; see Section 5.5). Some of the wastes that impact soil and groundwater, such as landfill leachate, may be a mixture of many compounds. Sometimes, **groundwater contamination** could be induced by unique circumstances. In Orange County, California, USA, highly purified wastewater has been used for groundwater replenishment (see Chapter 5 and 6 for more information) since 2008. Thanks to the advanced treatment technology, the product water met or exceeded drinking water standards. However, soon after the artificial recharge operations, the groundwater downstream of the recharge operation was found to contain elevated arsenic. After investigation, it was found that the highly purified recycled water was leaching the natural arsenic off the local aquifer because the water was corrosive due to lack of dissolved minerals. As a remediation measure, lime was added to the product water for pH adjustment and to add sufficient minerals to the water, and the arsenic issue was resolved.

7.2.2 Soil and Groundwater Contaminants

In Chapter 5, we discussed different types of pollutants that could impact surface water quality, including nutrients, pathogens, trash, sediment, toxic contaminants, and

---

[6] In the United States, these water quality standards are specified by the Safe Drinking Water Act as Maximum Contamination Levels or MCLs.

contaminants of emerging concerns (CECs). Some of these contaminants, under certain circumstances, could also impact groundwater quality. Some, such as trash and sediment, will be physically excluded so they cannot affect groundwater. Some compounds such as DDTs and PCBs, their extremely high hydrophobicity makes them strongly attached to soil particles, therefore their mobility in groundwater is quite limited. However, government regulations weigh more on toxicities than mobility. Pollutants with high toxicities can have very low allowed concentrations in groundwater (i.e., maximum contaminant levels [MCLs]). Although some can have low mobility or low solubility, and occur in groundwater at very low concentrations, they can still be pollutants of concern because observed concentrations can be orders of magnitude higher than MCLs. Table 7.1 lists **USEPA's** 126 priority pollutants.

Table 7.1 USEPA Priority Pollutant List

1. Acenaphthene	44. Methylene chloride	87. Trichloroethylene
2. Acrolein	45. Methyl chloride	88. Vinyl chloride
3. Acrylonitrile	46. Methyl bromide	89. Aldrin
4. Benzene	47. Bromoform	90. Dieldrin
5. Benzidine	48. Dichlorobromomethane	91. Chlordane
6. Carbon tetrachloride	49. (Removed)	92. 4,4-DDT
7. Chlorobenzene	50. (Removed)	93. 4,4-DDE
8. 1,2,4-trichlorobenzene	51. Chlorodibromomethane	94. 4,4-DDD
9. Hexachlorobenzene	52. Hexachlorobutadiene	95. Alpha-endosulfan
10. 1,2-dichloroethane	53. Hexachlorocyclopentadiene	96. Beta-endosulfan
11. 1,1,1-trichloroethane	54. Isophorone	97. Endosulfan sulfate
12. Hexachloroethane	55. Naphthalene	98. Endrin
13. 1,1-dichloroethane	56. Nitrobenzene	99. Endrin aldehyde
14. 1,1,2-trichloroethane	57. 2-nitrophenol	100. Heptachlor
15. 1,1,2,2-tetrachloroethane	58. 4-nitrophenol	101. Heptachlor epoxide
16. Chloroethane	59. 2,4-dinitrophenol	102. Alpha-BHC
17. (Removed)	60. 4,6-dinitro-o-cresol	103. Beta-BHC
18. Bis(2-chloroethyl) ether	61. N-nitrosodimethylamine	104. Gamma-BHC
19. 2-chloroethyl vinyl ethers	62. N-nitrosodiphenylamine	105. Delta-BHC
20. 2-chloronaphthalene	63. N-nitrosodi-n-propylamine	106. PCB-1242 (Arochlor 1242)
21. 2,4,6-trichlorophenol	64. Pentachlorophenol	107. PCB-1254 (Arochlor 1254)
22. Parachlorometa cresol	65. Phenol	108. PCB-1221 (Arochlor 1221)
23. Chloroform	66. Bis(2-ethylhexyl) phthalate	109. PCB-1232 (Arochlor 1232)
24. 2-chlorophenol	67. Butyl benzyl phthalate	110. PCB-1248 (Arochlor 1248)
25. 1,2-dichlorobenzene	68. Di-N-Butyl Phthalate	111. PCB-1260 (Arochlor 1260)
26. 1,3-dichlorobenzene	69. Di-n-octyl phthalate	112. PCB-1016 (Arochlor 1016)
27. 1,4-dichlorobenzene	70. Diethyl Phthalate	113. Toxaphene
28. 3,3-dichlorobenzidine	71. Dimethyl phthalate	114. Antimony
29. 1,1-dichloroethylene	72. Benzo(a) anthracene	115. Arsenic
30. 1,2-trans-dichloroethylene	73. Benzo(a) pyrene	116. Asbestos
31. 2,4-dichlorophenol	74. Benzo(b) fluoranthene	117. Beryllium
32. 1,2-dichloropropane	75. Benzo(k) fluoranthene	118. Cadmium
33. 1,3-dichloropropylene	76. Chrysene	119. Chromium
34. 2,4-dimethylphenol	77. Acenaphthylene	120. Copper
35. 2,4-dinitrotoluene	78. Anthracene	121. Cyanide, Total
36. 2,6-dinitrotoluene	79. Benzo(ghi) perylene	122. Lead
37. 1,2-diphenylhydrazine	80. Fluorene	123. Mercury
38. Ethylbenzene	81. Phenanthrene	124. Nickel
39. Fluoranthene	82. Dibenzo(,h) anthracene	125. Selenium
40. 4-chlorophenyl phenyl ether	83. Indeno (1,2,3-cd) pyrene	126. Silver
41. 4-bromophenyl phenyl ether	84. Pyrene	127. Thallium
42. Bis(2-chloroisopropyl) ether	85. Tetrachloroethylene	128. Zinc
43. Bis(2-chloroethoxy) methane	86. Toluene	129. 2,3,7,8-TCDD

Organics

Common organic contaminants of soil and groundwater are volatile organic compounds (VOCs). They are smaller organic compounds that are usually derived from petroleum products such as benzene, toluene, ethylbenzene, and xylene (**BTEX**); gasoline/diesel; fuel additives such as methyl tertiary butyl ether (**MTBE**) or other fuels or common industrial feed materials that are often stored in underground storage tanks (**USTs**). There are also many **VOCs** that are chlorinated, such as common solvents/degreasers trichloroethylene (**TCE**); perchloroethylene (**PCE**); among others.

Knowledge Box

MTBE: or methyl tertiary butyl ether, with a chemical formula of $C_5H_{12}O$, is a fuel oxygenate added to gasoline to reduce air pollution and to enhance octane ratings.[7] When it is released to the environment in pure form or as an ingredient of gasoline, its high-water solubility, chemical stability, and similar density with water make it difficult to contain and clean up. The persistence and mobility of MTBE and its wide uses have made it one of the most common contaminants in groundwater plumes, posing threats to many drinking water systems long after its banning in the 2000s. Even though MTBE's health effects are not clear, due to its significant impact on the taste of drinking water, it is regulated under one of the primary maximum contamination levels (MCLs) in California of 13 mg/L. There is no federal drinking water standard for MTBE in the United States.

Knowledge Box

TCE: Trichloroethylene has a chemical formula of C_2HCl_3, is another common soil and groundwater VOC pollutant. TCE is a volatile, colorless chlorinated hydrocarbon widely used as a solvent, degreaser, and cleaning agent. TCE is poorly soluble, 46% heavier than water, but is quite volatile. Many cities across the United States have TCE in their drinking water systems. In California between 2005 and 2008, TCE exceeded the MCL of 5 ppb in 133 water systems, the most for any VOC contaminants.

---

[7] Octane rating is a standard measure of the performance of an engine fuel. The higher the octane number, the more compression the fuel can withstand before detonating igniting so that the fuel can be ignited at the optimal and higher pressure for optimal performance. Gasoline with lower octane numbers may lead to the problem of engine knocking or premature ignition of fuel.

Inorganics

The most common inorganic groundwater contaminants include arsenic, lead and copper, nitrate, and perchlorate. Some, such as arsenic (a known carcinogen), are naturally occurring chemicals. However, most of the other inorganic contaminants are man-made or originated from human activities such as farming and industrial discharges. Arsenic could be used as pesticides, herbicides, or industrial uses such as wood preservatives. As a result, arsenic is widely detected in water systems, some with significant concentrations. According to the Safe Draining Water Information System (SDWIS), 86 drinking water systems in the United States exceeded the **maximum contamination level (MCL)** for arsenic of 10 µg/L in 2018.

Lead and copper may appear in the tap water due to corrosion of lead and copper materials in the drinking water distribution and plumbing systems. Lead is a toxic metal that can be harmful to human health, especially to younger children. The corrosion can be influenced by many factors such as water acidity, mineral content, temperature, and how long the water stays in the system. In many cases, there is a protective layer inside the water distribution system, mostly of a carbonate composition, that could protect the plumbing system from corroding. When this layer is impacted or removed, corrosion may occur and lead and copper may be leached out to the drinking water system. For this reason, USEPA has the Lead and Copper Rule (LCR) and established action levels for lead and copper to 15 ppb and 1.3 ppm, respectively.

Knowledge Box

Flint Water Crisis—The Flint Water Crisis is a monumental drinking water-related public health crisis that took place in 2014 in the City of Flint, Michigan, USA. The issue started when the City switched its drinking water source from treated water from Detroit to the Flint River. The switch was to lower the cost of its drinking water after a financial crisis that Flint had suffered in the past few years. Due to differences in water chemistry for these two sources, lead from aging pipes, some installed more than 100 years ago, started to be leached into the water supply, exposing over 100,000 residents to elevated lead levels in the drinking water. In addition to elevated lead levels, there were also taste, odor, color, and bacteria issues. Some city officials were criminally charged due to negligence and cover-up.

Nitrate and nitrite are important nitrogen compounds in the nitrogen cycle (Figure 2.8). They rarely exist in natural groundwater at high levels. However, sources such as fertilizers, septic tanks, and sewage could impact nitrate and nitrite in groundwater to levels unfit for drinking water use. High nitrate level in drinking water could cause methemoglobinemia, which is blood disorder in which an abnormal amount of methemoglobin[8] is produced. To prevent this health risk, an MCL for nitrate of 10 mg/L is set.

> ### Knowledge Box
> **Blue Baby Syndrome:** When nitrate-induced methemoglobinemia happens to infants, it is called 'blue baby syndrome' because the lack of ability for methemoglobin to transport oxygen by blood, causing the infant's skin to turn blue.

Perchlorate (ClO_4^-) is an oxidizing agent widely used in a number of industrial processes. For example, ammonium perchlorate (NH_4ClO_4), an oxidizer in rocket propellent, is the main source of perchlorate in drinking water in California and Nevada. The main health impact of perchlorate is on the thyroid gland, leading to reduced thyroid hormone production because perchlorate blocks the uptake of iodine, a critical component of this hormone. Reduced thyroid hormone could cause a range of health issues, including cardiovascular diseases, abnormal fetal brain development, and impairment of cognitive ability. Surprisingly, there is no federally-adopted MCL for perchlorate. In California, perchlorate MCL is 6 ppb.

Hexavalent chromium, or chromium-6, is the most common form of chromium salt. Its analogue trivalent chromium (chromium-3) is an essential element for humans. Chromium-6, on the other hand, is highly toxic and carcinogenic due to its oxidative properties. Chromium-6 is very soluble and can easily enter human body through drinking and inhalation, with inhalation posing the most significant and demonstrable carcinogenic risk. USEPA has an MCL of chromium-3 and chromium-6 of 100 µg/L, while California, the safe level for chromium-6 is somewhat controversial and is currently without an MCL due to a court ruling (see the Knowledge Box).

---

[8] Methemoglobin is a type of hemoglobin that can take up oxygen but do not readily release it as normal hemoglobin.

> ### Knowledge Box
>
> **Chromium-6**—The chromium-6 regulation in California, despite the famous work by Erin Brockovich, went through a bumpy road. USEPA's initial MCL for total chromium (i.e., the combined chromium-3 and chromium-6) was set at 50 µg/L but was raised to its current 100 µg/L in 1991. California did not follow USEPA's footstep and kept the 50 µg/L MCL. In 2011, California's Office of Environmental Health Hazard Assessment (OEHHA)[9] issued a public health goal for chromium-6 of a shocking 0.02 ppb. In 2014, California issued an MCL for chromium-6 of 10 µg/L. However, in 2017, a court invalidated this MCL for the reason that this limit failed to consider economic impact. After all the hoopla, this issue remains in limbo. This is a typical "California being California" case in the environmental field.

7.2.3 Human Health Risk Assessment

As discussed in Chapter 5, the human health or ecological risks of any chemicals are linked to the dose and length of exposure. Together, these risks will be the foundation in setting up the remediation targets for these chemicals at a contaminated site, as will be discussed in the next section. A human health risk assessment is the process of estimating the nature and probability of adverse health effects in the exposed individuals. It often involves building a conceptual site exposure model (CSEM) and the process includes the following steps. First, hazard identification is carried out to identify whether a chemical has the potential to cause harm; second, dose-response assessment to determine the relationship between exposure and adverse health effects; third, exposure assessment to determine the frequency, duration, and level of exposure to the chemical; lastly, risk characterization to quantify the health risk based on the above information as well as a standardized calculation scheme. The result of risk characterization is usually expressed in cancer risk. 1×10^{-4} and 1×10^{-6} excess lifetime cancer risk[10] are two thresholds that are often used as benchmarks for health risks. For example, if the risk is below 1×10^{-6}, there would be no further actions needed; if it is greater

---

[9] The Office of Environmental Health Hazard Assessment (OEHHA) is the lead agency in California for the assessment of health risks posed by environmental contaminants. Its mission is to protect human health and the environment through scientific evaluation of risks posed by hazardous substances. The Office is one of six state departments within the California Environmental Protection Agency (CalEPA).

[10] Excess lifetime cancer risk is the risk that one would get cancer due to exposure to a chemical for a lifetime over the baseline.

than 1×10^{-4}, response actions are definitely needed. In between these two thresholds, the need for action, and the type of actions, need to be evaluated. The exposure could be from ingestion, inhalation, or dermal contact. Aside from soil and groundwater, these exposures could be from air/soil vapor, surface water/drinking water, among others. The process of assessing risks and making decisions on remediation is called 'Risk-Informed Decision Making' or RIDM.

RIDM has both advantages and disadvantages. The strengths include that the conceptual site exposure model clearly lays out the technical and policy objectives upfront, so all stakeholders[11] would work on solutions from an early stage. The process also outlines a pragmatic tiered approach that helps prioritize the major issues and lend flexibility to issues that require more investigation, so there may be significant cost savings. Lastly, the risk-based approach allows the stakeholders to focus on results rather than politics and makes the process efficient and transparent. However, there are some drawbacks to this process, mostly because this is relatively new and technical, so some training is required. For the CSEM model, overly conservative assumptions could render the model impractical, and disputes on technical issues could slow down the process. The technical nature of this process also could make it challenging for the general public to comprehend.

7.3 Groundwater and Soil Remediation

Groundwater and soil remediation is one of the most important and prominent environmental issues. Similar to what the Cuyahoga River was to the Clean Water Act (see Chapters 3 and 7), the Love Canal disaster in New York and the Valley of the Drums in Kentucky were the triggers of the monumental **Superfund** law, which is officially known as the Comprehensive Environmental Response, Compensation, and Liability Act of 1980 (CERCLA). Please read Chapter 3 for more information about the Superfund law and environmental policy in general. In the United States, the Superfund law is administered by the USEPA to investigate and clean-up sites where the groundwater and soils are contaminated with hazardous substances as listed in Table 7.1. There are more than 40,000 Superfund sites across the United States, with 1,600 of them highly contaminated and are undergoing long-term remedial investigation and cleanups. Many (~30%) of the Superfund sites do not have responsible parties, or the responsible parties have bankrupt or ceased to exist. In these cases the USEPA will use federal funds (i.e., "Superfund") to pay for the

---

[11] For site cleanup projects, there usually are many diverse stakeholder groups. They could include regulators, responsible parties, property owners, banks, environmental groups, and the general public.

Figure 7.4 Crown Vantage Landfill Superfund Site in Alexandria Township, New Jersey, USA. The landfill is located near Delaware River and received hazardous wastes from several paper mills from 1930s to 1970s.

cleanup in a "use and replenish" mode, in which the cost recovery actions are usually taken toward the potential responsible parties to replenish the fund pool as much as possible.

Knowledge Box

Love Canal: Love Canal is a neighborhood in Niagara Falls, New York (Figure 7.5). The town started as a partially-built planned community envisioned by an entrepreneur Willian Love (hence the name). A canal ("Love Canal") was dug to introduce water features to the community but was unfinished due to economic downturn in the early 1900s. In the 1920s, the canal became a dump site and received municipal waste as well as thousands of tons of toxic chemicals (including dioxin, one of the most toxic organic chemicals) from a chemical manufacturer. In 1953, the property was sold to a local school district. Two schools were constructed and more residential developments were built nearby, some directly on top of the toxic dump. Since then, health issues began to appear in local residents. After numerous complaints, an investigation, and increasing public outcry, President Jimmy Carter declared a federal health emergency for Love Canal. The 1980 Superfund Act was passed specifically to address such a situation and it was the first Superfund site on the **National Priorities List**. It took the federal government until 2004 to complete the cleanup operations. The chemical company (now Occidental Petroleum) paid $129 million for damage and penalty in 1995.

Figure 7.5 Left: 1956 aerial photograph of Love Canal showing a school and houses built over and near a toxic landfill near Niagara Falls, New York. Right: current Google Earth® aerial of the same location.

Knowledge Box

The Valley of Drums: The Valley of Drums is a 9.3 hectare toxic waste site near Louisville in Kentucky, USA. The site received toxic wastes since the 1960s and waste-containing drums scattered across the area. USEPA investigated the area in 1979 and found extremely high levels of heavy metals, PCBs, and hundreds of other toxic chemicals in the soil, groundwater, and surface water. This site, similar to Love Canal, is known as one of the primary motivations for the passage of the Superfund Act. The cleanup operations began in 1983 and ended in 1990.

7.3.1 Underground Storage Tanks

VOCs and other contaminants as described in Section 7.2 above can get into soil and groundwater through many different ways, including accidental, and sometimes intentional dumping; leaking from pipeline for machineries, etc. The most common and most significant source is perhaps from underground storage tanks (USTs), see Figure 7.6. USTs are regulated in the United States by the USEPA to prevent the

Figure 7.6 A series of leaky underground storage tanks (USTs). Single-walled USTs could be leaking for years before the leak was detected.

contamination of groundwater and soil by the hazardous materials stored in the USTs. In 1984, U.S. Congress amended the Resource Conservation Recovery Act (RCRA) to include a section specifically for USTs. Around that time, there were more than 2 million USTs in the United States. In 1988, USEPA published initial underground storage tank regulations, including a 10-year phase-in period that required all operators to upgrade their USTs with spill prevention and leak detection equipment, to take financial responsibility for any releases or leaks, and to have UST insurance, a bond, or other abilities to pay for future cleanup.

In 2015, USEPA updated the UST and state program to increase the emphasis on properly operating and maintaining UST equipment. The revised minimum standards

Knowledge Box

UST—Underground Storage Tank (UST) is a tank (usually made of steel, aluminum, or composite materials such as fiber glass) and connected underground pipes used to contain regulated substances.[12] A tank does not need to be fully buried in the ground to be classified as a UST—only 10% of its volume (including the volume of underground pipes) needs to be underground. UST does not include farm or residential tanks of 1,100 gallons or smaller for motor fuel or heating oil or septic tanks.

---

[12] Regulated substances here mean substances are toxic, flammable, or have other properties that could pose threats to humans and the environment. As a result, there are regulations regarding their use, storage, transportation, and cleanup.

and new operation and maintenance requirements included secondary containment, operator training, among other things. The revision also increased the coverage of the regulation to emergency generator tanks, field constructed tanks, and airport hydrant systems.

When the contaminant plume hits the groundwater table, several processes will happen. For soluble contaminants, the migration will be controlled by groundwater movement as well as sorption/desorption of the contaminant with soil particles. The contaminant will also disperse to a larger plume with less concentration. For many heavy metals with several positive charges, they will often be sorbed strongly onto the soil particles and are generally not very mobile (with exceptions such as chromium-6 and Fe^{2+}). If the contaminants are lighter than water (such as some light nonaqueous phase liquid or **LNAPL**, including most VOCs), it will float above the water table and may form a "wedge" into the groundwater table. Such a plume will move with groundwater and travel downgradient. Dense, insoluble contaminants (such as dense nonaqueous phase liquid or **DNAPL**) will slowly sink to the bottom of the aquifer, making cleanup more difficult.

7.3.2 Soil Investigation

When a site is contaminated and requires cleanup, the first step is to conduct a contamination site assessment, which includes soil, vapor, and groundwater sampling and analysis, in order to know the contaminant source, type and the extent of the contamination in the soil and in the groundwater. Soil investigation is usually the first step. For volatile contaminants such as VOCs, soil vapor investigation is also required. Soil samples can be collected by a number of different methods, including coring (**Figure 7.7**), hand augering, or as simple as shoveling. The collected soil samples (**Figure 7.8**) can then be sent to a laboratory for analysis. In some cases where VOC contamination is the concern, onsite vapor testing (**Figure 7.9**) or soil vapor sampling by specialized equipment can be carried out. Soil and vapor investigation can also be done by passive method, which involves the emplacement of an adsorbent into the soil. With time, the adsorbent picks up the contaminants and can be analyzed.

7.3.3 Groundwater Investigation and Conceptual Site Model

Groundwater investigation requires more specialized equipment and methods. The standard method is to construct a monitoring well (**Figures 7.10** and **7.11**). If there are multiple layers of aquifers to be monitored, the monitoring well can be constructed in a way that different layers can be monitored separately. A typical structure

Figure 7.7 Geotechnical investigation and soil sampling by a drill rig.

of a groundwater monitoring well is shown in **Figure 7.11**. After construction, the groundwater can be monitored for water level, water chemistry, and contamination levels to assess groundwater impairment and to help design cleanup strategies. For most cleanup projects, multiple monitoring wells may be constructed. Based on water

Figure 7.8 Soil core samples for contamination site assessment.

Figure 7.9 Soil vapor testing by a portable device.

level and contamination pattern, a three-dimensional conceptual site model could be constructed. Since groundwater is not static and will carry the contaminant along its flow path, it is important to understand the groundwater hydrology by pumping the water out of the monitoring wells or specialized wells for groundwater testing. The

Figure 7.10 Groundwater monitoring well.

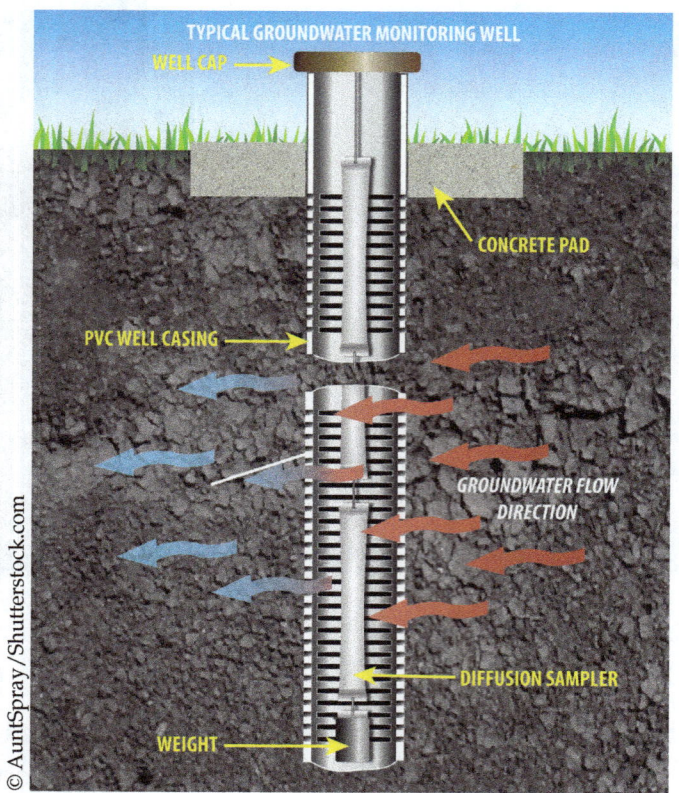

Figure 7.11 Groundwater monitoring well structure schematics.

groundwater production rate, soil and water testing results, and groundwater level recovery data will help investigators to understand the groundwater hydrology of the impacted site. This knowledge is critical to the design of an effective site remediation plan.

> ## Knowledge Box
>
> **Conceptual Site Model**—A conceptual site model (CSM) describes the sources and attributes of soil and groundwater contamination, local geology and groundwater hydrology, potential receptors,[13] and other factors that could impact the contaminant environmental fate and transport. A CSM should be developed as the first step of any investigative work at a contaminated site. The aforementioned soil, vapor, and groundwater sampling and investigation are all part of the CSM process. Document searching, interviews, and modeling could all be part of a CSM process. A CSM usually consists of the following components: hydrogeologic setting; source; contaminant transport and exposure pathways; and receptors.

---

[13] A receptor here means the subject (human; animal; plant) that, after intaking a contaminant or contaminants, may be impacted.

Table 7.2 Summary of Remedy Categories

Source Control
Treatment
Chemical, biological or physical means to reduce toxicity, mobility or volume of contaminated source media
Can be either in situ or ex situ
Examples include chemical treatment and in situ thermal treatment
On-site containment
Examples include the use of caps, liners, covers, and landfilling onsite
Off-site disposal
Includes excavation and disposal at an off-site facility
Monitored natural attenuation (MNA)
Reliance on natural processes
Natural recovery processes may include physical, chemical, and biological processes

7.3.4 Soil and Groundwater Remediation

After site investigation, remediation options will be evaluated and the best options will be implemented. Table 7.2 lists the general types of remedies, including source control; treatment; on-site containment; off-site disposal; and **monitored natural attenuation**.

In most cases, the pollution source(s) should be found and eliminated first. For example, the leaky underground storage tanks such as those shown in Figure 7.6 should be excavated and removed to prevent more fuels from leaking into the ground. Based on the nature, extent, and types of contaminants, the remediation options could be different. These methods can be classified into physical, chemical, biological, and passive methods. There are also novel methods that are in active research or trial period.

Physical methods are those methods where the contaminants are physically removed. For example, the contaminated soil can simply be excavated for disposal. VOC-contaminated groundwater can be pumped out and treated (see Figure 7.12). VOC-contaminated soil could be treated by soil vapor extraction, and VOC can be removed via filters containing granular activated carbon (GAC) or other materials that remove the VOCs from the vapor. For VOCs that absorb strongly to solid particles, in situ heating by electricity or natural gas could be used to free the VOCs from the soil particles for extraction and treatment. In some cases, air sparging is used to drive the air through the contaminated groundwater and soil, so that the air can strip the VOCs out. Extraction wells will be constructed nearby to remove the contaminated vapor from the air used for sparging. Figure 7.13 shows a soil and groundwater vapor extraction site impacted by trichloroethylene (TCE) in Tustin, California, USA.

Chemical methods for soil and groundwater remediation use chemicals to transform, destroy, or immobilize the contaminants. Common chemical treatment agents include oxidative chemicals such as hydrogen peroxide (H_2O_2), ozone (O_3), potassium permanganate ($KMnO_4$), and reductive chemicals such as ferrous chloride, zero valent iron.

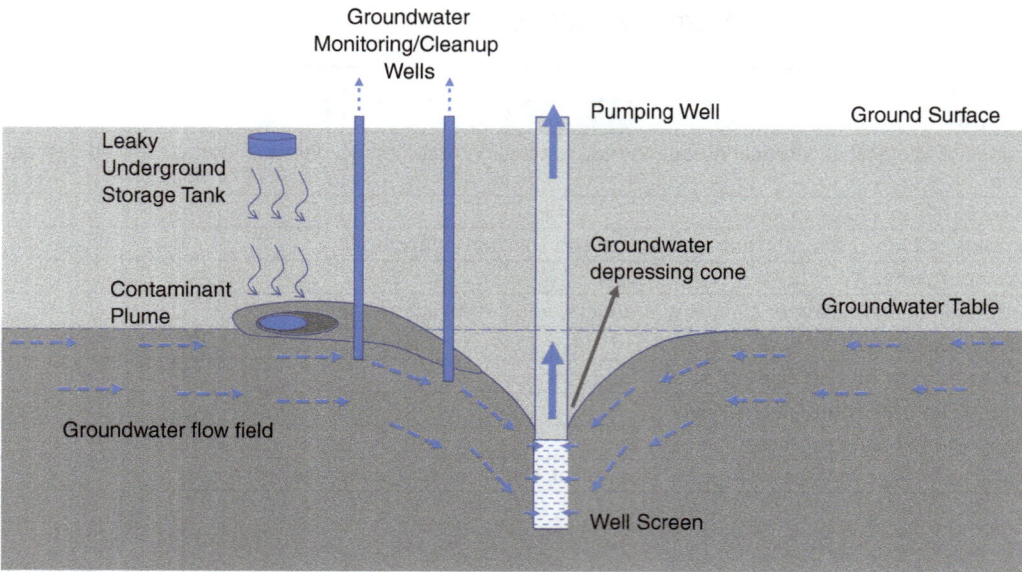

Figure 7.12 Groundwater monitoring and cleanup schematics. Monitoring wells will help establish the conceptual site model for the design of a cleanup plan, and cleanup wells can be constructed to intercept pollutant plumes for treatment before they reach the pumping/production well.

Source: Created by Jian Peng

Figure 7.13 Vapor extraction field for a site contaminated by TCE in Tustin, California.

These chemicals could be injected to the contaminated soil and groundwater, or be used to treat the excavated soil.

Biological methods are similar to chemical methods, but the processes are mediated or carried out by biological agents, mostly microbes or plants. Some natural or cultured bacterial species could use some contaminants as their energy source. By injecting these microbes into the soil and providing them with nutrients, these microbes could help reduce the levels of contaminants. The method that uses plants

Figure 7.14 Soil vapor barrier: two workers testing the air tightness of installed soil vapor barrier.

is called phytoremediation, where the plant could extract and accumulate the contaminants by the root system. The contaminant could either be evaporated into the air, degraded, or be harvested for disposal. As a result, the soil and groundwater are treated without conventional physical and chemical methods, which could be expensive and disruptive.

In many cases where human exposure is still a potential concern after site remediation, passive prevention could be implemented, such as soil vapor barrier, as shown in **Figure 7.14**. In some cases, passive methods could be used for treatment as well, such as permeable reactive barriers (PRB) where a trench is dug in the groundwater flow path. The trench is filled with materials such as oxidizers or zero valent iron (depending on the contaminants to be treated), so that the contaminated groundwater can be treated while flowing through the barrier. Occasionally, novel methods such as nanomaterials and nanotechnology could be used in site remediation as well. Some of these novel methods have been proven effective.

Table 7.3 listed some new techniques for groundwater remediation. Many are applicable for soil remediation as well. For details about these techniques, please refer to Miller (1997).

From 1982 to 2014, various treatment methods have been used on all of the 1,540 Superfund sites across the United States. 344 of them did not see active treatment. Instead, on-site containment, offsite disposal, monitored natural attenuation, or enhanced monitored natural attenuation were employed. Of the 1,196 site that underwent treatment, 275 had treatment at the source, 649 had treatment at both source and groundwater plumes, and 272 had treatment of groundwater only. With better source control and more stringent regulatory oversight, the number of active Superfund sites has been generally decreasing over the years after peaking in 1995.

Table 7.3 Groundwater Remediation Technology

In situ physical/chemical treatment techniques	Ex situ physical/chemical treatment techniques	In situ biological treatment techniques	Ex situ biological treatment techniques
Permeable reaction wall	Groundwater extraction and air stripping	Bioremediation	Bioremediation
In-well vapor stripping	Membrane filtration	Bioslurping	Aerobic biodegration
Catalytic decontamination	Solar detoxification	Bioventing	Biological aqueous treatment
Chemical treatment	Separator-filter-coalescer	In-well air stripping	Biodegradation
Cosolvent flushing	Polishing filter	Bio-fix beads	
Electrochemical reduction/immobilization	High energy electron irradiation	Biological treatment	
Pervaporation	Gas-phase chemical reduction	Biodegradation	
Precipitation/ filtration	Chemical treatment		
Dynamic underground stripping	Chemical fixation/solidification		
Oxidation	Oxidation		
Vacuum extration			
Air sparging			

Lastly, monitored natural attenuation, despite sounding passive and unexciting, can be the best option for a contaminated site because of cost-benefit considerations and availability of effective natural treatment options. If, for example, treatment is much more expensive than risk avoidance (such as properly relocating at-risk populations or protecting the perimeter of the impacted site), or the treatment itself is more environmentally damaging, it might be more sensible to have nature amends itself than to disturbing it even more.

Further Reading

USEPA. Underground Storage Tanks: Building on the Past to Protect the Future, USPEA Publication No. EPA 512-R-04-001, Washington, DC, 2004.

USEPA. Superfund Remedy Report, 15th Edition. EPA-542-R-17-001, USEPA Office of Land and Emergency management. Washington, DC, USA, July 2017.

USGS. A Groundwater Primer, US Government Printing Office, Washington DC, 1963.

Rong, Y. (eds.). Fundamentals of Environmental Site Assessment and Remediation. Boca Raton, FL: CRC Press, 2018.

USEPA USEPA Priority Pollutant List, 2020. Available at: https://www.epa.gov/sites/production/files/2015-09/documents/priority-pollutant-list-epa.pdf. Accessed May 3, 2020.

Miller, R.A. Analysis of information contained in the completed North American Innovative Remediation Technology demonstration projects. Groundwater Remediation Technologies Analysis Center. TI-97-01. Pittsburgh, Pennsylvania, USA, 1997.

Web Resources

California Department of Toxic Substances Control Guidance on Vapor Intrusion (great information with 7 videos): https://www.youtube.com/playlist?list=PL12aGMtO-GEbnlJJfvSMk8jH-ssZvDHLoO

USGS Webpage on Groundwater and aquifer: https://www.usgs.gov/special-topic/water-science-school/science/aquifers-and-groundwater?qt-science_center_objects=0#qt-science_center_objects

USEPA Safe Drinking Water Information System: https://www.epa.gov/ground-water-and-drinking-water/safe-drinking-water-information-system-sdwis-federal-reporting

USEPA Superfund: https://www.epa.gov/superfund

USGS National Water Information System: https://waterdata.usgs.gov/nwis

The Valley of the Drums: https://en.wikipedia.org/wiki/Valley_of_the_Drums

The Love Canal: https://en.wikipedia.org/wiki/Love_Canal

California Department of Toxic Substances Control: https://dtsc.ca.gov/

The story of Erin Brockovich (in addition to the movie): https://en.wikipedia.org/wiki/Erin_Brockovich

Questions and Exercises

1. What are the similarities and differences between CSEM and CSM?
2. How do you use the information developed in CSEM to determine the remedial options, as listed in Table 7.2?
3. If you were the project manager for a Superfund site with many stakeholders with very different views on treatment goals, do you want to bring them to the table early together, or do you work with separate stakeholder groups separately? How do you think the scientific team should play its role in the process?

4. On a site with contaminated groundwater with a potentially carcinogenic compound that will degrade completely in 10 years, what would be the most practicable remedy if the groundwater will not be used for drinking water purpose?
5. One of the soil/groundwater remediation goals is to keep the excess cancer risk to 1×10^{-6}, or one in a million chance of additional cancer during a lifetime of a person. Do you think this goal is reasonable? Should it be zero? Why?

Chapter 8

Ecology and Biodiversity[1]

Learning Outcomes	208
Key Concepts	208
8.1 Principles of Ecology	208
8.1.1 Concepts and Principles	208
8.1.2 Population Ecology	212
8.1.3 Ecosystem Health	212
8.2 Biodiversity	213
8.2.1 The Significance of Biodiversity	213
8.2.2 The Sixth Mass Extinction	214
8.3 Humans and the Ecosystem	217
8.3.1 Humans as Part of the Global Ecosystem	217
8.3.2 Ecosystem Services	218
8.3.3 Human Impact on Ecosystems	219
8.3.4 The Ecological Impact of Pollution	225
8.3.5 The Ecological Impacts of Habitat Loss	227
8.3.6 Ecological Impact of Hunting and Fishing	232
8.4 Genetically Modified Organisms	234
8.4.1 The Definition of GMOs	234
8.4.2 Benefits and Risks	236
8.5 Other Ecological Challenges	236
8.5.1 Endangered Species	236
8.5.2 Invasive Species	237
8.5.3 Deep Ecology: A Scientific Perspective	239
Further Reading	240
Web Resources	241
Questions and Exercises	241

---

[1] This chapter is greatly benefited from the careful review and edits by Dr. Peter Bowler of UC Irvine.

> **Learning Outcomes**
>
> - Understanding of the basic principles of ecology and the concept of an ecosystem
> - Understanding of the material and energy flow of within ecosystem
> - Understanding of the concept of biodiversity and its significance to ecological health
> - Appreciation of the fact that human is part of the ecosystem
> - Understanding of the disruptive force of human civilization to ecosystem health and biodiversity

> **Key Concepts**
>
> Ecology; ecosystem; biodiversity; food chain; food web; trophic pyramid; GMO; endangered species; invasive species; native species; deep ecology; mass extinction; The Sixth Mass Extinction; background extinction rate; ecosystem service; Ramsar; the International Union for the Conservation of Nature (IUCN)

8.1 Principles of Ecology

8.1.1 Concepts and Principles

Ecology is the scientific study of the distribution and abundance of living organisms and their relationships with both the biotic and abiotic components of their environments. Coined by the German zoologist Ernst Haeckel in 1866, the word ecology literally means "in house," and the scope of ecology as a scientific discipline has largely been true to its original meaning. Unlike common belief, ecology itself does not mean "green" or "sustainable."

Ecologists study populations[2] of organisms and groups of species in a habitat, and interactions among species are examined within an ecosystem. An **ecosystem** is a community of living organisms in conjunction with the nonliving components of their environment. These biotic elements are linked together through material and energy flows, as will be discussed below. Ecosystems are also controlled by external factors such as climate, geology, hydrology, and other environmental factors.

Ecology overlaps with other disciplines of science such as zoology, botany, microbiology, physiology, environmental science, hydrology, and many more, thus it is a truly

---

[2] In ecology, a population refers to a group of individuals that belong in the same species and live in the same area.

interdisciplinary science. The following list is a succinct summary of the key principles of ecological science.[3]

- Evolution organizes ecological systems into hierarchies of different dimensions. For organisms themselves, the hierarchies could be as individuals, populations, species, and higher taxa, such as genera, phyla and kingdoms. In an ecological context, there are individuals, populations, communities (within which different species coexist), ecosystems, biomes (continental scale groupings of ecosystems), and the overarching biosphere (all life on Earth). At each level within this hierarchy, ecological issues and study methods may be different (see **Figure 8.1**). An ecological unit could begin from an atom or a molecule and ecotoxicity could include cell organelles, cells, tissues, and organs. Most of ecological science begins with the organism and populations composed of many individuals of the same species. Different species within a localized geographic area form a biocenosis, an association of different organisms forming a closely integrated community. An ecosystem embraces all of the organisms and their physical environment. A biome (Latin name is *bioma*) is a naturally occurring large-scale community of plants and animals that have common characteristics for the environment in which they live. A term similar to biome is eco-region, a major ecosystem defined by distinctive geography. All of the biomes on Earth form the biosphere.

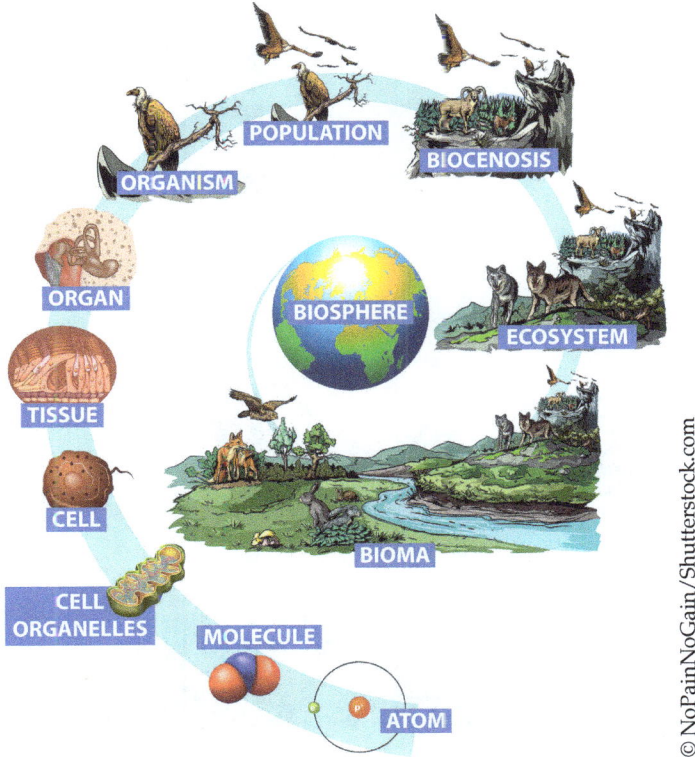

Figure 8.1 The hierarchy within ecological science.

---

[3] Revised based on the information provided by "The Kaspari Lab," https://michaelkaspari.org/2017/07/17/the-ten-principles-of-ecology/

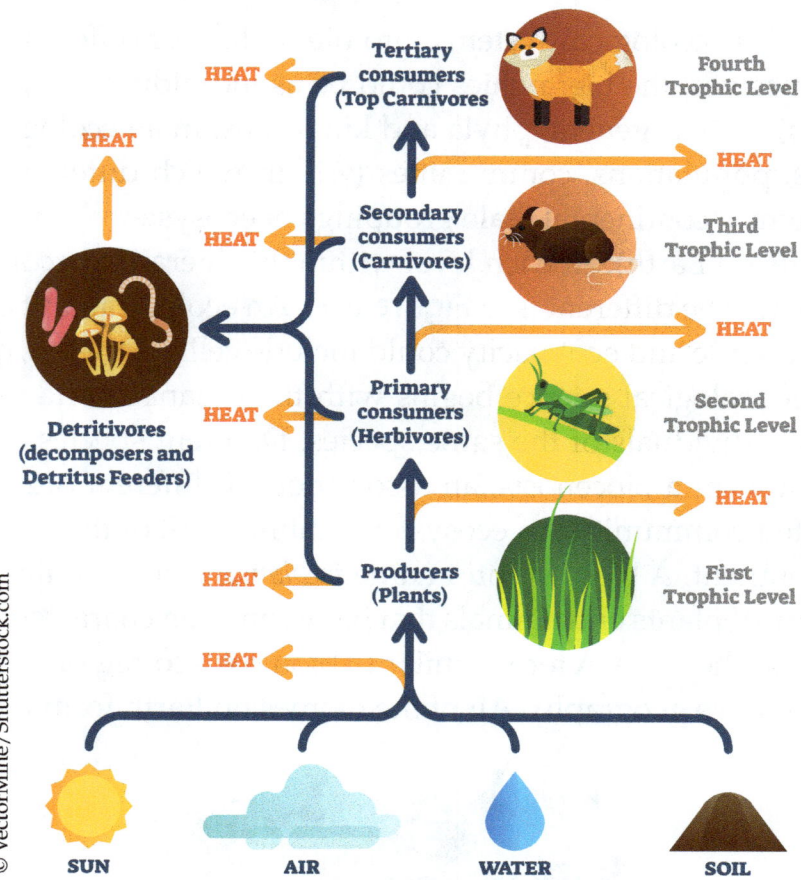
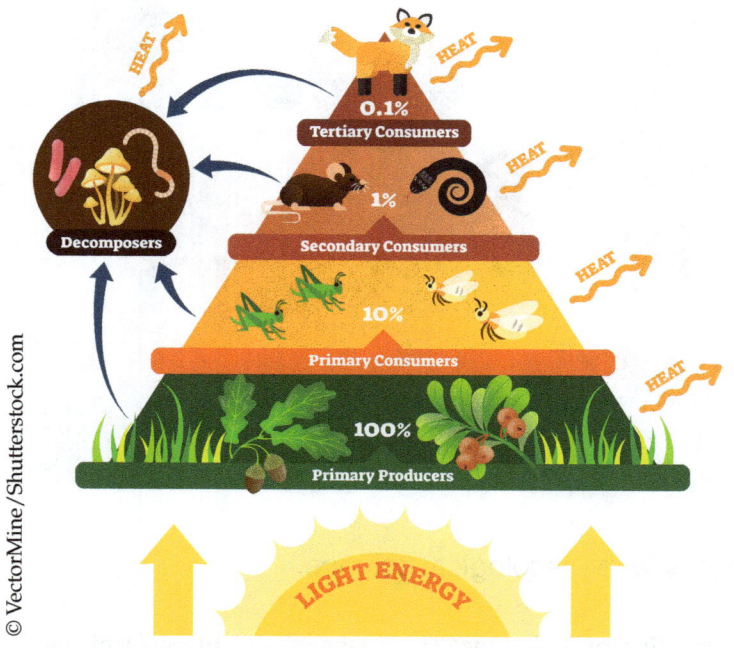

Figure 8.2 Food chain and energy flow in an ecosystem.

- There are continuous flows of energy and materials within an ecosystem. Solar energy is the ultimate source of energy in the biosphere (with few exceptions). Materials for the biosphere and ecosystem begin with inorganic matter such as carbon dioxide, water, and decomposed organic matter. The energy and materials then flow and expand through trophic levels and throughout the ecosystem. Energy enters the system primarily through photosynthesis by plants and other primary producers. The materials and energy then accumulate as it passes upward through the trophic levels to consumers of different hierarchies. At each level, some energy and material are lost to the environment, mediated by decomposers (mostly bacteria and fungi). These processes involve interactions between both the biotic and abiotic parts of the environment (**Figure 8.2**).
- While the energy will dissipate (but still be conserved), materials are constantly recycled within an ecosystem both through cycling with a trophic level and at different trophic levels. Individual organisms of a species or population could move in and out of an ecosystem. New species could enter to system, and existing species could move out or become extinct. In other words, physics and mathematics still work in ecological science!
- Organisms interact with each other in an ecosystem. The interaction could be predator–prey (such as in a **food web**, as shown in **Figure 8.3**), symbiosis,

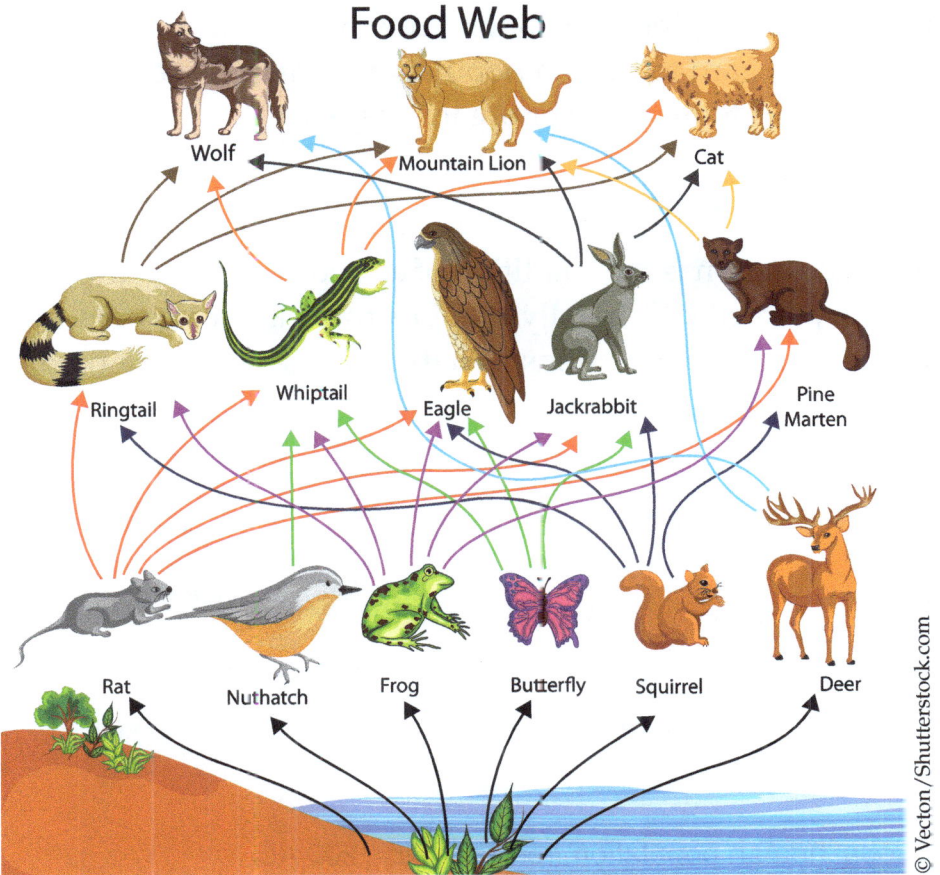

Figure 8.3 A food web example.

parasitism, or other forms. The species also competes for resources, including food and habitat. These interactions influence the population, diversity, and complexity of the ecosystem.

- Humans are one single but dominating species among millions of others on Earth. Humans are altering the Earth and reducing its natural resilience through population growth and technological advances that they are diminishing its biodiversity, climate stability, environmental quality, and sustainabilities for a diversity of ecosystems.

8.1.2 Population Ecology

Population ecology is the ecological study of how living and nonliving factors influence the density, dispersion, age structure, sex ratio, and size of a population. The size of the population could be affected by density-independent factors such as forest fires or other unpredictable events, or it could be affected by density-dependent factors such as limiting resources and carrying capacities. The latter is easy to comprehend because there are always limiting factors such as predation, diseases, and limited food availability that prevent the population to grow exponentially (as can be predicted easily for the case of unhindered reproduction). So in most cases, the population size can be predicted by an exponential model at the beginning of its growth phase, then it will level off due to limiting factors. Overall, the population growth curve will take an 'S' shape, as will be discussed later in regard to human population dynamics (see **Figure 8.4**). The same scenario can be applied to many other organisms as well.

8.1.3 Ecosystem Health

The health of an ecosystem is the condition of an ecosystem, but the benchmark for the condition can vary greatly. Generally, the health of an ecosystem can be shown by conditions such as viable populations of **native species**, natural biodiversity (see the

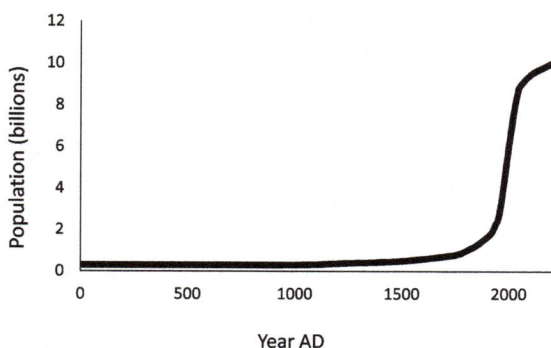

Figure 8.4 Global human population from 1 AD to 2200 AD.
Data source: United Nations. Created by Jian Peng

next section), maintenance of normal ecological and evolutionary processes; and resilience to external disturbances, including anthropogenic ones.[4] It can be affected by events or conditions such as fire, flooding, drought, climate change, extinctions, invasive species, and many anthropogenic impacts.

The concept of ecosystem health has been incorporated into some regulatory frameworks, including that for water quality management in the United States. For example, for sediment quality objective (SQO) assessment in bays and estuaries, benthic community health is evaluated in conjunction with sediment chemistry and toxicity. For the freshwater environment, the bioassessment process involves an assessment of the entirety of the ecosystem, including physical habitat, vegetation, hydrology, and benthic invertebrates, including their species abundance, diversity, and ratio between sensitive and tolerant species. Please refer to Section 5.5.6 for more information.

8.2 Biodiversity

8.2.1 The Significance of Biodiversity

Biodiversity is the variety and variability of life in various spatial scales (a region; an ecosystem; an ecoregion; or the entire biosphere). The significance of biodiversity can be seen by two United Nation (UN) actions. In October 2010, the United Nations designated 2011 to 2020 as the United Nations Decade on Biodiversity, and 2021 to 2030 is the United Nations Decade on Ecosystem Restoration. There are many different definitions for biodiversity. Generally speaking, biodiversity is the variety and variability of life within different ecological units as described above, from species, populations, habitats, ecosystems, or the global biosphere as a whole. The United Nations defines biodiversity as "the variety of life on Earth, it includes all organisms, species, and populations; the genetic variation within these; and their complex assemblages of communities and ecosystems" (UNESCO, 2010). Thus, biodiversity can be expressed in different levels such as genetic diversity, species diversity, and ecosystem diversity. Biodiversity could also be measured by many methods and at different levels. The most common measure of biodiversity is species richness in a given habitat, which can generally be measured by parameters such as Simpson's Diversity Index or Shannon's Diversity Index (see the Knowledge Box). Biodiversity is not distributed evenly on Earth in that it is very low in polar regions and is the richest in the tropics, where 90% of the species exist. This phenomenon is sometimes called 'latitudinal biodiversity gradient'. It is also often found that at a given region, the actual biodiversity is proportional to the size of the undisturbed habitat. The larger the habitat is, the higher biodiversity it can sustain. Therefore, habitat protection is key to preserving biodiversity.

---

[4] The notion that ecosystem health could be evaluated by its fitness to serve human needs is a controversial one.

> **Knowledge Box**
>
> **Measuring Biodiversity**—Biodiversity can be measured in different ways. Species richness, the number of species ("alpha diversity"), within a habitat, can generally be measured by indices such as Simpson's or Shannon's Index. Both indices are based on the number of different species, as well as the number of individuals within the same species in a habitat. On the ecosystem level, "beta diversity" is often used to compare two ecosystems or to determine changes in diversity over time. It is calculated by computing the differences between regional and local species biodiversity. The last term is "gamma diversity," or the total biodiversity, which is the sum of alpha and beta diversities. When discussing biodiversity, the scale or context of diversity must be specified.

8.2.2 The Sixth Mass Extinction

During the past 3.5 to 4 billion years when there has been life on Earth, there have been many **mass extinction** events (e.g., Bond et al., 2008). Since the Great Oxygenation Event of about 2.5 billion years ago, marking the profound impact of biota to the Earth's environment, there was a notable biodiversity "explosion" during the Cambrian period (i.e., "the Cambrian Explosion," see **Figure 8.5**), and five mass extinctions (**Figure 8.6**). The Odovisian–Siluvian extinction event occurred around 440 to 450 million years ago when 27% of all families, 57% of all genera and 60% to 70% of all species became extinct (for simplicity, we use 27%–57%–60% to denote the extent of this extinction). The others were Late Devonian extinction between 360 to 375 million years ago (19%–50%–70%), Permian–Triassic extinction at 252 million years ago (57%–83%–90%), Triassic–Jurassic extinction event around 201 million years ago (23%–48%–70%) that famously saw the

Figure 8.5 The Cambrian explosion took place around 540 million years ago when nearly all major animal phyla emerged. This extraordinary period was marked by rich fossil records, such as those for trilobites.

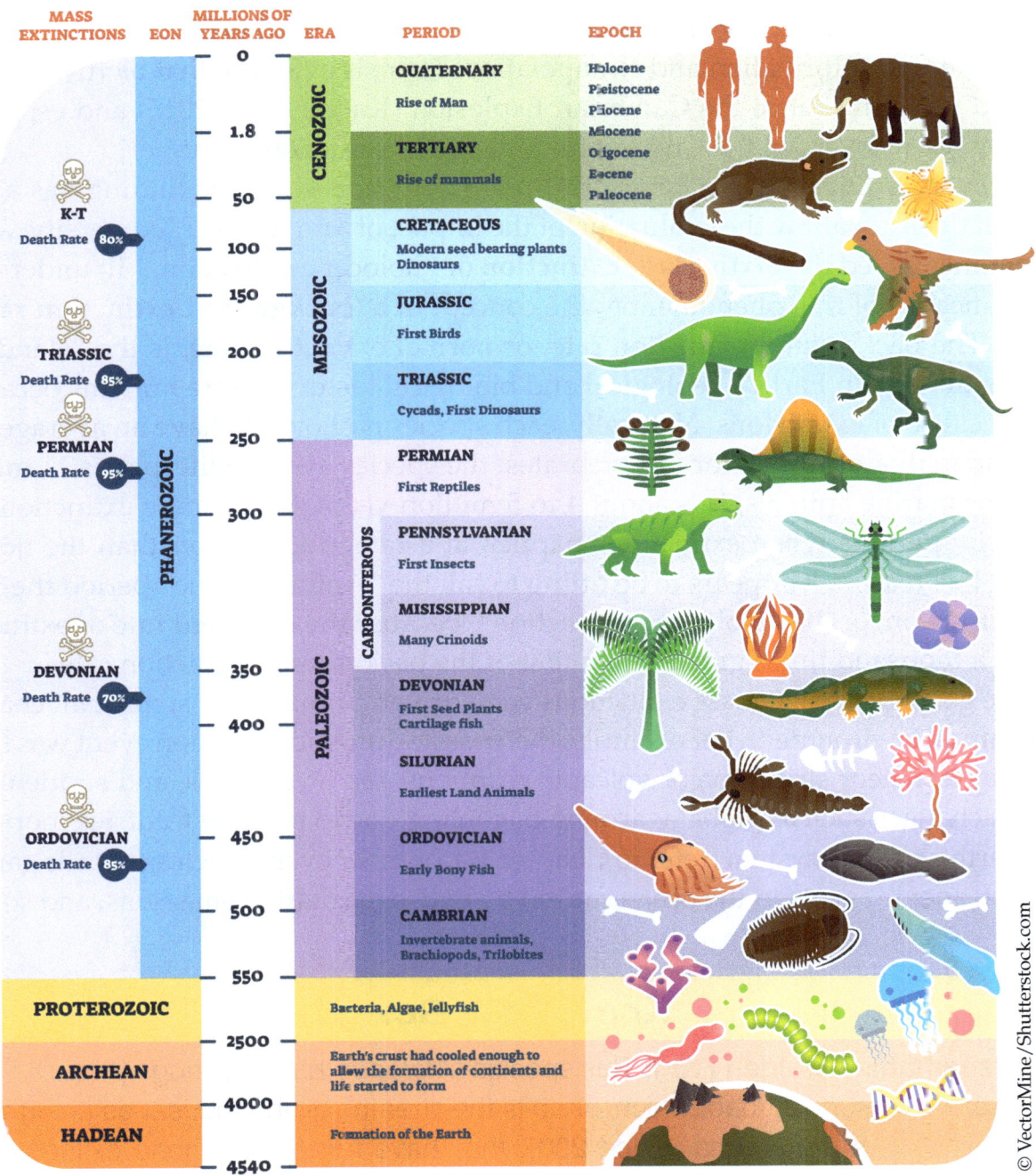

Figure 8.6 Geological history and the five mass extinctions (shown in the left column). We are probably experiencing the sixth mass extinction right now (Holocene).

extinction of dinosaurs, and lastly Cretaceous–Paleogene extinction event around 66 million years ago (17%–50%–75%).

Based on fossil record and extrapolation, it was postulated that 5 to 50 billion species, or more than 99% of the Earth's species were extinct long before humans emerged on Earth. Estimates of the number of the Earth's current species ranges from 10 million to 14 million, of which about 1.2 million have been documented and over 86% have not yet been described. In 2016, scientists reported that 1 trillion species are estimated to be on Earth currently with only one-thousandth of 1% described (Locey and Lennon, 2016) based on genetic information and extrapolation. There is evidence that biodiversity has increased with time since the Cambrian Explosion (Knope et al., 2015) and especially following the recovery of the Cretaceous–Paleogene extinction event.

However, as Elizabeth Kolbert (2014) points out, this scientific finding has a time scale that does not allow the evaluation of the impact of humans on biodiversity, which is sometimes called the **Sixth Mass Extinction** or "holocene extinction." To understand the significance of this phenomenon, the concept of a '**background extinction rate**' is needed. The background extinction rate, or normal extinction rate, is the "standard" rate of extinction in Earth's geological and biological history before humans became a primary cause of extinctions. Naturally, each species is shown to have an average species lifespan. For example, for invertebrates, the species average lifespan is 11 million years. For marine animals, it is about 4 to 5 million years. Major mass extinctions are marked by periods when extinctions happen at a rate much higher than the normal rate. For this reason, it appears to be fitting to call the "Anthropocene" period the Sixth Mass Extinction, or the "Holocene Extinction," because the observed rate of extinction is about a thousand times greater than that of the background extinction rate.

In the geological past, mass extinctions were driven primarily by significant changes in the natural environment. For example, the Triassic–Jurassic extinction event was likely caused by a meteor strike, major volcanic eruptions, sea level drops, and sudden temperature changes such as global warming or cooling due to positive feedback loop built in the Earth's climate system (refer to Chapter 11), or other events such as cosmic gamma ray burst, anoxic events in the ocean caused by hydrogen sulfide emissions, and so on.

Knowledge Box

The Panamanian Golden Frog: This species (*Atelopus zeteki*) of frogs (Figure 8.7) is actually a poisonous toad, which inhabits the streams by the thousands along the mountain slopes in Panama. Since 2007, they have functionally gone extinct with individuals living only in zoos in captivity around the world. It was found that a fungus called *Batrachochytrium dendrobatidis* is responsible for the extinction of Panamanian Golden Frog and many other amphibian species. Even though the fungus is natural, human traveling and commerce are believed to have enabled the fungus to infect most of the continents on Earth. The surviving Panamanian golden frogs in the zoos have to be carefully protected to avoid the deadly fungus.

Figure 8.7 Panamanian golden frog.

8.3 Humans and the Ecosystem

8.3.1 Humans as Part of the Global Ecosystem

The length of human history is a blink of an eye compared to the history of life on Earth (see Figures 1.12 and 8.6). Nonetheless, in such a short period of time, humans have altered the surface of the Earth in unprecedented ways. The greatest difference for humans from other species is that humans are intelligent beings and can use tools, machineries to improve their ability to live better and more comfortably. In the process, humans have changed the surface of the Earth. From the ecological standpoint, this is akin to a dominating species winning out against other species in the ecosystem. On the other hand, there are some fundamental differences between humans and other species. The single most difference is the dominant power of humans. Not only have we changed the surface of the Earth with a booming population that has strained many critical resources, humans have the capability to do extreme damage to the environment if our actions are conducted without thought or conscience about impacts upon nature and natural processes. We also have the ability to be ethical and restrain ourselves from destroying nature, and to restore areas historically damaged.

In the Earth's history similar events have happened in the past, such as the "Cambrian Explosion," about 541 million years ago (Figure 8.5) and the Age of Reptiles during the Mesozoic Era. During this remarkable period, dinosaurs dominated the Earth both on land, in the air, and in the oceans. Modern birds are the offspring of avian dinosaurs that survived the Cretaceous–Paleogene extinction event whereas all of the other dinosaur lines became extinct. It is remarkable to envision the Earth during that period when millions of dinosaurs of more than 1,000 species roamed on Earth. Some of these species were the largest land animals ever.

The emergence of humans as a new species on Earth was not a remarkable event in the beginning. However, with a fast-growing population of less than a million before 10,000 BC, to less than 200 million around 2000 years ago, to the current 7.7 billion, we are now the dominant species on the planet. It is projected that the global population will eventually stabilize at about 10 billion (Figure 8.4). Humans, for the most part, have lived in harmony with the environment and are part of the global ecosystem. However, the human dominance of the Earth looks and feels different. Many critical resources such as water and energy have approached breaking points.

It is also interesting to note that in terms of biomass, humans are less than livestock and much less than fish. There are orders of magnitude larger biomass for arthropods, fungi, bacteria, and plants, which are the largest living biomass on the planet (Bar-On et al., 2018; **Figure 8.8**), Humans comprise a meager 0.07 gigatons of carbon, compared with 450 for plants and 0.2 for viruses. However, despite the small mass compared with other species, humans are dominating the Earth and other species because of our intelligence and our thirst for resources.

8.3.2 Ecosystem Services

Ecosystem services are the benefits to humans provided by the natural environment by healthy ecosystems. These services include natural pollination of crops, clean air and water, noise reduction, climate mitigation, and even mental/physical well-being. This concept, despite being anthropocentric, has been used by environmentalists, scientists, and economists for a long time, was not well studied until the Millennium Ecosystem Assessment (MA) project conducted by the United Nations under the leadership of the then Secretary-General Kofi Annan. The findings of this project include: (1) humans have drastically changed the Earth's ecosystems to satisfy their need for food, water, and other materials necessary for the well-being; (2) while human's

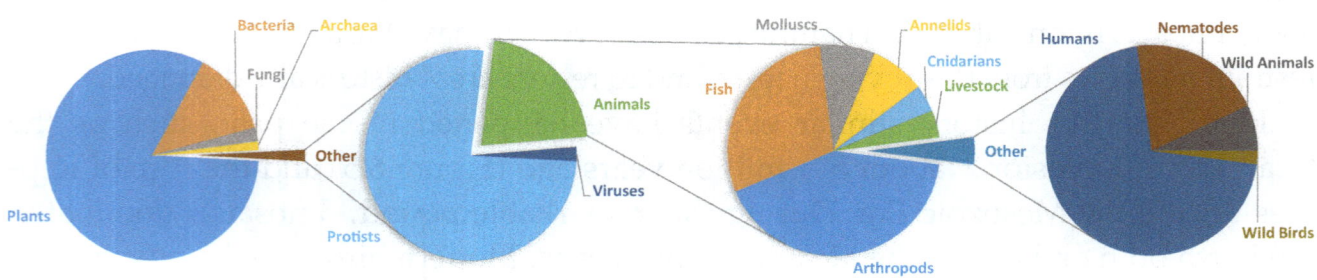

Figure 8.8 Biomass of different species (left: all biomass; right: animal biomass). Plants represent an overwhelming proportion of the Earth's biomass. In comparison, animals have smaller biomass than even bacteria and fungi.

Source: (Bar-On et al., 2018). Created by Jian Peng

well-being has improved significantly, the degradation of ecosystem quality and services have also worsened significantly; and (3) the conflicting needs between ecosystem protection and economic development can be assessed through the MA process to reduce the negative trade-offs and to provide positive synergies with other ecosystem services.

> ### Knowledge Box
>
> **The Catskill Story:** A positive story about ecosystem services can be found in New York City, where during the 1990s the quality of drinking water had fallen below federal and state standards. Agency authorities opted to restore the polluted Catskill Watershed that had previously provided the city with the ecosystem service of water purification, instead of resorting to downstream water treatment plants. The watershed restoration effort included elimination of sources of sewage and pesticides, and ecosystem restoration through means such as forestry management that helped raise the water quality above standard. This option cost between $1 and 1.5 billion, compared to an estimated $6 to 8 billion for cleaning using a water treatment facility (this reflects only the capital cost; annual operation and maintenance cost would be another $300 million). In this case, the value of the ecosystem services for a healthy Catskill Watershed is estimated to be at least $4.5 billion. If we count for other potential values such as clean air, biodiversity, and tourism, the value of the overall ecosystem provision will be even higher. This scenario is an example of positive synergy between human needs and ecosystem health.

8.3.3 Human Impact on Ecosystems

Human impact on the global ecosystem manifests in many different ways. International Union for Conservation of Nature (IUCN) Classification of Direct Threats includes 11 primary direct threats to conservation and ecosystems. They include: 1. Residential and commercial development; 2. Farming activities; 3. Energy production and mining; 4. Transportation and service corridors; 5. Biological resource usages; 6. Human intrusions and activities that alter, destroy, simply disturb habitats and species from exhibiting natural behaviors; 7. Natural system modifications; 8. Invasive and problematic species, pathogens and genes; 9. Pollution; 10. Catastrophic geologic events; and 11. Climate change. With the exception of #10, the others are all directly or indirectly linked to human activities. Table 8.1 lists these threats and subcategories of each. Due to the limited scope of this book, only a few of these factors will be presented.

Table 8.1 IUCN's Classification of Direct Threats

1. Residential and Commercial Development		
	1.1 Housing and Urban Areas	
		urban areas, suburbs, villages, vacation homes, shopping areas, offices, schools, hospitals
	1.2 Commercial and Industrial Areas	
		manufacturing plants, shopping centers, office parks, military bases, power plants, train and ship yards, airports
	1.3 Tourism and Recreation Areas	
		ski areas, golf courses, beach resorts, cricket fields, county parks, campgrounds
2. Agriculture and Aquaculture		
	2.1 Annual and Perennial Non-Timber Crops	
		farms, household swidden plots, plantations, orchards, vineyards, mixed agroforestry systems
	2.2 Wood and Pulp Plantations	
		teak or eucalyptus plantations, silviculture, christmas tree farms
	2.3 Livestock Farming and Ranching	
		cattle feed lots, dairy farms, cattle ranching, chicken farms, goat, camel, or yak herding
	2.4 Marine and Freshwater Aquaculture	
		shrimp or fin fish aquaculture, fish ponds on farms, hatchery salmon, seeded shellfish beds, artificial algal beds
3. Energy Production and Mining		
	3.1 Oil and Gas Drilling	
		oil wells, deep sea natural gas drilling
	3.2 Mining and Quarrying	
		coal mines, alluvial gold panning, gold mines, rock quarries, coral mining, deep sea nodules, guano harvesting
	3.3 Renewable Energy	
		geothermal power production, solar farms, wind farms (including birds or bats flying into windmills), tidal farms
4. Transportation and Service Corridors		
	4.1 Roads and Railroads	
		highways, secondary roads, logging roads, bridges and causeways, road kill, fencing associated with roads, railroads

	4.2 Utility and Service Lines	
		electrical and phone wires, aqueducts, oil and gas pipelines, electrocution of wildlife
	4.3 Shipping Lanes	
		dredging, canals, shipping lanes, ships running into whales, wakes from cargo ships
	4.4 Flight Paths	
		flight paths, jets impacting birds
5. Biological Resource Use		
	5.1 Hunting and Collecting Terrestrial Animals	
		bushmeat hunting, trophy hunting, fur trapping, insect collecting, honey or bird nest hunting, predator control, pest control, persecution
	5.2 Gathering Terrestrial Plants	
		wild mushrooms, forage for stall fed animals, orchids, rattan, control of host plants to combat timber diseases
	5.3 Logging and Wood Harvesting	
		clear cutting of hardwoods, selective commercial logging of ironwood, pulp operations, fuel wood collection, charcoal production
	5.4 Fishing and Harvesting Aquatic Resources	
		trawling, blast fishing, spear fishing, shellfish harvesting, whaling, seal hunting, turtle egg collection, live coral collection, seaweed collection
6. Human Intrusions and Disturbance		
	6.1 Recreational Activities	
		off-road vehicles, motorboats, jet-skis, snowmobiles, ultralight planes, dive boats, whale watching, mountain bikes, hikers, birdwatchers, skiers, pets in recreational areas, temporary campsites, caving, rock-climbing
	6.2 War, Civil Unrest and Military Exercises	
		armed conflict, mine fields, tanks and other military vehicles, training exercises and ranges, defoliation, munitions testing
	6.3 Work and Other Activities	
		law enforcement, drug smugglers, illegal immigrants, species research, vandalism
7. Natural System Modifications		
	7.1 Fire and Fire Suppression	
		fire suppression to protect homes, inappropriate fire management, escaped agricultural fires, arson, campfires, fires for hunting

(Continued)

	7.2 Dams and Water Management/Use	
		dam construction, dam operations, sediment control, change in salt regime, wetland filling for mosquito control, levees and dikes, surface water diversion, groundwater pumping, channelization, artificial lakes
	7.3 Other Ecosystem Modifications	
		land reclamation projects, rip-rap along shoreline, mowing grass, tree thinning in parks, beach construction, removal of snags from streams
	7.4 Removing / Reducing Human Maintenance	
		lack of mowing of meadows, reduction in controlled burns, lack of indigenous management of key ecosystems, ceasing supplemental feeding of condors
8. Invasive and Problematic Species, Pathogens and Genes		
	8.1 Invasive Non-Native / Alien Plants and Animals	
		feral horses, feral household pets, zebra mussels, Miconia tree, introduction of species for biocontrol
	8.2 Problematic Native Plants and Animals	
		overabundant native deer, overabundant algae due to loss of native grazing fish, plague affecting rodents, invasive grasses
	8.3 Introduced Genetic Material	
		pesticide resistant crops, hatchery salmon, restoration projects using non-local seed stock, genetically modified insects for biocontrol, genetically modified trees, genetically modified salmon
	8.4 Pathogens and Microbes	
		plague affecting rodents, Dutch elm disease or chestnut blight, Chytrid fungus affecting amphibians outside of Africa
9. Pollution		
	9.1 Household Sewage and Urban Waste Water	
		discharge from municipal waste treatment plants, leaking septic systems, untreated sewage, outhouses, oil or sediment from roads, fertilizers and pesticides from lawns and golf-courses, road salt
	9.2 Industrial and Military Effluents	
		toxic chemicals from factories, illegal dumping of chemicals, mine tailings, arsenic from gold mining, leakage from fuel tanks, PCBs in river sediments
	9.3 Agricultural and Forestry Effluents	
		nutrient loading from fertilizer run-off, herbicide run-off, manure from feedlots, nutrients from aquaculture, soil erosion

	9.4 Garbage and Solid Waste	
		municipal waste, litter from cars, flotsam and jetsam from recreational boats, waste that entangles wildlife, construction debris
	9.5 Air-Borne Pollutants	
		acid rain, smog from vehicle emissions, excess nitrogen deposition, radioactive fallout, wind dispersion of pollutants or sediments from farm fields, smoke from forest fires or wood stoves
	9.6 Excess Energy	
		noise from highways or airplanes, sonar from submarines that disturbs whales, heated water from power plants, lamps attracting insects, beach lights disorienting turtles, atmospheric radiation from ozone holes
10. Geological Events		
	10.1 Volcanoes	
		eruptions, emissions of volcanic gasses
	10.2 Earthquakes/Tsunamis	
		earthquakes, tsunamis
	10.3 Avalanches/Landslides	
		avalanches, landslides, mudslides
11. Climate Change		
	11.1 Ecosystem Encroachment	
		sea level rise (inundation of shoreline ecosystems, drowning of coral reefs), desertification (sand dune encroachment)
	11.2 Changes in Geochemical Regimes	
		ocean acidification, changes in atmospheric CO_2 affecting plant growth, loss of sediment leading to broad-scale subsidence
	11.3 Changes in Temperature Regimes	
		heat waves, cold spells, oceanic temperature changes, melting of glaciers/sea ice
	11.4 Changes in Precipitation and Hydrological Regimes	
		droughts, changes in timing of rains, loss of snow cover, increased severity of floods
	11.5 Severe/Extreme Weather Events	
		thunderstorms, tropical storms, hurricanes, cyclones, tornadoes, hailstorms, ice storms or blizzards, dust storms, erosion of beaches during storms

Figure 8.9 Agriculture and ecosystems: Soybean farm (left) and pig farm (right).

Agriculture and animal husbandry are two good examples of humans not only altering, but manipulating parts of an ecosystem so extensively that they are hardly ecosystems in the usual sense. See Figure 8.9 of a soybean field being actively managed by humans with tilling, pesticide and herbicide spray; and a large pig farm, where large number of pigs are raised in a crowded space in a highly industrialized manner. In both cases, they are not ecosystems because they are highly managed and they lack the biodiversity and interactions among different species and between species and its environment. Another example is mining operations. Large mining operations such as the Alberta oil sand exploration in Canada and open-pit mines such as the Earzberg mine in Austria result in complete habitat loss (Figure 8.10). Even after the mines cease operations, it will take decades for a healthy ecosystem to reestablish. Part of the reason may be due to pollution issues resulting from these operations, as will be discussed in the next section.

Figure 8.10 Mining operations have a heavy impact on the local ecosystem. Associated pollution could make their impact even worse. Left: Canada Alberta's oil sand operation. Right: Austria's Erzberg Mine.

Urbanization and associated deforestation and logging are other examples in which the human need for its own habitat is achieved by destroying the habitat essential to the survival of many other species. Commercial fishing is another illustration of anthropocentric technological prowess allowing them to hunt and harvest prey on a large scale. Humans are also impacting the ecosystem in ways that are more subtle. For example, the pollution of water, air, and soil; the overexploitation of water resources; genetically modified organisms (GMOs); and so forth. The following subsections examine some of these in greater detail.

8.3.4 The Ecological Impact of Pollution

Rachel Carson's "Silent Spring" is perhaps the best-known book on ecotoxicity and it raised international awareness about this serious issue. Rachel Carson, herself being an expert ecotoxicologist, told the seminal story of indiscriminate use of pesticides and their impacts to the ecosystem and human health. Due to her eye-opening research and books, laws were established regulating the use of many kinds of toxic compounds in the environment. Chapter 5 has more information about this book, and about the harmful impact that toxic chemicals have had upon ecosystems around the globe. There are a few examples of how chemical pollution could impact an ecosystem. An example of how chemical pollution has had a negative effect upon a key pollinator is the collapse of honey bee colonies. The majority of worker bees in a colony would suddenly disappear, causing the colony to be unsustainable. From 2005 to 2013, more than 10 million bee colonies declined and disappeared around the world. There are several theories on the possible causes of this phenomenon, and the leading theory is that pesticides such as neonicotinoids are to blame. Because honey bees are important pollinators, the collapse of bee colonies has damaged agricultural production. There are hundreds of billions of dollars equivalent of food and other agricultural products that require honey bees to pollinate.

Another famous example of ecological impact is that of DDT (dichlorodiphenyltrichloroethane) upon California Brown Pelicans along the Gulf Coast and around coastal Southern California. Pelicans are at the top of the **trophic pyramid** and feed on fish. In the Mississippi River watershed and along the Gulf Coast, DDT was extensively used as a primary pesticide. Along the California coast, thousands of tons of DDT were discharged to the coastal ocean by a DDT manufacturer (the largest in the world at that time) for decades. Through leaching and trophic transfer, there are increasing concentrations of DDT from water, sediment, phytoplankton, benthic invertebrates, and to organisms at higher trophic levels such as fish and birds. When fishes are contaminated by DDT and brown pelicans feed on them, DDT will further biomagnify as it passes through the **food chain.** One of the impacts experienced by brown pelicans was in the thinning of eggshells making them too thin for the eggs to hatch. By the 1960s,

Figure 8.11 Brown pelicans glide gracefully on the ocean, which is a common scene today. In the 1960s, they became an endangered species and nearly disappeared due to DDT contamination. Thanks to the ban of DDT in 1973 and other management measures, the pelican population has recovered and it is no longer an endangered species.

brown pelicans had nearly disappeared along the Gulf Coast and experienced nearly complete reproductive failure in Southern California. Because of this, they were declared an endangered species. With stricter regulations on DDT, especially the ban in 1973, the population of brown pelicans has gradually recovered to the point that they have been removed from the endangered species list (**Figure 8.11**).

> ## Knowledge Box
>
> **Bioaccumulation and Biomagnification:** bioaccumulation is the gradual accumulation of substances such as pesticides or other chemicals in the tissue of an organism faster than the rate the chemical is lost through metabolism and excretion. Biomagnification is the process where the concentration of a substance or chemical becomes greater with successively higher trophic levels. The case of the brown pelican is a good example of biomagnification. DDT passed from the primary producer/primary consumer to fish and then to the brown pelican. The concentrations of DDT increased by many orders of magnitude due to biomagnification, causing severe impacts to species at the top of the trophic level.

Other pollutants, such as nutrients, can cause severe ecological disruption. Due to large amounts of fertilizers in agricultural lands and home gardens in urban areas,

Figure 8.12 Fish kill at Fort Myers Beach, Florida, likely caused by toxic algae. Fish kills can also occur due to toxic chemical pollution or oxygen deficiency caused by eutrophication.

eutrophication has become one of the most common water quality challenges around the world. Even some large water bodies such as Chesapeake Bay (located at the east coast of the United States) and the Gulf of Mexico have suffered persistent eutrophication issues. In both cases, eutrophication has caused "dead zones" in the deeper waters, and at other sites, low dissolved oxygen makes the water uninhabitable. Sometimes, it may cause fish kills (e.g., **Figure 8.12**)

8.3.5 The Ecological Impacts of Habitat Loss

Habitat loss may be the single worst impact that humans bring to the natural world and to other species. Habitat loss can occur through direct destruction of habitat, fragmentation, or degradation. Because of dramatic increases in the human population and associated urbanization and the expansion of agriculture, the world's forests, swamps/wetlands, plains, lakes, and other habitats continue to disappear and to be replaced by residences, factories, and farmlands.

Wetland Loss

Wetlands are ecosystems characterized by the presence of water through runoff, periodic or seasonal flooding, or from groundwater that sustains them. Wetlands have characteristic hydric soils that are formed by alluvial processes, are wetted at least once a year, and have hydrophytic vegetation comprised of vascular species that are associated with wetlands. There are many categories of wetlands based upon the source of their water, vegetation type, substrate, periodicity of being flooded or inundated, and the flow and nutrient quality of their water. Surface water wetlands include bogs and seasonally wet vernal pools. Most wetlands have a distinct zonation of plant species from the adjacent uplands to the deepest occurrence of attached plants or algae. The pattern can include herbaceous emergent plants (characteristic of marshes), floating leaf species, and submergent plants. Bogs are surface water wetlands and fens are ground water depressional wetlands. Swamps are freshwater or marine wetlands with emergent trees or woody shrubs and marshes have emergent herbaceous vegetation. Wetlands can be supported by freshwater, brackish water, or salt water (such as coastal mangroves) and influenced by tides (salt marshes). Wetlands are globally important ecosystems that support biodiversity by providing critical habitats for a large number of species, many of which are endangered and rare species.

Wetlands provide water quality maintenance by filtering pollution through removal of sediment, producing oxygen, recycling nutrients, absorbing chemicals and nutrients, and serving as carbon sinks. The largest wetlands in the world include the Amazon River basin, the West Siberian Plain, and the Pantanal in South America (**Figure 8.13**). In the United States, a nationwide study conducted in 1989 showed that in the past 200 years, wetlands shrank from 392 million acres to 104 million acres, a 74% decrease (Dahl, 1990). California has sacrificed an estimated 91.5% of its original natural wetlands, a percentage of loss greater than that of any other state. According to **Ramsar**,[5] up to 87%

Figure 8.13 Pantanal wetlands in Brazil.

---

[5] Ramsar is the Ramsar Convention on Wetlands of International Importance especially as Waterfowl Habitat study and is an international treaty for the conservation and sustainable use of Wetlands. It is named after the city of Ramsar in Iran, where the Convention was signed in 1971. It provides protection to 2,331 wetlands around the world.

of global wetland resources have been altered and eliminated since 1700 and the rate of "take" for wetlands is three times faster than that of natural forests. The rate of wetland loss has been about 1 acre per minute for the past 200 years. Since 1970, 81% of inland wetland species and 36% of coastal and marine species have seen declines in population. Wetland protection and mitigation for historic losses is a critical but challenging endeavor.

> ### Knowledge Box
>
> **The Pantanal** (Figure 8.13): In Portuguese, Pantanal means wetland or swamp and it covers an estimated area of between 140,000 and 195,000 km^2. It is one of the world's largest surviving natural wetlands. It is located mostly in Brazil in the state of Mato Grosso do Sul with small areas in Bolivia and Paraguay as well. The Pantanal rests in a huge geological depression related to the formation of the Andes mountain range. It receives flood water from the Planalto highlands in Brazil and drains slowly to the Paraguay River to the south. There are many diverse ecosystems identified in this vast wetland with over 2000 plant species, 463 bird species, 269 fish species, 236 mammalian species, 141 reptile and amphibian species, and more than 9,000 species of invertebrates. The wetland provides essential sanctuaries for migratory birds and nursery grounds for many other species. In the past, the region was threatened by deforestation, gold mining, and cattle-ranching. It is now better protected through regulations and by the establishment of national parks and reserves, including one that was designated a Ramsar Site.

Deforestation

Habitat loss due to deforestation receives special attention because this activity severely damages the Earth's ecosystem in many significant ways. Forests, especially tropical rain forests, have one of the highest biodiversities of all ecosystems. Two best known tropical rain forests, one in Indonesia and another in Brazil's Amazon basin, are the sites where active and large-scale deforestation is taking place (**Figure 8.14**). In Indonesia, satellite studies show that more than 50% of protected lowland tropical rainforests in Kalimantan were cut down between 1985 and 2001 alone. One of the main drivers, in addition to logging for precious hardwood, is the lure of palm oil due to increasing global demand. Between 1985 and 2007, oil palm plantations saw a 10-fold increase, from 600,000 hectares to over 6 million hectares. Deforestation, together with other forms of habitat loss, is one of the main contributors to the Sixth Mass Extinction.

Figure 8.14 Deforestation in Borneo (left) and Amazon (right).

Tropical rainforests also contribute significantly to producing atmospheric oxygen and ameliorating the global greenhouse effect through the removal and reduction of carbon dioxide. By destroying these rainforests, humans are essentially digging their own grave.

Urbanization

Urbanization is the phenomenon of people migrating from rural areas to cities and surrounding developed areas. By itself, urbanization is not necessarily harmful to the environment and the ecosystem. This leaving of rural areas to cities is well known historically as a product of the Industrial Revolution as jobs were created by manufacturing and other industrial areas in cities. Urbanization is often viewed in a negative way because it often is driven by large increases in population. Urban growth however, focuses human presence in areas that are already developed and focuses dense human uses in areas no longer natural. This often requires mitigation that results in the restoration of habitats and the establishment of dedicated open space areas. The migration of residences and people from rural areas to urban areas does not necessarily cause net habitat loss. In fact, the opposite may be true. Here a more generic interpretation of urbanization will be used, in which the growth of urban areas and associated construction of roads, pipelines, parks, and other urban land uses can eliminate, segment, or degrade natural habitats—but on the other hand, it relieves rural areas from more intense anthropocentric utilization.

Urbanization has occurred throughout human history. Until several thousand years ago, most of the human population resided in villages where people have close bloodlines, close relationships, and communal behavior. Over time, cities began to emerge with urban cultural characteristics such as distant bloodlines, unfamiliar relationships, and competitive behavior. Entering the 20th century, urbanization around the globe accelerated with explosive population growth (Figure 8.4). Currently about 50% of the

Figure 8.15 Urbanization at Madrid and Shanghai.

world's population resides in cities. The UN projected that by 2050 about 64% of the population in the developing world and 86% of those in the developed world will be city-dwellers. With urbanization, large areas of the land will be completely covered by buildings, roads, parks with little or no open space for habitat (perhaps with the exception of some animals that are adapted to urban environments such as crows, racoons, rats, and in North America the coyote). Large cities such as Madrid and Shanghai (**Figure 8.15**) are such examples.

As urbanization and population increased, more roads are needed for commerce and connecting cities. Some of these roads have guard rails on the sides or at center dividers and segment the habitats for many wild animals. For the roads that allow wild animals to cross, vehicular traffic poses a severe threat to these animals. **Figure 8.16** shows two examples of such issues. Biological corridors over or under major highways are sometimes constructed to accommodate the movement of both resident and migratory wildlife.

Figure 8.16 Habitat segmentation by roadways is a significant problem when highways curtail free movement of animals (left) or cause road kills (right).

Figure 8.17 Sudan, the last male Northern White Rhino (1973–March 19, 2018). After its death, only two female northern white rhinos remain, thus this species is functionally extinct. Hunting for their horns, which in some Asian cultures are considered of medicinal value, is the main cause of the drastic decrease in their numbers and ultimately, their extinction.

8.3.6 Ecological Impact of Hunting and Fishing

Resource extraction operations such as commercial fishing/hunting and mining have caused severe ecological and environmental impacts in the past. A recent study by Ripple et al. (2019) found that many of the world's largest animals (megafauna) are being hunted nearly to extinction. Surprisingly, direct harvesting of megafauna for human consumption of meat or body parts is the largest individual threat to each of the categories examined. Several well-known examples include sharks that are hunted for their cartilaginous fins; rhinos that are hunted for their horns (**Figure 8.17**), reptiles for their eggs, meat and skin, and African elephants for their tasks.

Commercial fishing is a $185 billion per year industry and accounts for more than 171 million tons of fish a year, with over 90.9 million tons from capture and 110.2 million tons from aquaculture, including aquatic plants (United Nations FAO, 2016). Since the 1960s, the global per capita consumption of fish has doubled. China alone accounts for a third of the world's commercial fishing. Increasingly larger ships and more advanced technologies are being used, with large trawl nets capable of catching up to 100 tons of fish in a single cast (**Figure 8.18**). Some fishing methods, such as bottom trawl, can destroy deep-sea corals and sponge beds that will take centuries or longer to recover. Commercial fishing also leaves a significant quantity of abandoned fishing equipment including lines, nets, wires and cables that can cause severe harm or death to marine life. Due to overfishing and also probably global warming, habitat loss, and pollution, the global fish stock has steadily declined since the 1970s. By 2015, about a third of fish stocks were biologically unsustainable.

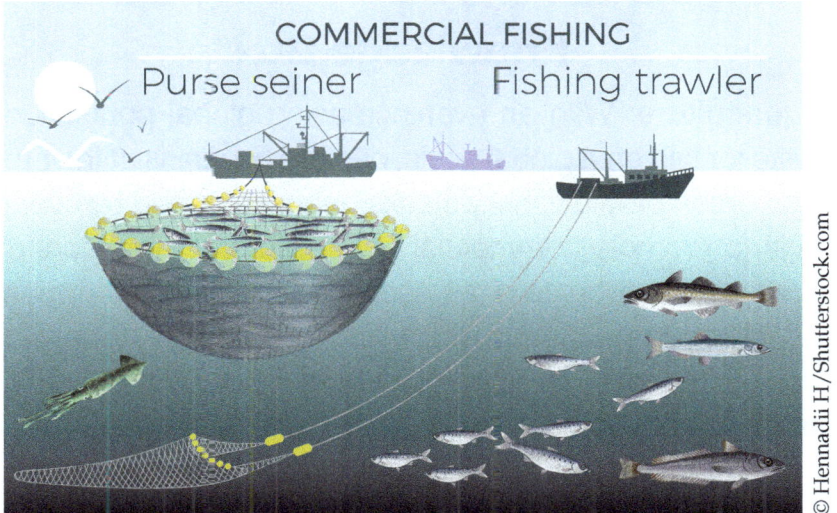

Figure 8.18 Purse seiner and trawler, two commonly used commercial fishing gears.

> ### Knowledge Box
>
> **Sustainable Fishing:** Sustainable fishing should ensure that fish stocks can replenish themselves so that their production can be maintained at a level that can both perpetuate healthy populations in the wild and allow regulated harvest. The concept of maximum sustained yield requires that a healthy natural population be maintained and levels of take be restricted to quantities that don't damage the species in the wild. This can be achieved by limiting the amount of fishing, avoiding fishing smaller fish, or limiting fishing periods to outside of the spawning season.

Overfishing is a classic "tragedy of the commons" phenomenon because most of the fishing takes place in international seas. In order to mitigate the problem, the United Nations Convention on the Law of the Sea Treaty deals with universal aspects of overfishing. Firstly, it requires that all coastal states ensure that the maintenance of living resources in their exclusive economic zones is of primary importance and that the fishery is not endangered by overexploitation. The same article addresses the maintenance or restoration of populations of species above levels at which their reproduction may become seriously threatened. Secondly, the coastal states: "shall promote the objective of optimum utilization of the living resources in the exclusive economic zone." Lastly, hunting for marine mammals should be prohibited, limited or carefully regulated.

> **Knowledge Box**
>
> **Sustainable Aquaculture:** With an ever-increasing global population and a steadily increasing appetite for fish on a global scale, capture fishery is at the breaking point with an increasing proportion of wild fishery stocks being damaged by over harvesting. One natural solution is to promote commercial aquaculture. As mentioned before, current global aquaculture still has room to grow. According to the UN, commercial aquaculture produced 110.2 million tons in 2016 (including 30.1 million tons of aquatic plants). With capture fishing reaching its maximum level since the 1990s, aquaculture has accelerated, but it needs to be done in such a way that it can feed the piscivorous people around the world and protect the marine and freshwater ecosystems it uses at the same time. An example that can serve as a cautionary tale is the rearing of Atlantic salmon in marine pens along the West Coast of North America. Escapees from the pens have established populations in many streams and rivers in British Columbia, both competing with native Pacific salmon and having a severe impact upon the freshwater habitats they have invaded. Most of the native anadromous salmonids die after spawning, thus providing nutrients to invertebrates that sustain young fish in the oligotrophic spawning habitats. The Atlantic Salmon does not necessarily die after spawning, and they can inhabit rivers and streams, thus reducing resources such as invertebrates that are critical to the survival of native species.

8.4 Genetically Modified Organisms

8.4.1 The Definition of GMOs

A **GMO** is an organism whose genetic material is altered using genetic engineering techniques. This is different from the common selective breeding method discovered by Mendel[6] and used later for cross-breeding, in which two subspecies are mated to achieve a combination of desirable traits from both parent lineages. A wide variety of organisms have been genetically modified, from animals to plants and microorganisms to render the organism some desirable traits, such as high yield; desirable color, taste, nutrition; resistance to diseases, pests or pesticides/herbicides; adaptability of harsh conditions such as drought or high/low temperature, and other factors.

To create a GMO, genetic engineers must isolate the gene they wish to insert into the host organism and combine it with other genetic elements, including a promoter and terminator region and often a selectable marker. There are a number of methods

---

[6] Gregor Mendel is a 19th century Moravian monk who did simple hybridization experiments with pea plants between 1856 and 1863 and established Mendel's Principles of Heredity. The published results, however, went unnoticed until 1900.

that are used to insert an isolated gene into the host genome using genome editing techniques. This approach was successfully established in the 1970s. A notable example of this technique is the Flavr Savr tomato that was developed in 1994 (see the Knowledge Box). The first genetically modified animal to be approved for food use was the AquAdvantage salmon in 2015. Many cash crops such as soybean, corn, canola, and cotton have significant portions of genetic modification.

> ## Knowledge Box
>
> **Flavr Savr:** Flavr Savr is a genetically modified tomato developed in 1994 and is usually considered to be the first commercially successful GMO. Developed by the California company Calgene, it was approved by the U.S. Food and Drug Administration (FDA) and is considered to be "as safe as tomatoes bred by conventional means." The genetic modification process added a gene to make the tomato more resistant to rotting, therefore they can be harvested when they are ripe, improving the flavor compared to conventional tomatoes, which have to be harvested early to avoid spoiling.

In 2015, 92% of corn, 94% of soybeans, and 94% of cotton produced in the United States were genetically modified strains. Around the world, GM soybean accounted for about half of all soybeans, mostly to make the soybean plant tolerate herbicides (mostly glyphosate, the most common herbicide) and to produce healthier oils. For example, a GMO apple could grow larger and have a better taste. It can also be made resistant to insects that damage non-GMO fruit. In comparison, a non-GMO organic apple, without the application of pesticides, may suffer from pest problems and may not be as appealing as the GMO variant (**Figure 8.19**). GMOs have also been used for biological control of disease vectors such as mosquitos.

Figure 8.19 A conceptual situation where a GMO apple looks and tastes better than an organic non-GMO apple suffering from pest issue.

8.4.2 Benefits and Risks

The ethics, safety, and ecological impacts of GMO have long been controversial topics. There is a scientific consensus that currently available food derived from GM crops poses no greater risk to human health than conventional food (Nicolia et al., 2013). However, GM food safety is a leading issue with critics and environmental conservationists. Many countries have adopted regulatory measures, mostly labeling requirements, to deal with these concerns. From the ecological point of view, GMOs, at a minimum, pose a threat to the genetic integrity of the natural environment due to the possibility of genetic contamination or competition with wild genotypes non-GMO species in the wild or interference with pollinators such as monarch butterflies.

8.5 Other Ecological Challenges

8.5.1 Endangered Species

According to Ceballos et al. (2015), about one fifth of the world's vertebrate species are threatened with extinction. The main causes of vertebrate biodiversity declines are overexploitation and habitat loss associated with an increasing human population (Hoffmann et al., 2010), as discussed earlier. An **endangered species** is a species that is likely to become extinct in the near future, either worldwide or in a particular political jurisdiction. Endangered species may be at risk due to factors such as habitat loss, poaching and invasive species. **The International Union for Conservation of Nature (IUCN)** Red List documents the global conservation status of many species, and various other agencies assess the status of species within particular areas. Many nations have laws that protect conservation-reliant species that, for example, forbid hunting, restrict land development, or create protected areas. Some endangered species are the target of extensive conservation efforts such as captive breeding and habitat restoration.

The duty to protect endangered and rare species lies in the intrinsic value of biodiversity and to some extent humans' responsibility to the natural world because many species are endangered because of human-induced impacts, as discussed before. Every year, considerable money is spent to protect endangered species around the world, with different degrees of success. The factors that harm biodiversity, such as urbanization, pollution, habitat loss, and so forth, are often beyond the ability and financial means of conservationists. The most common way to protect these endangered species is to create protected areas such as preserves or breeding grounds for these species. One such success story is the program used by China for giant pandas (**Figure 8.20**). A large swath of habitat was protected as habitat critical to survival of the species, and several breeding facilities were established. Thanks to coordinated efforts protecting foraging and breeding habitat, the wild panda population has increased from less than 1,000 in the 1970s to nearly 2,000 in 2016, enough to prompt IUCN to reclassify the

Figure 8.20 Giant pandas enjoying their time in a nature reserve (left) and in a breeding center in Sichuan, China.

species from "endangered" to "vulnerable." However, for every successful conservation story, there may be many unfortunate ones, such as that of the baiji, an extremely rare freshwater dolphin in China's Yangtze River (see the Knowledge Box), which was determined to be functionally extinct in the early 2000s.

Knowledge Box

Baiji: Baiji, or Yangtze River dolphin, is a sad story. It was an endangered and rare freshwater dolphin species native to Yangtze River, China. In Chinese, Baiji literally means 'white finned dolphin' and was nicknamed 'Goddess of the Yangtze' and considered a symbol of peace and prosperity in Chinese folklore. Due to pollution and other human intervention, possibly including the construction of the Three Gorges Dam, the population of baiji had decreased drastically in the late 20th century. Despite a Conservation Action Plan approved in 2001, multiple expeditions have failed to find any individuals. Consequently, baiji was declared to be functionally extinct on December 13, 2006 after a six-week survey of the Yangtze River by a large research team yielded no sighting of any Baiji.

8.5.2 Invasive Species

An **invasive species** is a non-native that spreads in a new habitat and causes various damages. The domination of an invasive species in a new habitat could be due to the absence of its natural controls (such as predators or other naturally limiting factors) or through a change in other environmental factors. This includes plant species labeled as exotic pest plants and invasive exotics growing in native plant communities. The term "invasive" is often poorly defined or very subjective. Sometimes it is used loosely for undesirable pest species. Notable examples of invasive plant species include the kudzu vine, Arundo, and Spanish mustard. Animal examples include New Zealand mud snail,

Figure 8.21 Invasive fauna and flora. Left: two nutria in a river; right: mustard along a trail in the Crystal Cove State Park in California.

quagga mussel, Asian carp and nutria (**Figure 8.21**). Invasion of long-established ecosystems by organisms from distant bio-regions is a natural phenomenon (e.g., from various natural dispersal processes), but has been accelerated by humans, from their earliest migrations through to the Age of Discovery, and now international trade.

Knowledge Box

Nutria: Nutria (Figure 8.21, left) is a large, herbivorous semiaquatic rodent with webbed feet and shaggy brown fur. It lives in burrows along waterways, and feeds on river plant stems. It is native to subtropical and temperate South America but was introduced to North America, Europe, Asia, and Africa by fur farmers. Its destructive burrowing and feeding habits make it a pest and an invasive species. The situation became quite dire when there were an estimated 20 million nutria in the 1950s in the bayous of Louisiana alone, destroying large areas of aquatic plants each day by feeding on their roots. It became such a problem that the government offered a generous $4 bounty on each nutria tail, and sponsored other programs to control it. The current nutria population is largely under control. Another well-known invasive species is the European black mustard (Figure 8.21, right) in California. This annual-flowering plant is the dominant inhabitant of many open spaces in coastal California, turning rolling hills into seas of yellow in the spring. According to folklore and confirmed by pollen records in the adobes of many Spanish missions, the mustards were brought to California by the Franciscan padres, who planted the mustards as a marker for the golden El Camino Real, the road that connects the California missions.

> **Knowledge Box**
>
> **New Pangaea**—While Pangaea was a supercontinent during the late Paleozoic and early Mesozoic era, when most of the major continents were fused into one major land mass and, presumably, land animals can travel unhindered. The concept of 'New Pangaea', however, refers to the modern world in which millions of people travel by air and across the oceans from continent to continent, virtually merging the continents into one and facilitating the proliferation of man invasive species.

8.5.3 Deep Ecology: A Scientific Perspective

Deep ecology is an ecological and environmental philosophy, rather than a science, that promotes the inherent worth of natural living beings and their natural habitats regardless of their instrumental utility to human needs. The implication of deep ecology on sustainability issues will be discussed in detail in Chapter 12. In this section, the scientific pespectives of deep ecology will be briefly discussed and related to and compared with traditional ecology. Deep ecology theory considers humans as a part of a natural ecosystem, and not a dominating or controlling factor within it. It recognizes intrinsic value in all species, regardless of their anthropocentric utility. According to this philosophy, humans are merely a part of a much larger whole, and should live and participate with nature as a context. Similar to the proclamation by the founding fathers of many western nations that "all humans are born equal," deep ecologists think that other animals, plants, and the natural world have intrinsic rights just as humans do. Other species have the right to exist and thrive without human intervention and regardless of their use to humans. Nature and natural species interrelationships are viewed as our evolutionary and present context, and humans should not interfere or destroy this philosophical and real foundation.

As discussed in Section 8.3, humans are indeed part of the global ecosystem. From the ecological perspective, humans are a dominant invasive omnivorous predator that stays solidly on the top trophic level in every ecosystem, exerting formidable and often damaging influence throughout the biotic and abiotic environment. However, in traditional ecological science, humans are given a special status and rarely discussed in the same context as other predators. When they are, the topics are often about the environmental and ecological impacts humans have caused, or ecosystem services that the natural world provides to humans (see Section 8.3.2). In popular culture, people often feel threatened or violated when wildlife attacks humans or 'invade' a community that was once their habitat. Above all, when considering the relationship between humans and the rest of the natural ecosystem, one needs to think deeply.

Further Reading

Nicolia, A., A. Manzo, F. Veronesi, D. Rosellini. "An Overview of the Last 10 Years of Genetically Engineered Crop Safety Research." *Critical Reviews in Biotechnology*, 34, no. 1 (2013): 77–88.

United Nations Food and Agriculture Organization. "The State of World Fisheries and Aquaculture 2018." 2018.

Hoffmann, M., et al. "The Impact of Conservation on the Status of the World's Vertebrates." *Science*, 330 (2010): 1503–09.

Ceballos, G., P. R. Ehrlich, A. D. Barnosky, A. García, R. M. Pringle, and T. M. Palmer. "Accelerated Modern Human-Induced Species Losses: Entering the Sixth Mass Extinction." *Science Advances*, 1 (2015): e1400253.

Ripple, W.J. et al. "Are We Eating the World's Megafauna to Extinction?" *Conservation Letters*, 12 (2019): e12627.

Locey, K.J. and J.T. Lennon. Scaling Laws Predict Global Microbial Diversity." *PNAS*, 113, no. 21 (2016): 5970–75.

United Nations. "What is Biodiversity?" (PDF). United Nations Environment Programme, World Conservation Monitoring Centre.

Staff (2 May 2016). "Researchers find that Earth may be home to 1 trillion species". National Science Foundation. Retrieved 6 May 2016.

McKinney, M.L. "How Do Rare Species Avoid Extinction? A Paleontological View." In: *The Biology of Rarity: Causes and Consequences of Rare-Common Differences*, Kunin, W.E. and K.J.Gaston (eds.), Netherlands: Springer, 2012.

The Kaspari Lab, 10 principles of Ecology: https://michaelkaspari.org/2017/07/17/the-ten-principles-of-ecology/.

IUCN, 2016. IUCN's Classification of Direct Threats (v2.0). https://cmp-openstandards.org/library-item/threats-and-actions-taxonomies/. August 23, 2019. Retrieved May 10, 2020.

Bar-On, Yinon M., Rob Philips, and Ron Milo. "The Biomass Distribution on Earth." *PNAS*, 115, no. 25 (2018): 6506–11.

Dahl, T.E. Wetlands Loss Since the Resolution, National Wetlands Newsletter, V.12, N.6. Washington DC: Environmental Law Institute, 1990.

Cepero, Almudena et al. "Holistic Screening of Collapsing Honey Bee Colonies In Spain: A Case Study." *BMC Research Notes*. 7 (September 2014): 649. doi:10.1186/1756-0500-7-649.

Bond, David P.G. and Paul B. Wignall. "The Role of Sea-Level Change and Marine Anoxia in the Frasnian–Famennian (Late Devonian) Mass Extinction" (PDF). *Palaeogeography, Palaeoclimatology, Palaeoecology*, 263, no. 3–4 (June 2008): 107–18.

O'Connor, Mary I., et al. "Principles of Ecology Revisited: Integrating Information and Ecological Theories for a More Unified Science." *Frontiers in Ecology and Evolution* (June 18, 2019), doi:10.3389/fevo.2019.00219.

Næss, Arne. "The Shallow and the Deep, Long-Range Ecology Movement. A Summary." (PDF). *Inquiry*, 16, no. 1–4 (1973): 95–100.

Leakey, R.E. The Origin of Humankind. New York, NY: Basic Books, 1994.

Meacham, M., C. Queiroz, A. V. Norström, and G. D. Peterson. "Social-Ecological Drivers of Multiple Ecosystem Services: What Variables Explain Patterns of Ecosystem Services Across the Norrström Drainage Basin?." *Ecology and Society*, 21, no. 1 (2016): 14. http://dx.doi.org/10.5751/ES-08077-210114.

Closs, G., B. Downes, and A. Boulton. Freshwater Ecology: A Scientific Introduction. Malden, MA: Blackwell Publishing, 2004.

Web Resources

World Wildlife Foundation: information about Borneo deforestation: https://wwf.panda.org/knowledge_hub/where_we_work/borneo_forests/borneo_deforestation/

The United Nations' proclamation of 2010 to 2020 as the decade on biodiversity: https://www.cbd.int/2011-2020/

Our Planet: From Deserts to Grasslands (and other episodes in the series): https://www.youtube.com/watch?v=XmtXC_n6X6Q

Population Ecology by Bozeman Science: https://www.youtube.com/watch?v=PQ-CQ3CQE3g

Energy Flow in Ecosystems, by Bozeman Science: https://www.youtube.com/watch?v=lnAKICtJIA4

From the Cambrian Explosion to the Great Dying, by PBS: https://www.youtube.com/watch?v=RDQa0okkpf0

Questions and Exercises

1. Do you agree that humans should consider themselves as an intrinsic part of the global ecosystem? Why or why not?
2. Study Figure 8.1 and understand how the living world can be "dissected" into smaller and larger units for analyses of ecological issues at different scales.
3. Why a more complex ecosystem is more resilient than a simpler ecosystem?

4. Study Figure 8.4. Which portion of the curve is exponential? What could be the causes of the leveling of the curve?
5. Should/could we protect all endangered species? Why or why not?
6. Do you support GMO? Why or why not?
7. Do you think deep ecology philosophy is valid or not? Do you believe that nature has its own intrinsic value?
8. Why is biodiversity important to ecosystem health? If you were asked to assess the health of an ecosystem by measurement of its biodiversity, what parameters would you use?
9. Identify any rare, threatened or endangered species at or near where you live. Learn about the factors that may have caused their decline, and think about what you could do to protect and restore them.
10. Identify any nonnative and invasive species at or near what you live. What brought them here? What are their environmental or ecological impacts? Are there any control measures implemented to control them?

Chapter 9

Solid Waste

Learning Outcomes	244
Key Concepts	244
9.1 Solid Waste Classification and Generation Rate	245
9.1.1 Overview	245
9.1.2 Solid Waste Generation in the United States	247
9.1.3 Solid Waste Generation around the World	247
9.2 Landfill	250
9.2.1 Overview and Waste Collection	250
9.2.2 Landfill Operation	251
9.2.3 Landfill Stormwater Management	252
9.2.4 Landfill Trash Management	253
9.2.5 Landfill Leachate Management	254
9.3 Solid Waste Recycling and Energy Recovery	257
9.3.1 Solid Waste Recycling	257
9.3.2 Waste to Energy	260
9.4 Hazardous Waste	263
9.4.1 Definition	263
9.4.2 Hazardous Waste Management	265
9.5 Plastics and Trash Pollution	266
9.5.1 Overview	266
9.5.2 Basel Convention	269
9.5.3 OECD Convention	271
Further Reading	272
Web Resources	273
Questions and Exercises	273

Learning Outcomes

- Understanding of the prevalence of waste generation
- Knowledge of the average amount of waste generated by each nation and around the world
- Understanding of the way how solid wastes are handled
- Understanding of the design principles of landfills, and potential environmental impacts of failing or improperly designed landfills
- Knowledge of the laws and regulations on solid waste management
- Knowledge of the laws and regulations on hazardous waste management

Key Concepts

Solid waste; waste management; cradle to grave; hazardous waste; landfill; bioreactor landfill; leachate; landfill gas; RCRA; NIMBY; Basel Convention; OECD Convention

We learned in Chapters 1 and 2 that Earth's geological history is defined in part by fossils, such as trilobites in the Cambrian period, amphibians in the Devonian period, and dinosaurs in the Jurassic period, and so on. Now that we are in Holocene, or so-called "Anthropocene" due to mankind's complete dominance over the Holocene Earth's environment, what are we leaving behind in the geological formations other than our skeletons? The answer is piles and piles of garbage in the form of **landfills**, open dumps, or worse, in the bottom of rivers, lakes, and the oceans. Have we wondered how the future generations will think about the legacy we leave for them in the geological record?

To help put the above thoughts in context, you can consider visiting a municipal landfill, preferably a large one where you can see the scale of the operation and amount of garbage generated on a daily basis (an example is shown in **Figure 9.1**). In the United States, most landfills are publicly owned and they host periodic public tours. When I was teaching environmental classes, landfills were one of the places where I always brought students for field trips. An average American generates more than 800 kg of garbage every year[1] and collectively waste more than $50 billion in recyclable materials to landfills. However, since everyone tosses the garbage in the garbage can, rolls the garbage can to the curb every week without thinking twice, the garbage is quickly out of sight and out of mind. It would be difficult to imagine the magnitude of

---

[1] Or, about 2.2 kilograms each day

Figure 9.1 An open garbage dump with a bulldozer.

the garbage issue without seeing in person how much is dumped every single day. Once you witness it, you might ask yourself—what could be done?

9.1 Solid Waste Classification and Generation Rate

9.1.1 Overview

Solid waste is a generic term for unwanted items and materials and is also often called garbage, refuse, trash, and other things. Since municipal solid waste, or MSW, represents the bulk of solid waste, this chapter will focus on MSW. The USEPA definition of solid waste includes any discarded items; things destined for reuse, recycle, or reclamation; sludges from wastewater treatment plants; and **hazardous wastes**. Encyclopedia Britannica simply defines solid waste as solid material that is discarded because it has served its purpose or is no longer useful. For MSW landfills, which accept the bulk of solid wastes, the wastes can come from homes, schools, hospitals, and businesses, and include various materials such as paper, food, plastics, metals, wood, yard trimmings, etc. Generally, solid wastes can be classified into the following:

- Biodegradable waste, such as food, yard trimmings, paper, etc., that are easily biodegradable
- Recyclable waste, such as paper, cardboard, glass, plastic bottles, aluminum cans that can be recycled to produce similar products
- Inert waste that are chemically stable and cannot be biodegraded, chemically treated/recycled, or incinerated, such as construction waste

- Electronic waste (e-waste), such as appliances, computers, TVs, etc.
- Composite waste, which is a mixture of different materials, such as clothing, packaging materials, waste plastics
- Hazardous waste, such as paints, chemicals, and other wastes containing hazardous and toxic materials (will be discussed specifically in Section 9.4), and biomedical waste

Unless recycled or incinerated, nearly all of the solid waste will go to landfills to be buried. **Figure 9.2(A)** shows the percentage of various wastes that are generated in the United States every year, totaling more than 250 million tons. **Figure 9.2(B)** shows the percentage of solid wastes ending up in landfills after some recyclable materials (paper, food, and yard trimmings) are removed for recycling/composting/reuse. Some, unfortunately, will end up in the environment, causing a number of issues. This issue was discussed in Chapter 4 by treating trash as a source of pollution in the water environment.

Even for properly disposed solid wastes, many metropolitan areas such as Greater Los Angeles are experiencing shortage of available lands for new landfills because the existing ones are all filled up. Due to environmental concerns, permitting new landfills is becoming increasingly difficult even if there were available lands. Therefore, cities are forced to transport MSWs farther and farther away, increasing the cost and causing more environmental impact (traffic, noise, greenhouse gases, etc.).

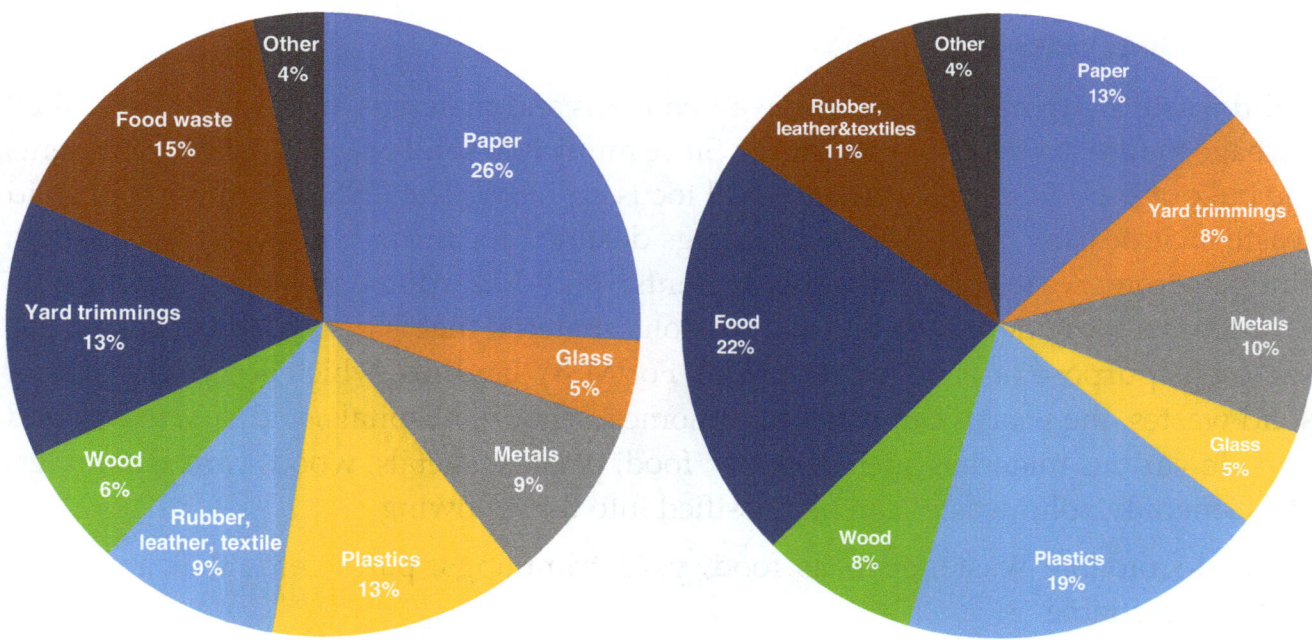

Figure 9.2 Municipal solid wastes in the United States, 2015. A (left): waste generated. B (right): waste landfilled after some materials (paper; yard trimmings and food waste) are recycled or reused.

Source: USEPA, 2018. Created by Jian Peng

9.1.2 Solid Waste Generation in the United States

People generate wastes every day. USEPA estimated that the national average rate of MSW generation was 1.22 kg/capita/day and a total of 80 million metric tons generated per year in the 1960s. Both increased in 2015, with the per capita rate at 2.42 kg/capita/day, and total national rate of 238 million tons. There are geographical differences in solid waste generation rates. For example, in 2010, Los Angeles residents generated an average 3.2 lbs/day (or 1.45 kg/day), while residents in the small town of Wilson in Wisconsin generated a mere 1 lb/day. A decade later, with population growth of about 10%, roughly 270 million tons of solid waste will be generated in 2020.

Figure 9.3 shows MSW generation rate as well as per capita generation rate for the United States from 1960 to 2015, as discussed above. As can be seen, there is a monotonic increase in the absolute amount of wastes throughout the years. However, the per capita generation rate has been flat since 1990. This trend is for the most part encouraging. However, as the nation with the highest waste generation rate per capita (see details below) around the world, the United States still has a long way to go.

9.1.3 Solid Waste Generation around the World

Based on the World Bank, the global average waste generation rate is about 0.74 kg, but the range could be from 0.11 to 4.54 kg (World Bank, 2020).[2] The average is significantly lower than that in the United States, which has one of the highest per capita waste

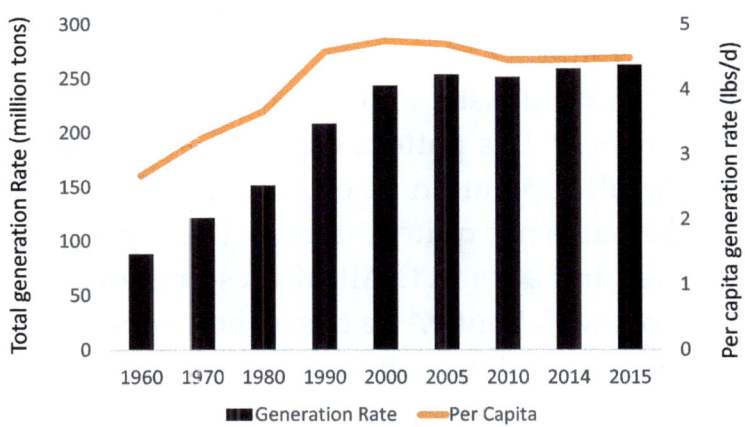

Figure 9.3 Municipal solid waste total and per capita generate rates in the United States, 1960 to 2015. Note the use of English units. Ton here is US short ton or 2,000 lbs.
Source: USEPA, 2018. Created by Jian Peng

---

[2] The discrepancy with the 4.48 kg/day as shown above is due to the difference in time. 4.48 kg/day was the data for the year 2015, and the World Bank data of 4.54 kg/day (United States has the highest waste generation rate in the world) is for 2017, indicating only a slight increase over that in 2015.

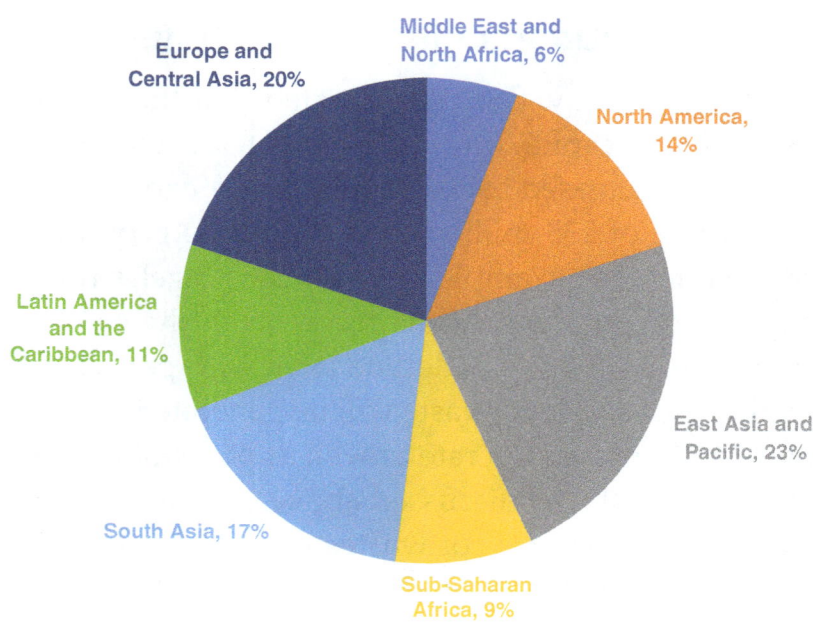

Figure 9.4 The global waste generation percentage by region.
Source: Created by Jian Peng

generation rate in the world. This is generally in line with the finding that more developed countries generate more waste per capita. For example, high-income countries account for 16% of the global population, but they account for 34% of the waste generation, or 683 million tons each year.

In terms of regional differences, the East Asia and Pacific region generates the most wastes, about 468 million tons a year. The Middle East and North Africa region generates the least, at 129 million tons a year. The rest of the regions, such as Sub-Saharan Africa; Americas, South Asia, and Europe are in between. This pattern is shown in **Figure 9.4**. However, this pattern could be misleading because of significant differences in the total population in each region. If calculated in per capita waste generation rate, the pattern is quite different. The top 13 nations with the most amount of MSW are listed in **Table 9.1**. All of these regions are expected to have increased waste generation, with South Asia to see the steepest increase (51.9%), from 334 million tons a year in 2016 to 466 million tons per year by 2030, and 661 million tons a year by 2050. The per capita waste generation rate projected for 2050 is shown in **Table 9.2** (World Bank, 2020). On a global scale, waste generation will increase from 2.01 billion tons in 2016 to 2.59 billion tons in 2030. By 2050, global waste generation will be an astounding 3.40 billion tons. At least for now, there seems to be no end in sight in this depressing trend.

Global waste composition is different from the United States (see **Figure 9.5**). Compared to the United States (Figure 9.2), there is clearly more food and green waste in global average waste composition (44% for the world, while the percentage for the

Table 9.1 Top 13 Nations with the Most Municipal Solid Waste in 2017 (Kaza et al., 2018)

Generation of municipal solid waste worldwide in 2017 (million tons)	
US	258
China	220.4
India	168.4
Brazil	79.89
Indonesia	65.2
Russia	60
Mexico	53.1
Germany	51.05
Japan	43.98
France	33.4
UK	31.57
Turkey	31.28
Pakistan	30.76

United States with food and yard trimmings adding up to 28%). There is also less paper in global waste composition (17% vs. 26%), less rubber and leather (2% vs. 9%). These data reflect differences in lifestyle, waste behavior and other socioeconomic factors in the United States from the global average.

There is a clear correlation between income level and per capita waste generation rate, with higher-income nations with high waste generation rates. This is not surprising because people with higher income consume more goods, including food and other nonperishable goods, and hence generate more wastes. Global per capita waste generation rates also positively correlate with urbanization rates. In terms of composition, it was found that higher income levels correspond to lower organic waste (food and

Table 9.2 Projection of Waste Generation Per Capita Worldwide in 2050 (kg/day) Compared to 2016

Projection of waste generation per capita worldwide in 2050 (kg/day)			
	2016	2050	% increase
North America	2.21	2.5	13.1%
Europe and Central Asia	1.18	1.45	22.9%
Latin America and the Caribbean	0.99	1.3	31.3%
Middle East and North Africa	0.81	1.06	30.9%
East Asia and Pacific	0.56	0.81	44.6%
South Asia	0.52	0.79	51.9%
Sub-Saharan Africa	0.46	0.63	37.0%

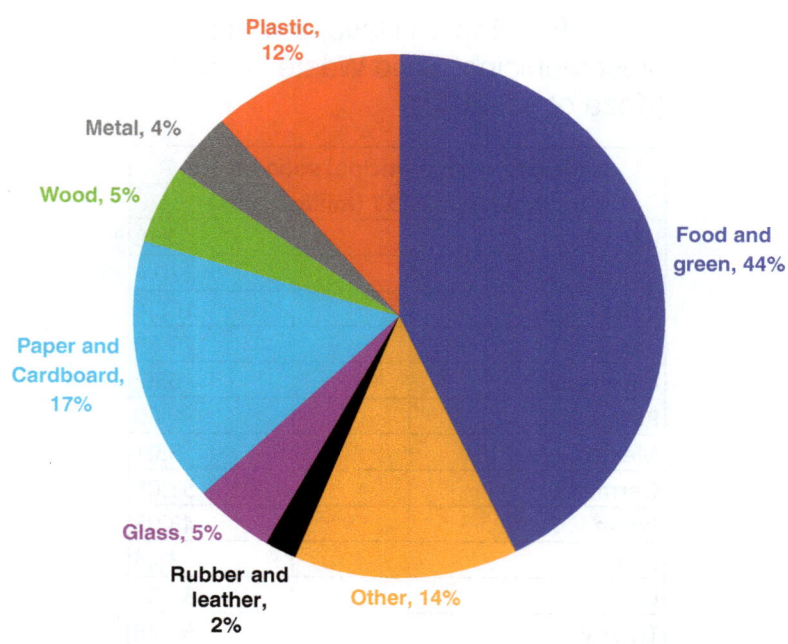

Figure 9.5 Global average waste composition.
Source: Kaza et al. (2018).

green waste). Higher income people also generate more paper, plastic, rubber, and leather wastes. This is consistent with the difference between the United States and the World as previously discussed, with the United States having a much higher per capita income than the global average.

9.2 Landfill

9.2.1 Overview and Waste Collection

A landfill is a site for the disposal of solid waste materials.[3] Similar to drinking water supply and wastewater treatment, garbage collection and landfill service is one of the key services a municipal government provides to its citizens. In developing countries, especially in the suburbs, wastes are often not collected. Rather, they are simply dumped in open dumps, pits, and other unsecured locations without proper management, creating potential sanitary and other environmental hazards. Globally, nearly a third of wastes are dumped at open dumps. For developed countries such as the United States, nearly 100% of wastes are collected and properly disposed of. In comparison, low income countries could see more than 90% of their wastes in open dumps.

---

[3] Except for hazardous wastes, which will be managed separately at specialized hazardous waste disposal sites.

Figure 9.6 A landfill in Irvine, California, USA. The tractor trailer trucks to the right are specialized garbage transportation vehicles that have mechanisms inside the trailers to push the garbage out. The bulldozers in the center are for spreading and crushing the garbage before covering it with compost, as shown in the foreground. This landfill receives an average 8,500 tons of municipal solid waste a day. It is expected to remain operational for 30 to 50 more years.

Even for the United States, solid **waste management** experienced gradual evolution over the years. Prior to the 1940s, solid waste was simply left in piles or thrown into pits. Some landfill sites are also used for waste storage, consolidation, transfer, and recycling purposes. Currently, most of the landfills in the United States are heavily regulated and operated by governmental agencies. **Figure 9.6** is such a landfill in operation in Irvine, California, USA. Currently, there are approximately 3,000 active landfills in the United States, and more than 10,000 inactive ones.

9.2.2 Landfill Operation

MSWs are usually collected weekly from households and businesses by commercial collectors. In most parts of the United States, the wastes are separated into garbage, recyclables, and green waste. These wastes will be treated separately. Recyclables will be transported to a sorting facility for sorting and recycling. Green waste is usually processed in specialized facilities for composting. General household garbage will be consolidated (if needed) and transported to a landfill directly without processing, as shown in **Figure 9.7**. A landfill usually charges individual waste transporters ("haulers") a fee based on weight, volume, or type of vehicle. A waste hauler company may have a contract with the landfill to pay a fixed monthly fee.

Management of a landfill involves control of odor, **landfill leachate**, stormwater, trash, and other issues. To deal with the odor issues, many landfills are required to operate on a very limited "active area" where the fresh garbage is dumped. Bulldozers and other heavy machinery will crush and spread the garbage to evenly distribute it across the surface. Once the active area is filled with garbage, cover materials such as

Figure 9.7 Garbage collection by a garbage truck as a general municipal service. The truck collects garbage from bins weekly in front of each residence, compacts the garbage to maximize the capacity. The garbage will then be dumped at a landfill either directly or after being consolidated into larger trucks at a municipal facility. Generally, no sorting or recycling will be performed on general household waste.

compost, dirt, or other materials will be used to cover it. For the active area that has not been filled overnight, large tarp will be used to cover the area to control odor and to prevent garbage from being blown away or scavenged by wildlife. Such a process can go on until the design height or the design life of the landfill is reached. Every year, a landfill would settle significantly due to decomposition of organic materials. This process usually prolongs the active life of a landfill. The released gas can also be harvested for beneficial use, as will be discussed later in this chapter.

9.2.3 Landfill Stormwater Management

Stormwater runoff from a landfill could contain elevated levels of many contaminants such as sediment, nutrients, bacteria, heavy metals, and other toxic materials. In California, stormwater for landfills is under stringent industrial stormwater regulation that requires best management practices to minimize stormwater exposure and to retain runoff on site. Stormwater runoff also needs to be monitored regularly to ensure the effluent limits are not exceeded. These requirements are generally specified in the facility's Stormwater Pollution Prevention Plan (SWPPP) that includes elements such as best management practices, maintenance plans, inspections, employee training, and reporting. To achieve compliance requirements, some landfills have achieved zero stormwater discharges by collecting stormwater runoff for treatment and reuse. Refer to Section 5.3 for a more detailed discussion on general stormwater management issues.

9.2.4 Landfill Trash Management

Trash problem for a landfill sounds strange but it could be a real problem for an urban area where there are residents living near the landfill. Wind could blow the trash away. Garbage trucks could litter some trash along the way and while dumping the garbage. Some birds could bring trash away from the landfill. In order to deal with the issue, in addition to covering the active area overnight (**Figure 9.8**), many landfills installed trash nets along the perimeters, especially in the downwind side, so wind-blown trash could be captured. Birds could be a nuisance because of trash and noise issues. They could be controlled by firing blank shots[4] or using falconry, as shown in **Figure 9.9**.

Figure 9.8 Compost and tarp cover of a landfill at the end of a work day. The light-colored tarp cover in the background is for the active area that will reopen the next day. The dark-colored compost cover in the foreground is for the inactive area.

---

[4] Blank shots, also called bird banger or pyrotechnic shots, are sometimes used to repel birds from a landfill. However, it could be a source of nuisance to local residents, including pets. In a complaint filed by local resident near a landfill in California, the resident claimed that his dog developed depression due to the frequent blank shots. As the result, the landfill started using falconry for bird control.

Figure 9.9 A falconer and a trained falcon at a landfill in Irvine, California, USA.

> ### Knowledge Box
>
> **Public Relations Management for Landfills:** Management of public relations with local residents is an important responsibility. Sometimes this issue could become a sensitive topic. For example, the residences close to a landfill will have a disclosure in the property title document acknowledging that the property owner knows, and the future buyer will be notified, that there is a landfill nearby. Landfill odor could be a subject of frequent complaints. Heavy traffic by garbage trucks on a daily basis could be a nuisance. Landfill management has to deal with all of these complaints carefully and professionally.

9.2.5 Landfill Leachate Management

Landfill leachate is a common and difficult issue. Much of the MSW is organic material with various water contents (Figures 9.2 and 9.5). When the materials degrade, gases (mostly carbon dioxide and methane) and liquid are produced through a complex process. The liquid, which is called landfill leachate, can also contain precipitation or local groundwater. Due to its complicated sources, landfill leachate could contain high organic matter, suspended solids, nutrients (nitrogen and phosphorus), salts, and metals.

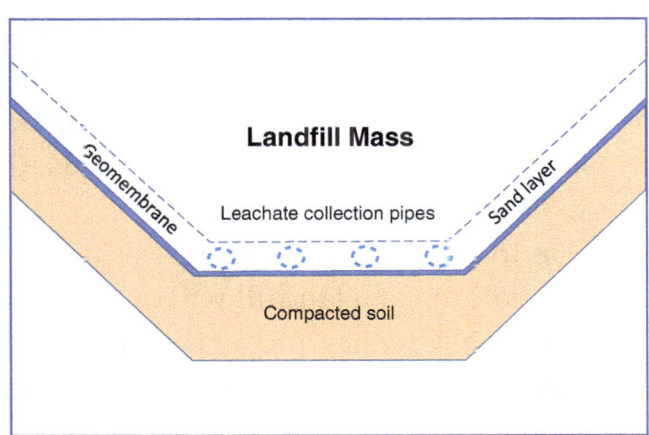

Figure 9.10 Landfill management measures to control leachate. Left: A landfill in construction where a geomembrane is being laid down. Right: a cross section of a typical landfill, with compacted soil, geomembrane, sand layer, and leachate collection pipes.

Microbes could thrive in the leachate as well. Due to presence of heavy metals and some organic pollutants such as VOCs, landfill leachate is usually considered a significant pollutant. In many places, landfill leachate has caused soil and groundwater pollution.

To prevent leachate pollution of soil and groundwater, modern landfills are usually lined and compartmentalized. The lining materials can be a low-permeable geomembrane combined with a clay layer, as shown in **Figure 9.10**. This lining will prevent the leachate from leaving the landfill site and facilitate leachate collection and treatment (**Figure 9.11**). In most cases, a sand layer will be placed on top of the impermeable layer, with perforated pipes embedded in the layer to collect leachate for treatment. Many modern landfills are compartmentalized where multiple cells, each with geomembrane seal on all sides, are stacked on top of each other to enhance water- and air-tightness. When a landfill goes out of commission, a final, permanent geomembrane

Figure 9.11 A landfill leachate collection pond in Sweden.

can be put on top of the landfill. If sealed properly, a decommissioned landfill can be used to build parks or golf courses, while the gas and leachate collection systems can remain operating for decades afterwards. Unlike leachate, which is mostly hazardous waste, landfill gases could often be used for energy recovery. This topic will be discussed in the next section.

Due to settling and decomposition of abundant organic-rich materials in the landfill, the volume of a landfill will decrease with time, prolonging the lifetime of the landfill. Sometimes, water and air could be pumped into the interior of a landfill to accelerate its degradation. In this case, the landfill is called a **bioreactor landfill**. There are three types of bioreactor landfills. An aerobic bioreactor landfill has its leachate removed from the bottom layer and recirculated back into the landfill. Air is injected into the landfill with vertical or horizontal wells to keep the landfill waste mass aerobic to accelerate waste stabilization. An anaerobic bioreactor landfill is one similar to an aerobic bioreactor landfill but without addition of air. Therefore, the landfill waste mass remains anaerobic, and biodegradation produces **landfill gas** that is mostly methane that can be captured for reuse and energy recovery. The third type of bioreactor landfill is hybrid aerobic/anaerobic bioreactor landfill, which injects air periodically to create alternating anaerobic/aerobic conditions. In a hybrid bioreactor landfill, the organics in the top layer will degrade rapidly, and landfill gas can be collected from lower sections.

Compared to a traditional "dry tomb" landfill, a bioreactor landfill has many advantages. First, decomposition and biological stabilization take a few years for a bioreactor landfill compared to a few decades for a traditional landfill. The addition of water and air in both aerobic and anaerobic conditions reduce waste toxicity and mobility. The recycling and reuse of leachate drastically reduced treatment and processing cost of leachate. Moreover, increased biodegradation rate results in higher landfill mass density and corresponding 15% to 30% increase in landfill space, thus increasing the utility and efficiency of the landfill. Bioreactor landfills also see a significant increase in landfill gas, which can be captured and used for energy recovery. Lastly, bioreactor landfills require less post-closure care and associated long-term environmental risks and operating costs.

Bioreactor landfills have some special considerations compared to traditional landfills. First, a bioreactor landfill will generate more landfill gases than a traditional landfill during its operating period. The landfill gases, which are mostly methane and carbon dioxide (both are significant greenhouse gases), will need enhanced management (capture and reuse) to avoid adverse environmental impacts. These landfills also emit more odor and may require special odor management. They may be physically unstable due to rapid degradation and decrease in size, increased moisture and density. This instability could cause the liner system to fail and require more attention. Bioreactor landfills could also see surface seeps or landfill fires due to increased moisture and landfill gases. However, these issues could be managed safely. Therefore, bioreactor landfills generally are engineered systems that have higher initial costs and require additional monitoring and control. Over time, they require no more management than a traditional landfill.

9.3 Solid Waste Recycling and Energy Recovery

9.3.1 Solid Waste Recycling

Someone's waste could be a treasure for others. An observation of the typical composition of MSW, as shown in Figure 9.5, can show that much of the waste can be recycled, including plastics, metals, paper and cardboard, glass, rubber, etc. Some materials such as food and green waste can be reused in other ways, including composting and energy recovery, and will be discussed in the next section. The data for MSW generation, recycling, composting, and energy recovery from 1960 to 2015 are compiled and plotted on **Figure 9.12**. From this figure, it can be seen that the waste generation rate has flattened around 250 million tons since 2000 despite significant population increase. Enhanced recycling seems to be the most important contributor, assisted by combustion with energy recovery as well as composting. These combined factors have essentially kept the landfilling and other disposal flat since the 1980s. Overall, the trend is encouraging.

However, what has been done is far from enough. In 2015, there was still 125 million tons of MSW that ended up in landfills. A significant portion could have been recycled (paper, 13%; metals, 9%; plastic, 19%; etc.), composted (yard trimming, 8%), or used for

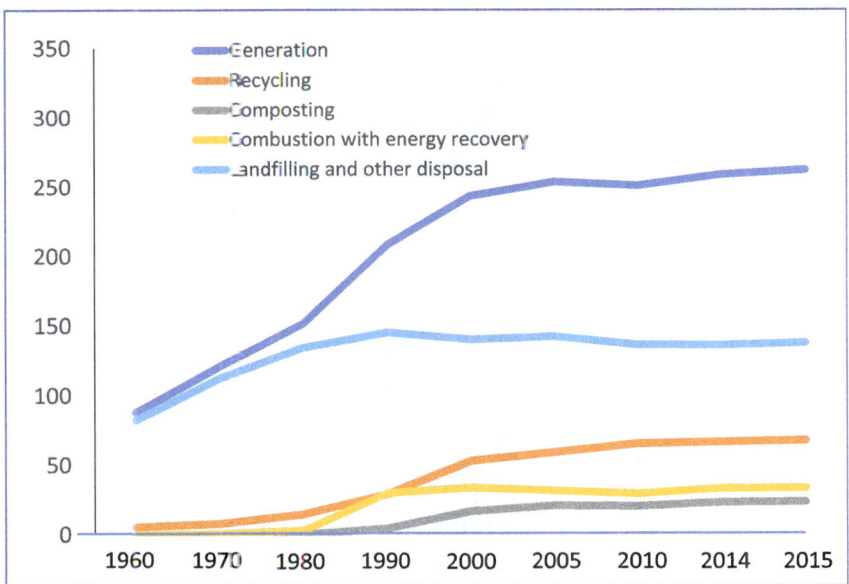

Figure 9.12 Trends in municipal solid waste generation, recycling, composting, and energy recovery.
Source: Created by Jian Peng

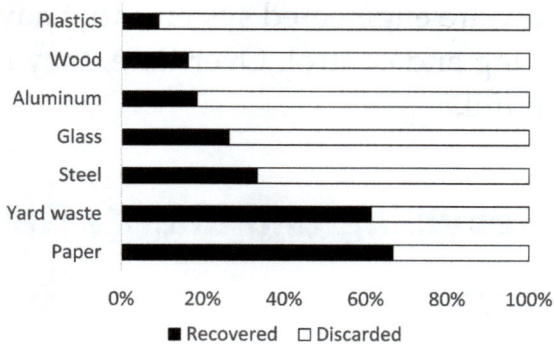

Figure 9.13 Landfilled solid waste composition in the United States in 2015 (left), with the percent recovered for each category on the right.
Source: Created by Jian Peng

energy recovery (paper, 13%; plastic, 19%; wood, 8%; food, 22%, etc.), as shown in **Figure 9.13**. In many cases, the challenges lie in the fact that waste classification is lagging behind. For most households in the United States, there are still only trash cans and recycling cans. For single family residences, there will be a separate bin for yard waste. While the green wastes can be composted for beneficial uses (fertilizers, mulch, landfill cover, etc.), the lack of further waste classification is limiting further reduction of landfilled MSW and improvement of recycling and energy recovery. For example, if there is only one bin for recycled materials, paper, plastics, glass bottles, cardboard will all be mixed together. Usually they will be transported to a sorting facility, but a significant portion of the "recyclable" materials are discarded. Other than plastic bottles, many other plastic containers and packaging materials are not recyclable. This is the key reason why plastics have the lowest recovery rate of less than 10% (Figure 9.13).

On the other hand, the general (i.e., nonrecyclable) trash can will receive food waste, which has high organic matter and is ideal for energy recovery by composting and fermentation. However, for an average household, general waste materials such as packaging, wood, nonhazardous electronic waste, glass, pottery, and a very wide range of miscellaneous items are also thrown away in the trash bin. These materials dilute the energy density of the trash, making it undesirable for energy recovery via composting and fermentation. As will be discussed in Chapter 12, environmental, social, and economic aspects should all be considered in order to make a sound management decision.

Therefore, it is important to have improved trash classification and sorting at its sources in order to improve the degree of reuse and decrease the amount of trash that goes to landfills. As shown in an example in Sweden (**Figure 9.14**), many materials such as plastic, glass, metal, organics/food waste, e-waste, and paper could be recycled much more efficiently if they are sorted at the household or individual business level.[5]

---

[5] Improved waste sorting may not be easy to implement, especially at household level. Better classification will reduce the amount of each type of recyclable material, potentially making the collection and transportation costs too high to be economical. Community-based recycling centers should work better, but improved public education is needed.

Figure 9.14 Waste classification in Sweden, where the waste is collected in six different bins, with five of them recyclable.

Based on the information shown in Figures 9.2 and 9.5, up to 78% of trash that goes to a landfill could have been recycled. Note, however, that a portion of plastics cannot be readily recycled, such as plastic bags, packaging materials, foams, and so on. Even for a progressive country such as Sweden, currently only 47% of all plastics is recycled (the need to reduce plastic waste and trash is discussed in the next section and Chapter 12). Much of the plastic waste enters the environment and causing a range of issues, including the proliferation of microplastics in the environment (Section 5.5.2).

Knowledge Box

Solid waste and sustainability: Solid waste issue is significant in many environmental and sustainability aspects. The popular 3-R, i.e., reduce, reuse, recycle, is not nearly enough. Many believe that the first step should be 'avoid', where the materials that will end up in a landfill should be avoided at its source (see Chapter 12 about the concept of 5-Rs, or refuse, reduce, reuse, recycle, and rot). Recycling is often incomplete and costly not only in financial terms, but also in terms of water and carbon footprint. When life cycle cost is considered, including that associated with production, transportation, and landfilling, the trash that ends up in landfills in shocking quantities represents a significant "waste" (pun intended) of the limited resources of our society and our planet.

There are other issues that have prevented better trash classification. For example, unlike general recyclables and general trash, glass, metals, and other materials are not regularly or evenly generated by each household or business. Therefore, using a fleet of trucks to go door to door to collect these recyclables is not economically or environmentally viable. Community-based recycling centers may work better in this case. Regardless, an improved trash classification will improve trash recycling, reuse, and energy recovery. At the very least, every household should have four types of waste bins: food waste, recyclables, garbage, and hazardous waste.

> ### Knowledge Box
>
> **Food waste:** Food waste is increasingly viewed as a resource that should be diverted from landfills due to its high energy content. In a circular economy, food waste could be converted via technology such as anaerobic digestion into biogas (methane) for energy production, and the remaining materials can be used as organic fertilizer. Food waste could be codigested with other biomasses. In the United States, it is often collected from restaurants together with fat, oil and grease to be sent to biosolid digesters at local wastewater treatment plants for biogas production and energy recovery. Treatment plants that implement such measures can produce enough energy to cover a significant portion of its operation needs.

> ### Knowledge Box
>
> **Sweden Recycling Revolution:** Sweden is one of the leading countries in recycling and circular economy (for more information about circular economy, see Chapter 12). In 2017 per capita trash generation 473 kg/year, but 50% of it was turned into energy, 85% of bottles and cans was recycled, and 69% of all packaging was recycled. For 2020, the nation's goal for food waste is to use 50% for natural fertilizer, and another 40% for energy recovery. Sweden is aiming for a zero-waste society. It is no surprise that there is a saying "Recycle like the Swedes" (see Figure 9.14).

9.3.2 Waste to Energy

In addition to collecting leachate as shown in Figures 9.10 and 9.11, landfill gases that are mostly carbon dioxide and methane could also be collected. Unlike leachate,

Figure 9.15 A landfill gas collection system in Sunnyvale, California, USA with an information bulletin board.

which is a waste, landfill gas could be reused for energy recovery if the methane gas is of sufficient amount and level to make it economically viable to generate electricity. **Figure 9.15** shows a landfill gas collection and cogeneration system for energy recovery in California.

Energy recovery could also be achieved by incineration. If MSW is properly sorted in a way that facilitates a high-energy portion of the waste stream to be separated as "refuse-derived fuel," or RDF, which contains about 12 to 16 million joules per kilogram, incineration of RDF can be a profitable and environmentally friendly method of treating MSW. The sorting could be done by a rotating screen or trommel, a shredder, and a blower to collect only the "light" fraction (paper, plastics, wood, textiles, etc.) for combustion in an incinerator. If properly designed, this will result in a higher percentage of energy recovery than the landfill gas energy recovery due to its higher efficiency. This will also reduce transportation and landfill operation cost for trash, because the weight and volume of the residue after incineration are a small fraction of those for the feed materials.

Knowledge Box

RDF—A full-scale RDF plant has been operational in Ames, Iowa, USA since 1975. The Southeastern Virginia Public Service Authority has an RDF plant as part of its comprehensive integrated waste disposal system to reduce waste in the landfill and recover a significant portion of energy in the waste stream. It processes about 83% of all of the waste through the RDF plant, generating more than 200 megawatts of electricity. This plant also drastically extended the life expectancy of local landfills. **Figure 9.16** shows another RDF plant.

Figure 9.16 A power plant using refuse-derived fuel (RDF) produced from municipal solid waste.

Yard trimmings, or "green waste," are shredded grass and other plant materials from household or business landscaping operations. In the United States, about 60% of yard trimmings were recovered as of 2015 (Figure 9.13). Still, yard trimmings occupied 8% of the volume of waste that went to landfill. The best way to handle this waste is composting, as shown in **Figure 9.17**. The finished product, usually after about 40 days, could be used for in many purposes, such as gardening, landfill cover, soil amendment, and so on. Composting could be done both commercially and on household level (Figure 9.17).

Figure 9.17 A commercial composting facility (left) and a household composter (right).

9.4 Hazardous Waste

9.4.1 Definition

Hazardous waste is waste with properties that make it dangerous or capable of having a harmful effect on human health or the environment. Hazardous waste could be deemed a subset of solid waste, such as the hazardous household waste as shown in **Figure 9.18**, or it could be a subset of hazardous or toxic materials and could be generated from at least as many sources as hazardous materials are produced and used. Common hazards by these wastes include ignitability, reactivity, corrosivity, and toxicity.

Almost universally, hazardous wastes are treated and regulated separately from the other solid waste categories. For example, in the United States, the Resource Conservation and Recovery Act, or **RCRA** of 1976, was established to set up a framework for the proper management of hazardous waste. Under RCRA, a hazardous waste is defined through a rather convoluted four-step process, as shown in **Figure 9.19**. The waste has

Figure 9.18 Examples of household hazardous waste. In the United States, the local waste management agencies collect these wastes from residents with no cost. These hazardous wastes will then be sorted, consolidated, and sent to hazardous waste site for disposal.

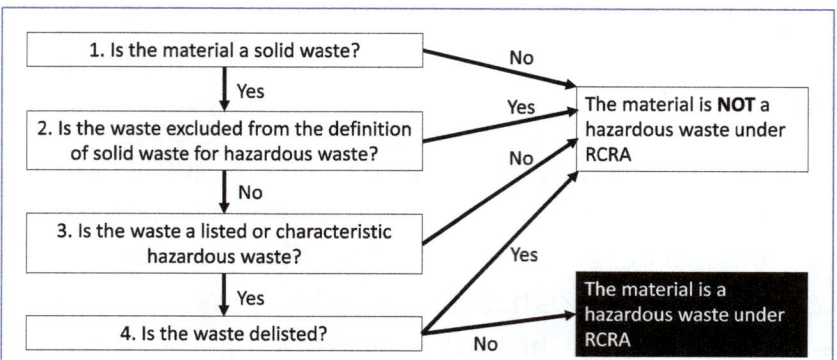

Figure 9.19 RCRA four step definition of hazardous waste (USEPA, 2020).
Source: Created by Jian Peng

to be a solid waste that is not excluded from the definition. There are seventeen (17) seemingly hazardous materials excluded from RCRA, such as household hazardous waste, petroleum contaminated media and debris from underground storage tanks, as shown in **Table 9.3**. This odd list is the result of many lawsuits, lobbying, and negotiations. For the excluded solid wastes that are not excluded from these 17 categories, they need to be specifically listed or characterized to be hazardous waste in order to be considered hazardous waste. The listing encompasses four lists (so-called F, K, P, and U lists) that contain hundreds of very specific and detailed descriptions of types of waste. There are 149 entries in the F and K lists, and 724 entries in the P and U lists.

The so-called characteristic wastes are those that have four characteristics mentioned above, including ignitability, corrosivity, reactivity, and toxicity. In addition,

Table 9.3 List of Solid Wastes that are Excluded from Being Considered Hazardous Waste Under RCRA (USEPA, 2020)

Solid Wastes Which Are Not Hazardous Wastes
Household Hazardous Waste
Agricultural Waste
Mining Overburden
Fossil Fuel Combustion Waste (Bevill)
Oil, Gas, and Geothermal Wastes (Bentsen Amendment)
Trivalent Chromium Wastes
Mining and Mineral Processing Wastes (Bevill)
Cement Kiln Dust (Bevill)
Arsenical-Treated Wood
Petroleum Contaminated Media & Debris from Underground Storage Tanks
Injected Groundwater
Spent Chloroflurocarbon Refrigerants
Used Oil Filters
Used Oil Distillation Bottoms
Landfill Leachate or Gas Condensate Derived from Certain Listed Wastes
Project XL Pilot Project Exclusions
Project XL Pilot Project Exclusions

mixed radiological **waste** is also considered hazardous waste under RCRA. For energy-related radiological waste, more discussion can be found in Chapter 10.

The Love Canal and the Valley of Drums incidents, as mentioned in Chapter 7, are both infamous incidents related to hazardous waste. They are the two key incidents that prompted the United States government to establish the Comprehensive Environmental Response, Compensation, and Liability Act (CERCLA), or the Superfund Act.

9.4.2 Hazardous Waste Management

The USEPA, under RCRA, mandates a so-called **Cradle-to-Grave** approach to hazardous waste management. Under RCRA, there is a comprehensive regulatory program to ensure that hazardous waste is managed safely from the time it is generated (cradle), transported, treated, and stored, until it is disposed of (grave). Hazardous waste generators are the first link (i.e., the "cradle") in the hazardous waste management process. All generators must first determine whether their waste is hazardous (see Figure 9.19). If so, generators must ensure and fully document that the hazardous waste that they generate is properly identified, managed, and treated prior to recycling or disposal.

Transportation of hazardous waste is also highly regulated because the transportation process involves public roads, highways, rails and waterways. The United States Department of Transportation has a comprehensive set of regulations on hazardous waste transportation. For this purpose, the Federal Motor Carrier Safety Administration has a division specifically for hazardous waste.[6]

The recycling, treatment, storage and disposal of hazardous wastes (i.e., the last step in the "cradle-to-grave" process) are also critical steps that ensure the proper handling of hazardous wastes. To the extent possible and practicable, some wastes are recycled if it can be done safely and effectively. Recycling of hazardous wastes is an environmentally responsible way to manage hazardous wastes because it reduces the consumption of raw materials and the volume of waste materials that must be treated and disposed, which are expensive processes both economically and environmentally.

Those that cannot be recycled will be treated, stored, or disposed of. Storage of hazardous materials might cause spills, leaks, fires, and contamination of soil and drinking water. Therefore, USEPA has a set of regulations specifically for hazardous material storage.

Lastly, treatment storage and disposal facilities (TSDFs) provide temporary storage and final treatment of disposal for hazardous wastes. Similar to other hazardous waste

---

[6] Website: https://www.fmcsa.dot.gov/regulations/hazardous-materials/how-comply-federal-hazardous-materials-regulations.

facilities, TSDFs are highly regulated as well and have to follow facility management standards, specific provisions, and implement additional precautions to protect soil, groundwater, and air quality.

> ### Knowledge Box
>
> **Hazmat:** Hazardous materials, usually called hazmat in short, are different from hazardous waste. According to the Institute of Hazardous Materials Management, a hazardous material is any item or agent (biological, chemical, radiological, and/or physical), which has the potential to cause harm to humans, animals, or the environment, either by itself or through interaction with other factors. Hazardous materials are defined and regulated in the United States primarily by laws and regulations administered by the U.S. Environmental Protection Agency (EPA), the U.S. Occupational Safety and Health Administration (OSHA), the U.S. Department of Transportation (DOT), and the U.S. Nuclear Regulatory Commission (NRC). Each has its own definition of a "hazardous material."

9.5 Plastics and Trash Pollution

9.5.1 Overview

As shown in Figures 9.2 and 9.5, plastic is one of the major components (13%) in solid waste generated, and is the largest component of landfilled solid waste (19%) due to the low percentage of recovery or recycling (>10%). The main reason is much of the plastics are nonrecyclable packaging materials, plastic bags, and other items that cannot be economically recycled. The proliferation of plastics and lack of education and government intervention have caused plastic pollution around the world. In 2017, it was estimated that the global production of plastic waste was 4.9 billion tons, and about half of the plastics ever made are either in the landfill or are polluting the environment. **Figures 9.20** and **9.21** show two examples of plastic pollution at Bali, Indonesia, and Fullerton, California, USA, respectively.

Plastic is roughly 80% of the trash that impacts waterways and causes nuisance and ecological impact. On a global scale, plastic is also the main culprit of the notorious "ocean garbage patches" (**Figures 9.22** and **9.23**). There are five oceanic garbage patches, with the Great Pacific Garbage Patch (Figure 9.23) the biggest of all, spanning 1,760,000 square kilometers. 80% of this garbage is from land discharges and 20% is

Figure 9.20 Plastic pollution at Kuta Beach, Bali, Indonesia.

Figure 9.21 Trash, most plastic cups, near a storm drain outfall at Fullerton Creek, Fullerton, California, USA.

from ships. For the Great Pacific Garbage Patch, 99% of the garbage is plastic, numbering 1.8 trillion big and small pieces. Nearly half are discarded fishing gear.

Despite its severe pollution impacts, trash is not always regulated as a pollutant, but California, USA is an exception. Please refer to Section 5.5 for more information about regulations that target trash as a pollutant. Due to the concerns for plastic and trash pollution, many places have passed laws and regulations prohibiting common plastic items that are found in waterways. They include plastic bags, straws, foams,

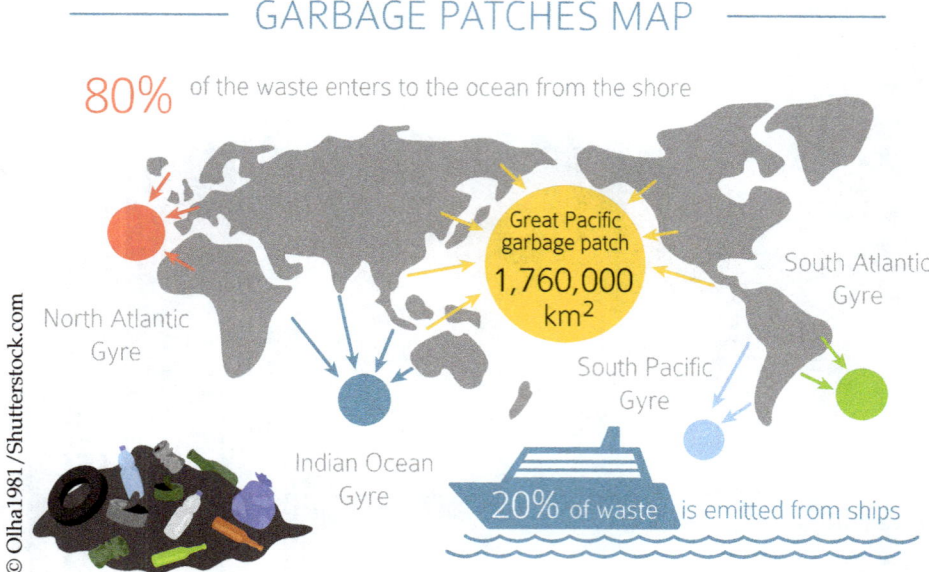

Figure 9.22 The global ocean garbage patches.

Figure 9.23 The Great Pacific garbage patch.

and the like. Various packaging materials, many of which are plastic-based (especially polystyrene foams), are also a significant source of plastic pollution in the environment and in the waterways. To curb the proliferation of plastic wastes from packaging materials, increasing pressure is being put on manufacturers to minimize the use of packaging materials, or to encourage the use of reusable packing materials. By controlling these plastic wastes from their sources, it is hoped that plastic pollution can be managed more sustainably at lower environmental, social, and economic costs. More in-depth discussions on plastic and other solid waste and sustainability issues are presented in Chapter 12 near the end of this book.

9.5.2 Basel Convention

The issue of both hazardous wastes and plastic wastes in the 1970s to 1980s has raised public awareness of these issues. The increasing amount of hazardous waste, rising disposal costs, economic gap between rich and poor nations, and a general **NIMBY** (not in my backyard) attitude among the industrialized nations have resulted in significant transboundary transport and disposal (often in improper ways) of hazardous and other wastes. It was found in the late 1980s that Africa, Eastern Europe, and other developing countries had been the disposal sites for toxic wastes imported from abroad. To deal with this international crisis, on March 22, 1989, the **Basel Convention** on the Control of Transboundary Movements of Hazardous Wastes and Their Disposal was adopted by the Conference of Plenipotentiaries in Basel, Switzerland. The Convention became effective in 1992 and was amended several times afterwards. As part of the requirements, member nations report their annual rate of generation of hazardous and other wastes, see **Figure 9.24** for the data for 2018.

The overarching objective of the Basel Convention is to protect human health and the environment against the impacts of hazardous wastes. The definition of "hazardous waste" here is fairly broad (i.e., different from that in the United States as defined by the Toxic Substances Control Act; Resource Conservation and Recovery Act; and other regulations) and includes domestic waste and many hazardous wastes, as shown in **Tables 9.4a to c**. There are three main goals for the Basel Convention:

- The reduction of hazardous waste generation and the promotion of environmentally sound management of hazardous wastes, wherever the place of disposal;

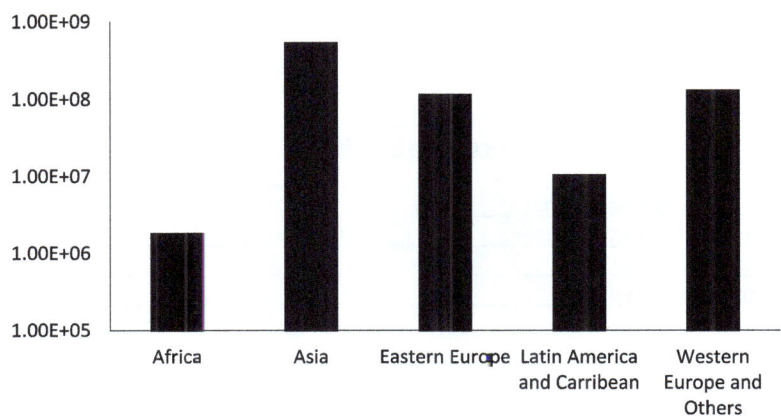

Figure 9.24 Total amount of generation (tons) of hazardous and other wastes in 2018.
Data source: http://ers.basel.int/eRSodataReports2/ReportBC_Generation2.htm. Created by Jian Peng

Table 9.4a Basel Convention Categories of Wastes to be Controlled: Waste Streams
The table is called 'Annex' because the information is contained in the annex (or appendix) of the Basel Convention protocol (The Secretariat of the Basel Convention, 2018)

	Annex I: Basel Convention Categories of Wastes to be Controlled
	Waste Streams
Y1	Clinical wastes from medical care in hospitals, medical centers and clinics
Y2	Wastes from the production and preparation of pharmaceutical products
Y3	Waste pharmaceuticals, drugs and medicines
Y4	Wastes from the production, formulation and use of biocides and phytopharmaceuticals
Y5	Wastes from the manufacture, formulation and use of wood preserving chemicals
Y6	Wastes from the production, formulation and use of organic solvents
Y7	Wastes from heat treatment and tempering operations containing cyanides
Y8	Waste mineral oils unfit for their originally intended use
Y9	Waste oils/water, hydrocarbons/water mixtures, emulsions
Y10	Waste substances and articles containing or contaminated with polychlorinated biphenyls (PCBs) and/or polychlorinated terphenyls (PCTs) and/or polybrominated biphenyls (PBBs)
Y11	Waste tarry residues arising from refining, distillation and any pyrolytic treatment
Y12	Wastes from production, formulation and use of inks, dyes, pigments, paints, lacquers, varnish
Y13	Wastes from production, formulation and use of resins, latex, plasticizers, glues/adhesives
Y14	Waste chemical substances arising from research and development or teaching activities which are not identified and/or are new and whose effects on man and/or the environment are not known
Y15	Wastes of an explosive nature not subject to other legislation
Y16	Wastes from production, formulation and use of photographic chemicals and processing materials
Y17	Wastes resulting from surface treatment of metals and plastics
Y18	Residues arising from industrial waste disposal operations

Table 9.4b Basel Convention Categories of Wastes to be Controlled: Wastes Having as Constituents (i.e. wastes having any of the following consituents). This is also part of Annex I

	Wastes Having as Constituents:
Y19	Metal carbonyls
Y20	Beryllium; beryllium compounds
Y21	Hexavalent chromium compounds
Y22	Copper compounds
Y23	Zinc compounds
Y24	Arsenic; arsenic compounds
Y25	Selenium; selenium compounds
Y26	Cadmium; cadmium compounds
Y27	Antimony; antimony compounds
Y28	Tellurium; tellurium compounds
Y29	Mercury; mercury compounds
Y30	Thallium; thallium compounds
Y31	Lead; lead compounds
Y32	Inorganic fluorine compounds excluding calcium fluoride
Y33	Inorganic cyanides
Y34	Acidic solutions or acids in solid form
Y35	Basic solutions or bases in solid form
Y36	Asbestos (dust and fibres)
Y37	Organic phosphorus compounds
Y38	Organic cyanides
Y39	Phenols; phenol compounds including chlorophenols
Y40	Ethers
Y41	Halogenated organic solvents
Y42	Organic solvents excluding halogenated solvents
Y43	Any congener of polychlorinated dibenzo-furan
Y44	Any congener of polychlorinated dibenzo-p-dioxin
Y45	Organohalogen compounds other than substances referred to in this Annex (e.g. Y39, Y41, Y42, Y43, Y44)

Table 9.4c Basel Convention Categories of Wastes to be Controlled: Wastes Requiring Special Consideration (the table is called 'Annex' because the information is contained in the annex (or appendix) of the Basel Convention protocol (The Secretariat of the Basel Convention, 2018)

	Annex II: Basel Convention Categories of Wastes Requiring Special Consideration
Y46	Wastes collected from households
Y47	Residues arising from the incineration of household wastes
Y48	Plastic waste

- The restriction of transboundary movements of hazardous wastes except where it is perceived to be in accordance with the principles of environmentally sound management; and
- A regulatory system applying to cases where transboundary movements are permissible.

Knowledge Box

The United States and Basel Convention: Surprisingly, the United States is not a signatory of this Convention due to a lack of implementing legislation (it did sign the Convention in 1990). However, because the United States is a member of the Organization for Economic Co-operation and Development (OECD[7]), a 34-member organization consisting of western developed countries, and OECD has specific agreements governing waste movements among member countries, the United States is still largely in compliance with Basel Convention stipulations.

9.5.3 OECD Convention

The OECD has established another international convention to control export and import of wastes which may pose a risk to human health and the environment. Different from the Basel Convention and the European Union regulations that concern both recovery and disposal, the OECD requirements only concern movements of

---

[7] The OECD (Organization for Economic Co-Operation and Development) is an international organization established in 1960 to assist Member countries in achieving sustainable economic growth, employment, and an increased standard of living, while simultaneously ensuring the protection of human health and the environment. OECD Member countries concern themselves with a host of international socio-economic and political issues, including environmental issues. OECD member countries include: Australia, Austria, Belgium, Canada, Chile, the Czech Republic, Denmark, Estonia, Finland, France, Germany, Greece, Hungary, Iceland, Ireland, Israel, Italy, Japan, Luxembourg, Mexico, the Netherlands, New Zealand, Norway, Poland, Portugal, the South Korea, the Slovakia Republic, Slovenia, Spain, Sweden, Switzerland, Turkey, the United Kingdom, and the United States.

wastes destined for recovery operations within the OECD area. The United States is required to comply with **OECD Convention** because it is an international agreement that created binding commitments on the United States. In addition to OECD convention, the United States also established regulations[8] for the transboundary movement of hazardous wastes between United States, Canada and Mexico. Overall, the United States handles anything related to hazardous wastes in a strict and responsible manner.

Further Reading

Kaza, Silpa, Lisa Yao, Perinaz Bhada-Tata, and Frank Van Woerden. 2018. What a Waste 2.0: A Global Snapshot of Solid Waste Management to 2050. Urban Development Series. Washington, DC: World Bank. doi:10.1596/978-1-4648-1329-0. License: Creative Commons Attribution CC BY 3.0 IGO. Weblink: https://openknowledge.worldbank.org/handle/10986/2174.

Humes, E. Garbology: Out Dirty Love Affair with Trash. New York, USA: The Penguin Group, 2013.

USEPA, 2006. Industrial Stormwater Fact Sheet Series. Sector L: Landfills and Land Application Sites. USEPA Office of Water. EPA-833-F-06-027.

USEPA. 2012. Hazardous Waste Listings: A user-friendly reference document. Weblink: https://www.epa.gov/sites/production/files/2016-01/documents/hw_listref_sep2012.pdf.

USEPA. 2018. Advancing Sustainable Materials Management: 2015 Fact Sheet. Weblink: https://www.epa.gov/sites/production/files/2018-07/documents/2015_smm_msw_factsheet_07242018_fnl_508_002.pdf.

The Secretariat of the Basel Convention (SBC), 2018. Basel Convention on the control of transboundary movements of hazardous wastes and their disposal. Protocol on Liability and Compensation for damage resulting from transboundary movements of hazardous wastes and their disposal. Texts and Annexes. United Nations Environment Programme.

OECD, 2009, Guidance Manual For The Implementation Of Council Decision C(2001)107/Final, As Amended, On the Control Of Transboundary Movements Of Wastes Destined For Recovery Operations.

---

[8] This regulation can be found at 40 CFR Part 262, Subpart H, which can be access here: https://www.law.cornell.edu/cfr/text/40/part-262/subpart-H

U.S. Environmental Protection Agency. Landfill bioreactor performance: second interim report: outer loop recycling & disposal facility—Louisville, Kentucky, 2007. EPA/600/R-07/060.

Web Resources

World Bank, 2020, Trends in solid waste management, https://datatopics.worldbank.org/what-a-waste/trends_in_solid_waste_management.html. Accessed May 17, 2020.
USEPA website for municipal solid wastes: https://archive.epa.gov/epawaste/nonhaz/municipal/web/html/
OECD Resource productivity and waste website: http://www.oecd.org/env/waste/
The Swedish Recycling Revolution: https://sweden.se/nature/the-swedish-recycling-revolution/
A NYU student was able to drastically reduce her trash generation: https://www.youtube.com/watch?v=nYDQcBQUDpw
USEPA Hazardous Waste Website: https://www.epa.gov/hw/learn-basics-hazardous-waste
IHMM Definition of hazardous materials: https://www.ihmm.org/about-ihmm/what-are-hazardous-materials
The Love Canal Tragedy: https://en.wikipedia.org/wiki/Love_Canal
Global plastic waste estimate: https://cosmosmagazine.com/society/global-plastic-waste-totals-4-9-billion-tonnes

Questions and Exercises

1. Take a look at your waste basket and conduct an inventory of your garbage. Which of these items can be dealt with by each of the 5-Rs (refuse, reduce, reuse, recycle, rot)?
2. Are you generating garbage at a rate higher or lower than the national average? By looking at your answer to question 1, what could be done to bring the generation rate below the national average?
3. Is it possible to have a zero-waste lifestyle?
4. What do you think of the "polluters pay" principle, in which the generator of waste, recyclable or not, would pay for the recycling and disposal costs? Should they also be responsible for environmental impacts?

5. Why is improved waste classification one of the keys to better solid waste management? List at least three reasons.
6. How does a bioreactor landfill operate? What are the advantages and challenges of operating such a landfill?

Chapter 10

Energy and Sustainability

Learning Outcomes	276
Key Concepts	276
10.1 Global Energy Consumption and Trends	277
10.1.1 Energy Sources	277
10.1.2 Trends in Global Energy Consumption	278
10.2 Conventional Energy	281
10.2.1 Overview of Conventional Energy	281
10.2.2 Crude Oil	282
10.2.3 Coal	285
10.2.4 Natural Gas	288
10.3 Renewable Energy	289
10.3.1 Overview of Renewable Energy	289
10.3.2 Solar Energy	291
10.3.3 Wind Energy	294
10.3.4 Other Renewable Energy Sources	297
10.4 Nuclear Power	300
10.4.1 Overview	300
10.4.2 Advantages of Nuclear Power	303
10.4.3 Challenges of Nuclear Power	304
10.5 Energy and Sustainability	308
Further Reading	311
Web Resources	311
Questions and Exercises	312

Learning Outcomes

- Appreciation of the need for energy for everyday life and economy
- Understanding of the significance of energy industry to the generation of greenhouse gases and recognize the need to curb climate change in the energy sector
- Knowledge of different types and sources of energy, both conventional (including fossil fuel energy) and renewables
- Knowledge of energy generation process through conventional and non-conventional process
- Knowledge of each major renewable energy type and its characteristics, including both advantages and issues
- General knowledge of the future changes in global energy portfolio in both quantity and composition, especially with regard to renewable energy

Key Concepts

Conventional energy; fossil fuel; Renewable energy; energy consumption; alternative fuel; solar energy; wind energy; geothermal energy; hydropower; nuclear power

Of many fundamental environmental issues that Earth is facing, nothing is more worrisome than global warming and climate change, as will be discussed in Chapter 11. However, the most important driving force behind global warming and climate change is the ever-increasing consumption of **fossil fuels** (such as coal, petroleum, and natural gas) to produce more and more energy required by the increasing global population as well as economic growth. It is predicted that by 2050, global **energy consumption** will rise more than 50% from the current level, to about 260 quadrillion watt hours from 160 quadrillion watt hours in 2010. This significant increase, much of it is expected to be "fueled" by fossil fuels, will inevitably bring more pressure on greenhouse gas emission and potentially cause a series of global changes. This chapter will dive deep into the history, present and future of energy sources, energy consumption by different nations and different sectors, and the trend in different energy sources, both **conventional energy** (fossil fuel based) and **renewable energy**. By the end of this chapter, the reader will get a good grasp of the status and trends of global energy issues. This knowledge will be critical to understanding the materials of the last two chapters on climate change (Chapter 11) and sustainability (Chapter 12).

10.1 Global Energy Consumption and Trends

10.1.1 Energy Sources

By definition, energy is a property that performs work or heats an object. Cars, trucks, airplanes, and machinery need energy from fuels or from electricity to work. Electricity, in turn, can be produced from various sources, most of which (at least for now and in the near future) are based on various fuels, such as coal, petroleum, natural gas, and so on. Humans need energy to function and get the energy from food, which turns into energy after digestion and assimilation.[1] Plants need **solar energy** for photosynthesis in order to grow and reproduce. Some microbes can harvest chemical energy to reproduce and propagate. In this chapter, we will focus on energy on the macro scale that works outside biotic systems.

There are different types of energy such as kinetic, potential, elastic, chemical, radiant, and thermal energy. Other than the famous mass-energy equivalence equation,[2] energy is conservative. It can be passed from one object to another, or from one part of the system to another, but the total amount of energy has to be conserved at all times.

Earth's surface, including the hydrosphere, atmosphere, and biosphere, receive most of its energy from the sun. With 174 petawatts of incoming solar radiation at the upper atmospheric, a portion (~30%) is reflected back to space, the rest (~70%) is absorbed by the Earth, with a total amount of about 3.85×10^{24} joules per year (see the Knowledge Box). This portion of solar energy essentially drives nearly all processes on the Earth's surface (with exceptions such as volcanoes; plate tectonics; geothermal vents; etc.), including the cycling of water, air, and the biogeochemical cycles of various elements (see Chapter 2). The only exception is the residual heat from the time when the Earth was formed, as discussed in Sections 1.4 and 1.5. Of the materials that formed the primordial Earth from the solar nebula, there are many radioactive elements such as potassium and uranium that are still decaying slowly today, generating considerable energy in the form of heat within Earth's interior.

---

[1] For humans and most animals, energy comes from adenosine triphosphate, or ATP, which is an organic compound that provides energy for most cell activities such as muscle movement, neural signal transfer, and so on.

[2] The famous mass-energy equivalence equation $E = mc^2$ is the key equation in special relativity discovered by Albert Einstein. Clearly, special relativity is a little beyond the scope of this book.

> **Knowledge Box**
>
> **Solar Energy Budget:** Of the 100% incoming solar energy, about 30% is reflected back to space (roughly 20% reflected by the clouds, 6% scattered from the atmosphere, and 4% reflected by the Earth's surface). The rest 70% is absorbed by the Earth, with 19% absorbed by the air and the clouds, and 51% absorbed by the solid Earth. To maintain energy balance without causing global warming, nearly all of this energy eventually escapes back to space in the form of long-wave radiation, as will be discussed in Chapter 11.

Despite the vast amount of energy gifted to us from the sun, very little of the solar energy is being harvested, as will be elaborated later in this chapter. This is due to the higher cost of solar energy compared to traditional energy sources or other renewable energy sources, such as **wind energy**. Instead, we use a combination of several sources of energy. They include conventional energy based on fossil fuels such as crude oil, natural gas, and coal, and renewable energy sources such as solar, wind, tidal, **geothermal energy**, **hydropower**, biofuel,[3] etc. Terms such as "green energy" or "clean energy" are similar to renewable energy but they are sometimes used loosely,[4] so these terms will not be discussed in this chapter. Energy production from unconventional fuel or from recycled or reused wastes such as biogas, biofuel, landfill gas, trash, etc. is covered briefly in Chapter 9 (solid waste) and will not be discussed in detail here. Nuclear energy and hydropower will be discussed in Section 10.4.

10.1.2 Trends in Global Energy Consumption

Global per capita energy consumption has been steadily increasing since the 1970s (Figure 10.1) from 56 gigajoules to over 80 gigajoules per year. To put this in context, 80 gigajoules is roughly equivalent to the energy released by burning 1,920 kilograms

---

[3] Biofuels such as firewood, and ethanol produced from corn or algae are a type of renewable energy because the source materials (trees, algae, and crops) can be readily renewed in a short-time period.

[4] For example, hydropower is a "clean" energy, but with ecological impacts (will be discussed later in this chapter). Nuclear energy could be called a "clean" energy if the issues with nuclear wastes could be ignored. On the other hand, even solar and wind energy generate wastes, sometimes significant amounts of waste, so there is no "clean energy" that is absolutely clean.

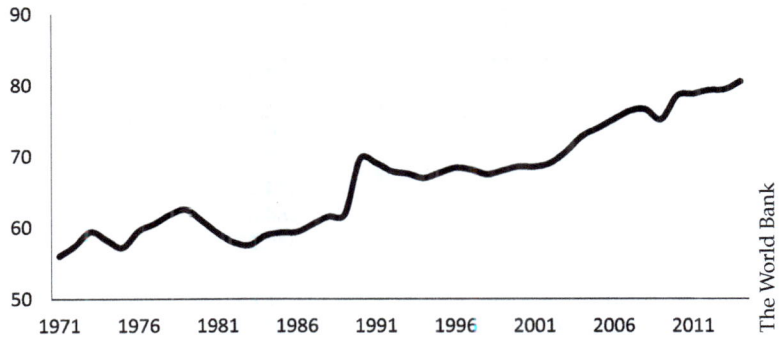

Figure 10.1 The global trend in per capita energy consumption 1971 to 2014 in gigajoules.
Source: Created by Jian Peng

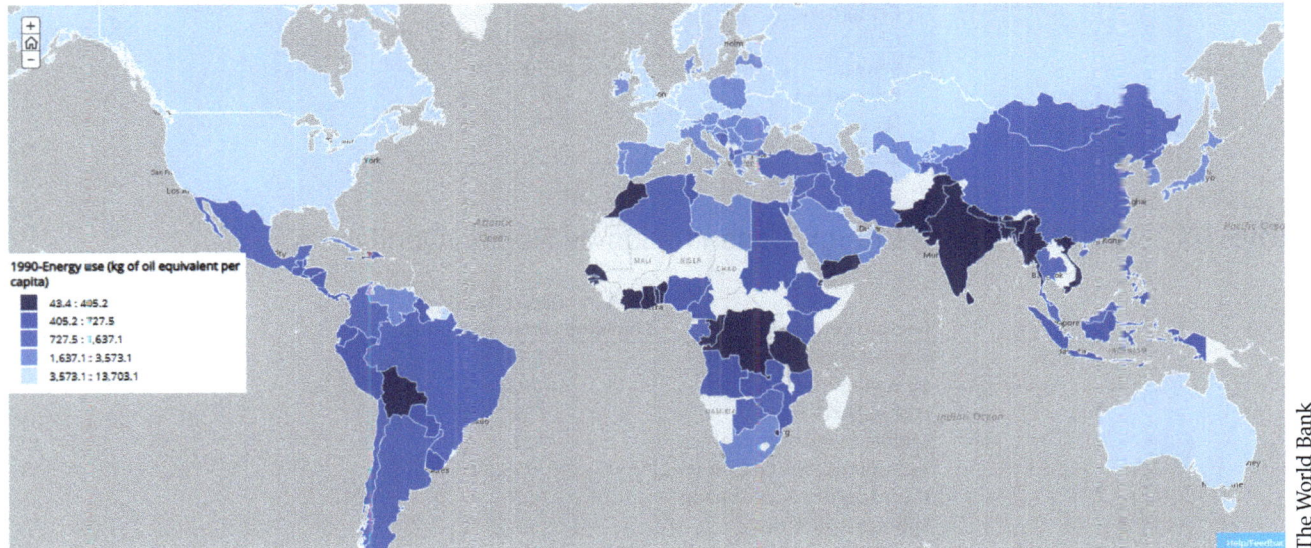

Figure 10.2 Energy consumption per capita (plotted based on data from the World Bank).

of oil.[5] However, there are large geographical differences, as shown in **Figure 10.2** where developed countries and those at higher latitude tend to consume much more than developing countries. The difference between the countries with the highest and lowest per capita energy consumption can be more than two orders of magnitude (see **Table 10.1**). **Table 10.2** summarizes the average per capita energy consumption for different regions and continents around the world. It should be noted that North America and European Union per capita energy consumptions have been largely flat, and those in the developing countries, especially East Asia and Pacific, led by China, have been increasing rapidly in recent years.

---

[5] In fact, this unit, termed "kilogram of oil equivalent," is actually the unit used by the source of this data (The World Bank).

Table 10.1 2014 Per Capita Energy Consumption—Top 10 and Bottom 10 Nations

Country	2014 Per Capita Energy Consumption in gigajoules
Qatar	750.96
Iceland	750.69
Bahrain	444.00
Kuwait	384.60
Canada	330.92
United Arab Emirates	320.47
United States	291.65
Saudi Arabia	289.35
Finland	260.34
Oman	253.10
Congo	16.31
Sudan	15.96
Myanmar	15.48
Tajikistan	14.24
Cameroon	14.04
Ghana	13.90
Senegal	11.70
Bangladesh	9.61
Niger	6.29
South Sudan	2.78

Table 10.2 2014 Per Capita Energy Consumption by Region/Continents

Region/Continent	2014 Per Capita Energy Consumption in gigajoules
North America	296
European Union	129
East Asia & Pacific	79
Middle East & North Africa	62
Latin America & Caribbean	57
Latin America & Caribbean	55
Sub-Saharan Africa	29
South Asia	24
World Average	**81**

Looking into the future, the projected global energy production until 2050 (**Figure 10.3**) shows a general increasing trend. However, the trend has a clear distinction between developed countries (represented by OECD nations) and non-OECD nations, with the former barely increasing, and the latter more than doubling during this 40-year period. For developing countries, all four major categories of energy consumption (industrial; transportation; commercial, and residential) will see increases. Industrial energy consumption, which is used for manufacturing and other fundamental industrial activities, is currently the largest portion of overall energy consumption by developing countries. In comparison, for OECD countries, the proportion of industrial power consumption is now less than 50% and will remain flat through 2050. Even for developing countries, the relative proportion of industrial energy consumption will decrease due to faster increases in transportation, commercial and residential energy consumption levels.

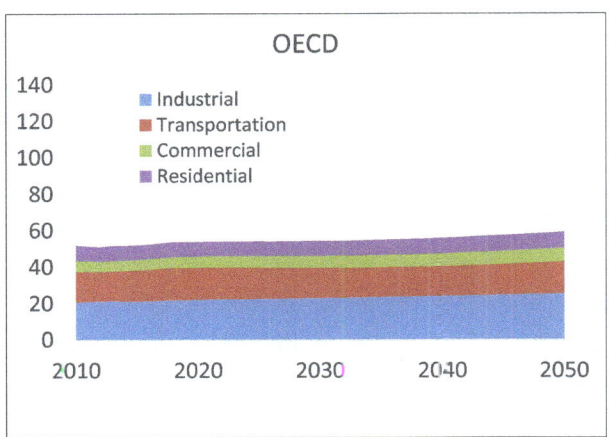 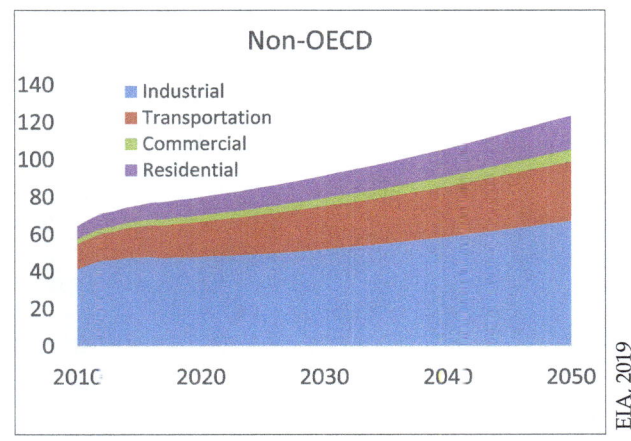

Figure 10.3 Projection of global energy consumption by sectors. Left: OECD rations. Right: non-OECD Nations. Unit: quadrillion watts hour per year.

10.2 Conventional Energy

10.2.1 Overview of Conventional Energy

For simplicity, conventional energy here refers to energy produced from fossil fuels, which has been the predominant energy source for the last century and will remain so for the next few decades, if not longer. Global consumption of crude oil was 80 million barrels a day in 2018 and is projected to reach 100 million barrels a day in 2050. Until the late 1960s, about 95% of the world's energy was from fossil fuel. The 1973 oil crisis significantly reduced the world's reliance on oil and gas, but the level hovered around 80% for the next 45 years (**Figure 10.4**). As shown in **Figure 10.5**, the net increase of overall global energy consumption is achieved by other energy sources, such as hydropower, **nuclear power**, then increasingly by renewable energy. Historically, consumption of fossil fuels, especially crude oil did not take off until after World War II when countries around the world started to rebuild. Relatively speaking, coal has been an important energy source since the Industrial Revolution and remains a major fuel in the global energy portfolio. Currently, coal occupies 40% of global energy and electricity production. Despite the continued decrease in coal consumption in the United States and other developed countries due to pollution concerns as well as booming natural gas uses, overall coal consumption globally has actually been increasing due to increasing use in China[6] and other countries. Going forward, the coal industry will face increasing challenges due to pollution concerns and greenhouse gas emission under increasingly stringent national and global regulations.

---

[6] China alone accounts for about half of global coal consumption.

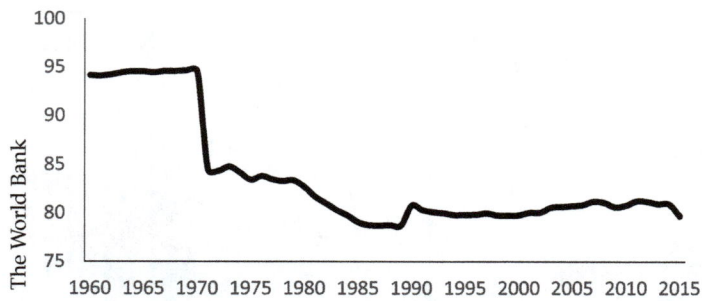

Figure 10.4 Percentage of fossil fuel (coal, oil, petroleum, and natural gas) in total energy consumption, 1960 to 2015. The key driver here is crude oil price.

Source: Created by Jian Peng

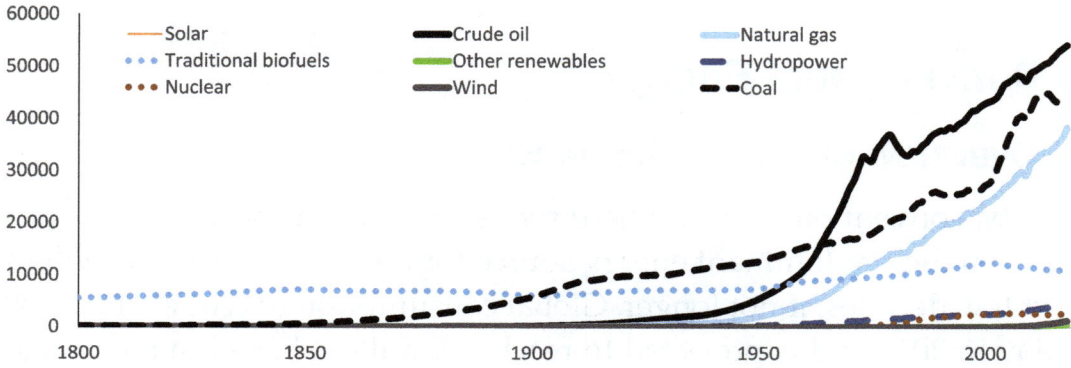

Figure 10.5 The world energy portfolio from 1800–2018 in terawatt hour.

Data source: https://ourworldindata.org/energy. Plot created by Jian Peng

10.2.2 Crude Oil

People started using crude oil, or petroleum, for various uses several thousand years ago in the Middle East and in China. Distillation of petroleum to use its various components started from as early as the 9th century, and became an industry when the world's first oil refinery was built in 1856 in Europe. Since the 1960s, crude oil has passed coal and became the top energy source in the world.

According to the U.S. Energy Information Administration, the global crude oil reserve is sufficient to sustain the world's demand for liquid fuels through at least 2050, depending on oil prices, development of alternative and renewable energy sources, and other factors. Crude oil is a naturally occurring liquid found in geological formations. It is formed from burial of organic materials such as zooplankton and algae under anaerobic conditions over thousands of years or longer. After burial of these materials, increased pressure and temperature over millions of years turn these organic materials into liquid form through a series of complex reactions. The liquid could stay in place, or migrate to a reservoir elsewhere. Oil wells can be drilled into the reservoir to extract the oil by oil rigs either on land or in the ocean (**Figure 10.6**).

Figure 10.6 Top: an oil rig platform in the ocean. Bottom: an oil pump on land.

> ### Knowledge Box
>
> **Fracking:** Hydraulic Fracturing, or fracking, is a technique to extract natural gas or crude oil from shale and other forms of "tight" rocks, which are impermeable and difficult to extract oil and gas from. Using large amounts of water, chemicals, and sand at extremely high pressure can fracture these formations and release oil and gas for extraction. The technique started as early as 1862 as so-called "shooting the well" using explosives. In the 1940s, explosives were replaced with high pressure blasts of liquids. This technique greatly increased production of the oil and gas industry, but has caused many environmental issues.

After crude oil is extracted, it is transported to oil refineries (**Figure 10.7**) to be converted to many petroleum products that people use every day. Most of the refineries focus on producing transportation fuels. On average, a 42-gallon barrel of crude oil produces 19 to 20 gallons of gasoline, 11 to 12 gallons of diesel, and 4 gallons of jet fuel. More than a dozen other products are also produced by oil refineries as by-products.

In addition to producing fuels for vehicles, to heat buildings, and to produce electricity, petroleum can also produce critical materials such as plastics, polyurethane, solvents, and hundreds of other raw and finished products that are critical to many industry sectors. However, production, refining, and use of petroleum products may cause a

Figure 10.7 An oil refinery with multi-level distillation towers that separate crude oil into different products. A flare burns excess process gas.

Figure 10.8 The site of Deepwater Horizon oil rig, owned by British Petroleum (BP), in the Gulf of Mexico after a blowup that killed 11 workers and spilled 780,000 m^3 of crude oil. It is the largest oil spill in U.S. history and resulted in a fine of up to $20 billion to BP and other firms.

number of environmental issues. Recently, extensive use of hydraulic fracturing (fracking, see Knowledge Box) has caused significant environmental concerns. Fracking uses high pressure to fracture deep geological formations in order to release oil and gas. This process produces large amount of wastewater, generates earthquakes, and so on. In addition to the most salient issue of producing greenhouse gases, the petroleum industry is also challenged by high-profile environmental disasters such as Exxon-Valdez oil spill and Deepwater Horizon disaster (**Figure 10.8**).

10.2.3 Coal

Coal, a combustible black or brownish-black sedimentary rock with a high amount of carbon and hydrocarbons, is the oldest fossil fuel. It is formed by the burial of plants that lived millions of years ago. The pressure and heat then turn the plant materials into coal. Depending on the carbon content and other properties, coal can be classified into anthracite, bituminous coal, subbituminous coal, and lignite, with decreasing carbon content and poorer performance as fuel. Lower-grade coals also tend to give out more pollutants when burning. Despite decades of mining and consumption, there is still more than one trillion tons of coal reserves, of which 22% is in the United States, followed by Russia (15%), Australia (14%), China (13%), and India (10%).

Worldwide coal production has been steady in recent years at about 8 billion tons per year. Therefore, the world's reserve can sustain more than 100 more years of consumption.

OECD countries will see a steady decrease of coal consumption from 2018 through 2025 before holding steady through 2050 (**Figure 10.9**). China has been the largest coal producer and consumer, but its coal production will see a slight decrease. However, India will see significantly increase, more than compensating for the decrease in China. Other non-OECD nations also see an increase in coal consumption. As a result, after a brief dip in coal consumption around 2020 and shortly after, it will increase slowly through 2050.

Coal is often referred to as the dirtiest fossil fuel among the three, with good reason. Mining of coal could cause significant environmental damage, especially for surface mines (also called strip mines). In the United States, mountaintop removal and valley fill mining in the Appalachian Mountains has affected many areas (e.g., **Figure 10.10**). Burning coal will also produce a number of air pollutants in addition to carbon dioxide. These pollutants include sulfur dioxide, nitrogen oxides, particulates, mercury and other heavy metals, among others (see Chapter 3 for more details). In addition, fly ash after coal burning could be a significant pollutant and is usually sent to landfill (**Figure 10.11**).

To combat these issues, the coal industry has found several ways to reduce sulfur and other impurities from coal. Nowadays most power plants use flue gas desulfurization equipment to clean sulfur from the smoke before it leaves smokestacks. There are a number of technologies that can be used to remove other impurities from coal before it is burned. There is also equipment to remove nitrogen oxides, mercury, and other impurities. However, dealing with carbon dioxide emission has been challenging, and

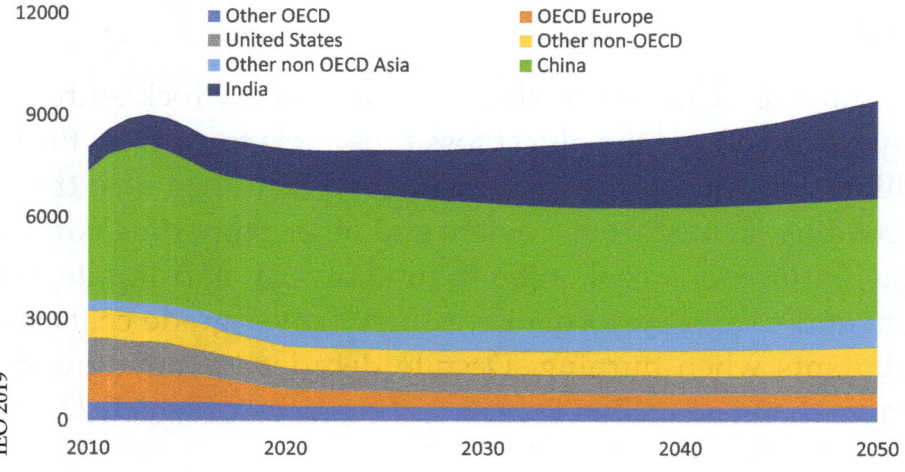

Figure 10.9 Projected global coal consumption to 2050. Unit: million tons.

some attempts have been made to purify the gas for beneficial use or injection into deep geological formation. Some of these cutting-edge technologies will be described in Chapter 12.

Figure 10.10 Mountaintop removal for coal mining in central West Virginia.

Figure 10.11 Coal-burning factories in China.

> ### Knowledge Box
>
> **The 1973 Oil Crisis:** The 1973 oil crisis began in October 1973 when the members of the Organization of Petroleum Exporting Countries (OPEC) proclaimed an oil embargo targeting nations supporting Israel during the Yom Kippur War. The initial nations targeted were Canada, Japan, the Netherlands, the United Kingdom the United States, Portugal, Rhodesia and South Africa. The embargo forced the price of crude oil up by nearly 400% and caused an oil crisis.

10.2.4 Natural Gas

Natural gas is mostly methane and is usually collected during the exploration of coal and oil, especially the latter. Impurities in natural gas, such as carbon dioxide, water vapor, and heavier condensates need to be removed before it can be commercially used. Sometimes sulfur compounds need to be removed as well to minimize air pollution. Compared to crude oil and coal, natural gas has enjoyed a more steady and faster increase over the years. Despite being one of the major types of fossil fuel, natural gas has not suffered the same level of scrutiny or bad press as crude oil and coal have. This is mainly due to the fact that natural gas is the cleanest burning fuel because of its purity and complete burning with little or no residue or pollutants other than carbon dioxide. This positive image has contributed to the increased use of natural gas in recent years and continued increase is expected in the future. At the current rate of consumption, natural gas can last just a few decades for the United States.[7]

> ### Knowledge Box
>
> **Flare or Release**—Figure 10.7 is a common process of burning off excess petroleum gas, mostly methane, in a refinery. Some oil fields do the same thing when they have more natural gas than they can safely handle. Is this good or bad? Should the natural gas be released directly into the atmosphere? The answer lies in the fact that methane is a much more potent greenhouse gas than carbon dioxide. Flaring the excess natural gas, rather than releasing it directly into the atmosphere, is actually an environmentally better way. Of course, the energy industry is working hard to recover as much process gas or excess natural gas as possible, and you will see less and less flares with technology improvement.

---

[7] There is considerable uncertainty in the estimate on how long natural gas will last. For the United States, the latest confirmed reserve is 500 trillion cubic foot and annual consumption is about 31 trillion cubic foot. However, the confirmed reserve keeps increasing, but consumption also has been increasing.

10.3 Renewable Energy

10.3.1 Overview of Renewable Energy

By definition, renewable energy is energy that can be replenished *naturally* and *on a human timescale*. While the "natural" aspect is easy to understand, the latter part of the definition needs some explanation, i.e., the renewal has to be fast enough to be meaningful. This is because fuels such as coal, crude oil, and natural gas are also "renewable" in a sense that they are naturally forming around the Earth as part of the global carbon biogeochemical cycle (see Chapter 2). However, humans are exploring the existing resources so fast that natural replenishment is nearly negligible. On the other hand, biofuels such as firewood and crop wastes, even though they could be significant carbon sources to the atmosphere, are renewable energy sources because they can be replenished quickly.[8] On a year over year basis, there is no net carbon input to the atmosphere from these sources. A wood-fire stove and fireplace as shown in **Figure 10.12** is actually a good example of renewable energy at work. In fact, this renewable energy has been utilized by humans since at least a million years ago for heating and cooking.

In this section, the focus will be on solar, wind, hydropower, and tidal power, or the "clean" renewable energy sources. To put renewable energy in context with other energy sources, **Figure 10.13** shows the consumption of different types of energy sources in 2018 and projected composition in 2050, when renewable energy as a whole will surpass fossil fuels (petroleum, coal, natural gas) and nuclear power and become the largest energy source. Compared to its humble start as shown in Figure 10.5, renewable energy is expected to become increasingly important with time.

Overall, the amount of renewable energy will see a drastic increase of nearly five-fold from 2010 to 2050, driven mostly by solar and wind energy in both absolute

Figure 10.12 Using firewood for heating and cooking.

---

[8] This assumes that the firewood is harvested in an environmentally sustainable manner, not through deforestation or over-exploitation.

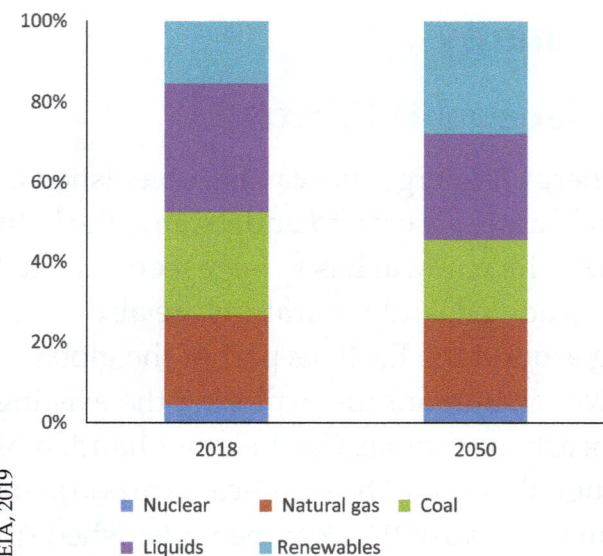

Figure 10.13 Global energy source projection. Renewable energy is projected to become the leading source of primary energy consumption by 2050.

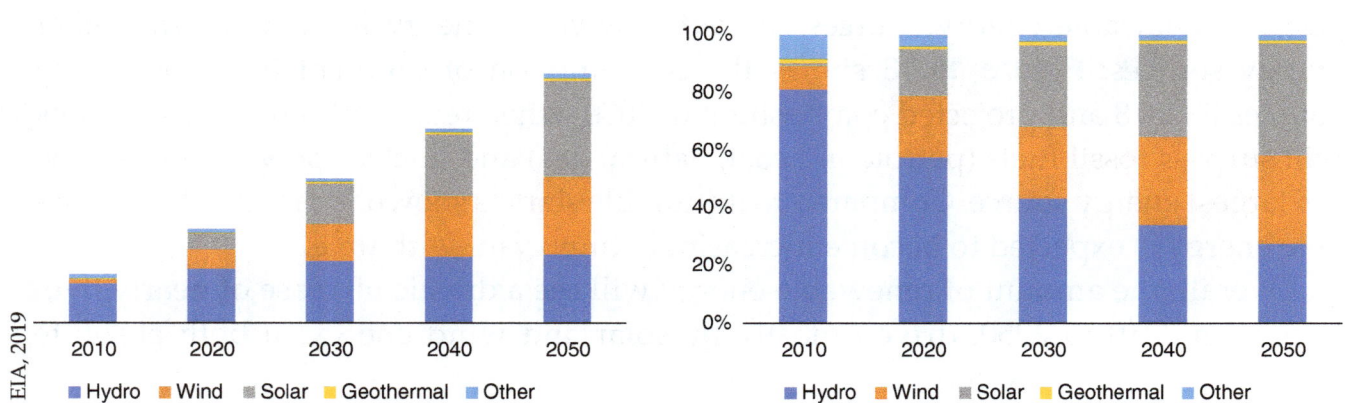

Figure 10.14 Global electricity production by renewable sources, 2010 to 2050 projection. Left: amount of production in TWh. Right: percentage composition.

Source: Created by Jian Peng

amounts as well as relative proportions compared to other renewable energy sources such as hydropower, geothermal, and other renewable sources, as shown in **Figure 10.14**. Compared to the data in Figure 10.13, this fivefold increase in the amount of renewable energy does not result in a proportionate increase in the weight of renewable energy in the overall energy portfolio because the other conventional energy sources will see increases as well, albeit smaller compared to that for renewable energy.

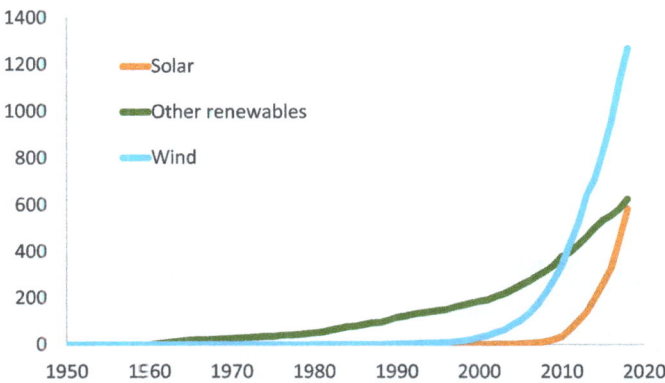

Figure 10.15 Renewable energy production 1950 to 2018 (unit: terawatt hour). This figure is based on the same data used for Figure 10.4 but is plotted with different scale for clarity.

Source: Plotted by Jian Peng with data source from https://ourworldindata.org/energy.

10.3.2 Solar Energy

The vast amount of solar energy hitting the Earth dwarfs the amount of primary energy used by the entire human civilization, which is currently about 5.39×10^{20} joule per year, or about only 0.1% of the solar energy (as mentioned in Section 10.1, the total solar energy reaching the Earth is 3.85×10^{24} joules per year). In other words, the sun gives the Earth more energy in 1 hour than mankind uses in one entire year.[9] However, only a tiny portion of the available solar energy is being used (**Figure 10.15**, which is a blow-up figure for the renewable energy sources in Figure 10.4). Similar to other renewable energy sources, solar energy has certain limiting factors that prevent it from becoming the dominant energy source over fossil fuels despite the vast gift from the sun. One factor is pricing, which remains uncompetitive compared to most conventional energy sources. However, the cost has been decreasing steadily and significantly. Since 1976, the solar photovoltaic (PV) module price dropped from about $100 per watt produced to less than $1 per watt, a two orders of magnitude drop. At the same time, solar PV production increased from less than 1 megawatt in 1976 to more than 100,000 megawatts in 2016. Concurrent with the price drop, solar energy enjoys fast growth, and the growth will continue for the next few decades. In 2016, the percentage of solar energy was only 1.3% of total energy. It is anticipated to grow to the largest energy source (16% for photovoltaics[10] and 11% for concentrated solar power[11]) by 2050. Therefore, there is great potential for solar energy to grow.

---

[9] Another shocking aspect is that 1 year worth of solar energy that the Earth receives will be more than the energy stored in crude oil, coal, natural gas, and mined uranium combined.

[10] This is so-called solar panel that converts solar energy to electricity.

[11] This system uses mirrors or lenses to concentrate solar power to a central location for conversion into power or electricity.

Figure 10.16 Solar panels.

Not all of the solar energy that reaches Earth can be harvested. It is estimated that only 1.557 to 49.837×10^{21} joules of solar energy could potentially be harvested, or less than 1% of the incoming solar energy. Still, this amount is many times more than the total world energy consumption. Development of affordable, inexhaustible, and clean solar energy technologies would have tremendous long-term benefits, including sustainability, less pollution, resource conservation, among others.

Solar energy can be harvested in many different ways. The most common ways include photovoltaic panels (**Figure 10.16**), which convert solar energy directly into electricity with up to 20% efficiency, and concentrator solar power generators (**Figure 10.17**), which concentrate solar power to generate steam or molten salt to power high-efficiency turbines (~50% efficiency). Solar energy could also be harvested by low-tech options including simple solar water heaters for household uses (**Figure 10.18**), cooking, or simply using sunlight to disinfect drinking water.

Solar energy is not free of environmental impacts. Solar panels have a lifespan of 25 to 30 years. Waste solar panels can be a significant source of solid waste. Moreover, both photovoltaic and concentrator solar plants require very large areas of open space to be economically viable. These plants, by shielding most of the sunlight and by disturbing the soil substrate during installation, could cause significant ecological impacts.

Solar energy also has other issues inherent with the timing and location of the power generated. While smaller rooftop solar panels can be readily integrated into the local electricity grid large-scale solar farms generate electricity in the middle of the day when the demand is low. There may also be significant seasonal variations in solar radiation and hence the output power of the solar energy plant changes seasonally as well. Solar panels cannot generate power at night or with cloud cover. The power also needs to be

Chapter 10   Energy and Sustainability   293

Figure 10.17   Concentrator solar power plants.

Figure 10.18   Solar water heater

transferred, sometimes over long distances, to where it is needed. Therefore, backup power generation systems (usually fossil-fuel-powered) and efficient energy storage systems are needed for solar power plants.

> ### Knowledge Box
>
> **Energy Storage:** The difference in timing between solar energy production and end-user demand requires ways to store the energy for use at a different time. There are many ways to do so, and some have seen commercial applications. The most straightforward way is to store the power in rechargeable batteries. The energy could also be stored in phase-changing materials such as paraffin wax, or materials with high heat capacity such as water or molten salt. Pumped-storage hydroelectricity could be used as well, where excess solar energy is used to pump water to a higher elevation to be converted back to electricity using a method similar to hydropower to put electricity back to the grid. There are also some innovative methods such as hydrogen or methanol fuel production, but more research and development are needed before they see commercial uses.

10.3.3 Wind Energy

Wind energy also originates from solar energy, which creates temperature and pressure differences in different areas on Earth, causing the air to flow from higher pressure areas to lower pressure areas. Wind power has traditionally been used in the past for simple mechanical applications such as milling and pumping (e.g., **Figure 10.19**). It is one of the cleanest types of energy with a very low footprint on the environment. Currently wind energy is mostly harvested by modern wind turbines, which is increasingly becoming part of the scenery in many parts of the world (e.g., **Figure 10.20**). As one of the fastest growing fields in renewable energy, wind energy has been outpacing solar energy by a rough 2:1 ratio over the last 10 years.

> ### Knowledge Box
>
> **Windmills in Netherland** (Figure 10.19)—The windmills in Netherland were built to pump water out of the land in southern Netherlands (mostly "Holland"), which is very low-lying and some parts are even below sea level or below high tidal levels. Canals were dug to drain the land, but the drained soil started settling, much like California's Bay-Delta area. Starting from the 1730s, many windmills were built to pump water from low-lying fields into the canals. Currently most of the work is done by pumps, and the windmills became historical landmarks.

Figure 10.19 Windmills in Kinderdijk, southern Netherlands.

Figure 10.20 Wind turbines in Gansu Wind Farm, Jiuquan, Gansu Province, China.

A wind farm is a group of wind turbines in the same area. Sometimes there may be hundreds of individual turbines in one single farm. Unlike a solar farm where the land is generally of very limited use, the land between the turbines may be used for agricultural or other purposes. A wind farm could be located offshore as well, but installation and maintenance could be challenging.

> ### Knowledge Box
>
> **Gansu Wind Farm**, located in Jiuquan, Gansu, China, is the largest wind farm in the world with several thousand turbines (Figure 10.20). Built on the Gobi Desert with very high winds that contribute to the genesis of the loess plateau, it has a design capacity of 20 gigawatts. However, due to its remote location, this monumental project has not been able to operate to its full capacity due to a lack of transmission and other infrastructure.

Most of the large wind turbines have a horizontal axis wind turbine with three blades (usually made of carbon or glass fibers) on top of a tall tubular tower. These turbines usually have a wind sensor to make sure that the turbine directly faces the wind. Some commercial wind turbines could have blades as long as 80 meters (**Figure 10.21**), generating up to 8 megawatts of electricity in one single turbine. Because larger turbines usually have smaller unit cost per power generated, wind turbines are getting larger with time.[12]

Figure 10.21 An oversized truck transporting a single wind turbine blade.

---

[12] A "mid-sized" 1.5-megawatt wind turbine is 80 meters high. The blades and rotor weigh 22 tons. The concrete base uses 26 tons of reinforced steel and more than 450 tons of concrete.

Figure 10.22 An idealized schematic where wind and solar energy solutions are enhanced by energy storage.

In a wind farm, individual turbines are interconnected with a medium voltage (often 34.5 kV) power collection system and communications network. In general, a distance of seven times the rotor diameter of the wind turbine is set between each turbine in a fully developed wind farm. To minimize transmission loss of electricity, at a substation, this medium-voltage electric current is increased in voltage with a transformer for connection to the high voltage electric power transmission system.

Similar to solar power, wind power may cause some environmental impact as well. It has been reported that the blades of wind turbines could kill birds. The sheer size of a typical wind turbine could have visual impacts as well. It is also a challenge to recycle the blades because they are made of thermosetting composite materials such as glass fiber or carbon fiber. In Germany, wind turbine blades are commercially recycled as an **alternative fuel** mix for a cement factory. The most common way of disposing of wind turbine blades is still landfilling. Fortunately, these blades are nontoxic.

Similar to solar power, wind power is also intermittent, and there could be significant time difference between generation and usage. Similarly, transmission and storage capacity (Figure 10.22) are important considerations in the planning of a wind power project.

10.3.4 Other Renewable Energy Sources

Tidal Energy

Tides are caused by gravitational interactions among the Earth, the moon, and the sun. Historically, there were tide mills in Europe and North America as far back as the Middle Ages or earlier, where incoming tidal water was trapped in tidal lagoons. When it went out, the tidal flow drove the water wheel to mill grains. Therefore, unlike wind and solar power, tidal power generation is predictable. Tidal power stations are more costly to be built, and can only be built where tidal patterns and geographic and geological conditions are favorable to harness the tidal power, for example, where a tidal lagoon can be built or created. With new turbine technology and other new designs, more tidal power energy could potentially be harnessed for beneficial uses in the future.

> ### Knowledge Box
>
> **Tidal Power:** The world's first large-scale tidal power plant was the Rance Tidal Power Station in France (Figure 10.23). It was built in 1966 and has an installed capacity of 240 megawatts. Currently the largest tidal power station in terms of output is the Sihwa Lake Tidal Power Station in South Korea (built in 2011) with an installed capacity of 254 megawatts.

Geothermal Energy

Geothermal energy is residual thermal energy generated during Earth's formation and stored in Earth's interior, and from the continuous decay of radionuclides. Hot springs (such as the famous Old Faithful in Yellowstone National Park, Wyoming, USA, see Figure 10.24) have been used by humans since prehistoric times, but it was not until recently when geothermal energy was used to produce electricity.

Figure 10.23 Rance Tidal Power Station in France.

Figure 10.24 Old Faithful Geyser in Yellowstone National Park, Wyoming, USA.

The enormity of Earth's thermal energy is manifested by large-scale events such as volcanic eruptions and plate tectonics, as well as smaller events or phenomena such as hot springs and geysers. However, only a tiny fraction of this energy could be harnessed and a geothermal power plant has to be built on a favorable location (i.e., nearly always at the tectonic plate boundaries or near a volcano) to be economically viable. Currently, only a small amount (11,700 megawatts) of geothermal power is produced, and more geothermal energy (28 gigawatts) is being used directly as heating. Currently, the United States, Philippines, Indonesia, Mexico, Italy, New Zealand, Iceland, and Japan are the top nations with significant geothermal power capacity. Of these nations, Iceland has the highest percentage (30%) of its electricity from geothermal power.

Geothermal power plants (**Figure 10.25**) usually use superheated steam with temperatures as high as 350°C to drive steam turbines for electricity generation. Since no fuel is used, environmental impacts from geothermal power plants are minimal. Even though geothermal energy is not strictly renewable in the conventional sense, it is generally classified as a renewable, or at least a clean, energy source because the available energy from Earth's interior is essentially unlimited.[13]

Other sources of renewable energy include wave energy, and energy from biofuels and many waste-to-energy sources as discussed in Chapter 9. Hydropower is a

---

[13] For example, Earth has a total potential heat energy of 10^{31} joules, which is 100 billion times the 2010 global energy production.

Figure 10.25 A geothermal power plant at Reykjanes Peninsula, Iceland.

renewable (renewable by the hydrological cycle driven by solar energy) and clean energy source (no carbon emission), but due to potentially significant ecological and other environmental impacts, hydropower has been increasingly scrutinized (see Chapter 6 for more details). With a global energy production of 4,198 terawatt hours in 2018 (Figures 10.5 and 10.14), hydropower generates more than twice as much energy as solar, wind, and other renewable energy sources combined. Due to environmental and ecological impacts, its portion in the global energy portfolio has been decreasing and will decrease further beyond 2050 (Figure 10.14).

10.4 Nuclear Power

10.4.1 Overview

Nuclear power relies on the nuclear energy released from nuclear reactions, either nuclear fusion (i.e., two or more atoms fuse to form one atom) or nuclear fission (i.e., one atom splits into two or more smaller atoms). Currently, all nuclear power plants use nuclear fission to produce power, with controlled nuclear fusion technologies still under development. On a per-weight or per-volume basis, nuclear fuel has extremely high energy density,[14] making it a favorable energy source in many applications and locations where conventional fossil fuels may not be feasible or economically viable.

---

[14] For example, one kilogram of gasoline can power a car for 20 kilometers. One kilogram of nuclear fuel, on the other hand, can power the same car from the Earth to the moon and back.

The process diagram of a typical nuclear power plant is shown in **Figure 10.26**. Basically, nuclear reaction takes place in the reactor where the nuclear fuel, usually uranium-235 (U235), is bombarded with neutrons. The U235 atoms then fission into two smaller atoms and more neutrons, which in turn drive more fission reactions of other U235 atoms, causing a chain reaction. These reactions release an enormous amount of energy in the form of heat. The chain reaction can be accelerated, maintained, slowed down, or shut down by controlling the number of neutrons using neutron-absorbing materials such as silver and boron. In Figure 10.26, control rods

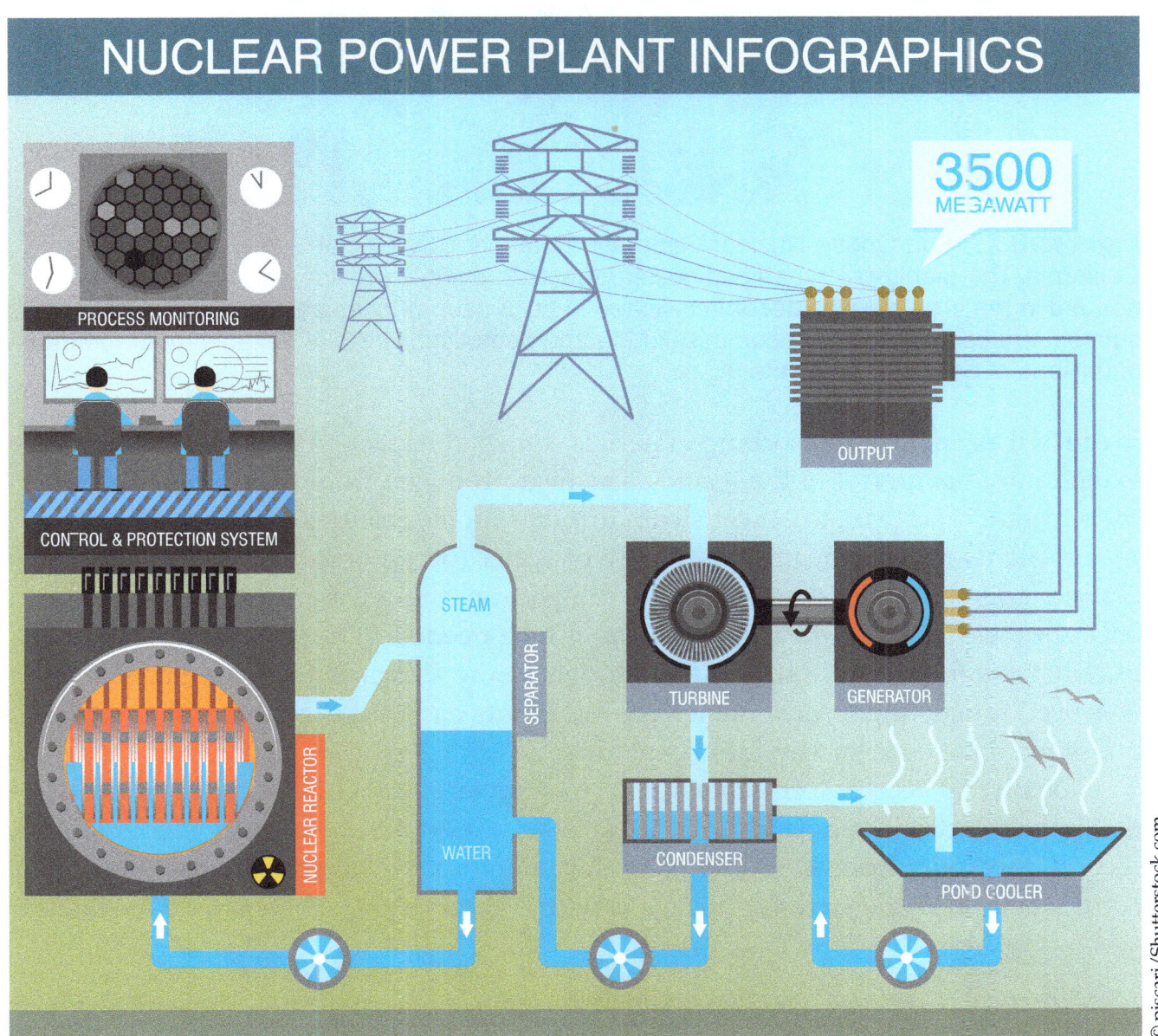

Figure 10.26 Process diagram of a typical nuclear power plant. While the large cooling towers could be the most prominent signature of nuclear power plants. If the plant is built near a large water body (e.g., an ocean), cooling could be achieved using ocean water without a cooling tower.

Figure 10.27 San Onofre Nuclear Generation Station (SONGS) at San Onofre Beach, California. The plant, which had two nuclear reactors, used to provide 40% of the energy needed in the area but was decommissioned in 2013 after mechanical failures and subsequent controversies. Decommissioning and demolition will take decades and hundreds of millions of dollars to complete.

made of these neutron-absorbing materials control the reaction rate by the depth they extend into the interior of the reactor. The heat produced from the nuclear reaction is used to generate steam to power steam turbines for electricity generation. For nuclear power plants that are near a large body of water (ocean or a large lake or river), as many plants do, cooling water may be drawn from these water sources and released back. In these cases, cooling towers may not be needed. **Figure 10.27** shows such a nuclear power plant in California, USA, where the seawater from the Pacific Ocean was used for cooling.[15] Most nuclear power plants have a design lifespan of 40 years, but many have been successful in extending it.

Globally, nuclear power generated 2,563 terawatt hours of electricity, or 10.3% of the global electricity demand, in 2018. It is an important low-carbon energy source, second only to hydropower in terms of the amount of power generated (Figure 10.5). As of March 2020, there are 441 nuclear reactors[16] in the world. There are also 54 nuclear

---

[15] The San Onofre Nuclear Generation Station shown in Figure 10.27 has been decommissioned since 2013 due to mechanical issues as well as public outcry since as early as 1977. The protests grew stronger after the Fukushima Daiichi Disaster in 2011.

[16] This number is larger than the number of nuclear power plants because some plants have more than one reactor. The best economic efficiency could be achieved by having 6 reactors in one single plant. Many plants have two reactors, as did the San Onofre Nuclear Generation Station in California, USA (Figure 10.27—the reactors sit inside the spherical domes).

power reactors under construction and 109 reactors planned, with a combined capacity of 61 and 119.5 GW, respectively. There are an additional 360 reactors being proposed with a combined capacity of 361 GW.

10.4.2 Advantages of Nuclear Power

Nuclear power uses nuclear fuels that have an average energy density about 10,000 times higher than conventional fossil fuel. Therefore, nuclear power requires much less transportation cost. For the same reason, nuclear fuel is long-lasting with one single fuel pallet lasting many years (usually 5 years) without changing. For this reason, many nuclear-powered aircraft carriers and submarines can cruise for years without refueling. The nuclear power generation process is relatively simple and operation costs are low. Nuclear power plants generate the same amount of energy with much smaller footprint and hence less ecological impact.

One of the most important advantages of nuclear power over fossil fuel power is that it generates no carbon emission. This is an important issue given that climate change and global warming receive increasing attention around the globe. Nuclear power plants also generate no air pollutants such as sulfur oxide, nitrous oxide, or fine particles. There are new technologies that allow high efficiency reactors; reprocessing and reuse spent fuel; seawater uranium mining; and nuclear fusion energy, etc. (see the Knowledge Box), that can potentially make nuclear energy a virtually inexhaustible energy source for the foreseeable future.

Knowledge Box

Future of Nuclear Energy: Nuclear power technology has improved significantly over the years. Among these improvements include spent fuel reprocessing technology that separate U235 from spent fuel to make new fuel. Uranium could be extracted from seawater, which contains nearly inexhaustible amount (4.5 billion tons) of uranium. Some new reactors ("breeder reactors") can use the more abundant U238 as fuel. If both breeder reactor technology is used and seawater can be used to source uranium, nuclear power can be considered a sustainable energy source that can last 5 billion years (McCarthy, 2006).

Contrary to common knowledge, nuclear power is actually much safer than most of the other energy options as shown in **Figure 10.28**. Fossil fuel power generation is not risk free from both operations as well as from air pollution and carbon dioxide emission. Based on the study by Kharecha and Hansen (2013), nuclear power actually

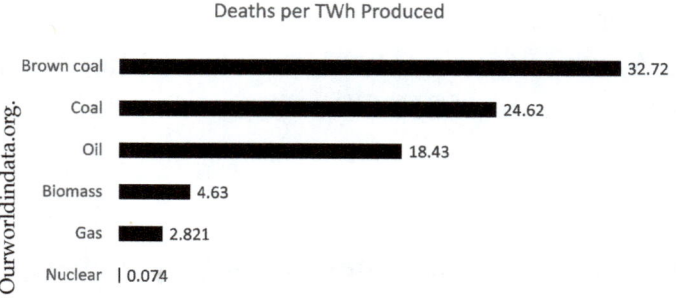

Figure 10.28 Death rates from energy production per terawatt-hour (TWH).

has prevented an average of 1.84 million air pollution-related deaths and 64 gigatons of carbon dioxide-equivalent greenhouse gas emissions that would have resulted from fossil fuel burning.

10.4.3 Challenges of Nuclear Power

Safety and associated public perception issues are perhaps the greatest challenges faced by the nuclear power industry. Despite being the much cleaner option to fossil fuel, as well as millions of prevented deaths, nuclear power always has a stigma as being dirty and dangerous. There are indeed quite a few high-profile incidents that have been blamed for this perception. Three Mile Island incident, where a partial meltdown of a nuclear reactor at the Three Mile Island Nuclear Generating Station, Pennsylvania, USA, took place in 1979 (Figure 10.29). This is by far the most significant nuclear power plant-related accident in U.S. history. The accident caused long-lasting, nationwide antinuclear sentiment until today, and prompted new regulations for the nuclear industry in the United States.

The Chernobyl Disaster took place on April 26, 1986 at the Chernobyl Nuclear Power Plant, near the City of Pripyat, Ukraine (formerly part of the Soviet Union). The disaster started when a safety test failed and caused an uncontrolled nuclear chain reaction. The extreme heat and ensuing high pressure steam caused steam explosion and reactor core fire releasing a large amount of airborne radioactive contamination that impacted large areas of the Soviet Union and Europe. The disaster killed 2 staff, hospitalized 134 others, of which 28 died soon after and 10 more died of radiation-induced cancer. The impact to other areas, while uncertain, is considered significant too.[17] It is the worst nuclear disaster in human history (Figure 10.30).

---

[17] Based on Peplow (2006), the estimated deaths as a result of Chernobyl Disaster ranges from 4,000 to 16,000, depending on models and assumptions.

Figure 10.29 Three Mile Island Nuclear Generating Station. After the accident, the power station used only one nuclear generating station on the right. The other station to the left has not been used since the 1979 accident.

Figure 10.30 Chernobyl nuclear reactors. After the disaster happened in 1986, the city and the surrounding area have been a ghost town.

Fukushima Daiichi nuclear disaster took place on March 11, 2011 when the Tohoku earthquake and tsunami flooded the plant and stopped the coolant system. The resultant three nuclear meltdowns and explosions released significant amounts of radioactive contamination to the Pacific Ocean (**Figure 10.31**). Due to the swift evacuation and dilution of the radioactive waste by the Pacific Ocean, there has been no confirmed health impact or fatalities. However, this disaster remains the second worst nuclear power plant disaster in the world after Chernobyl.

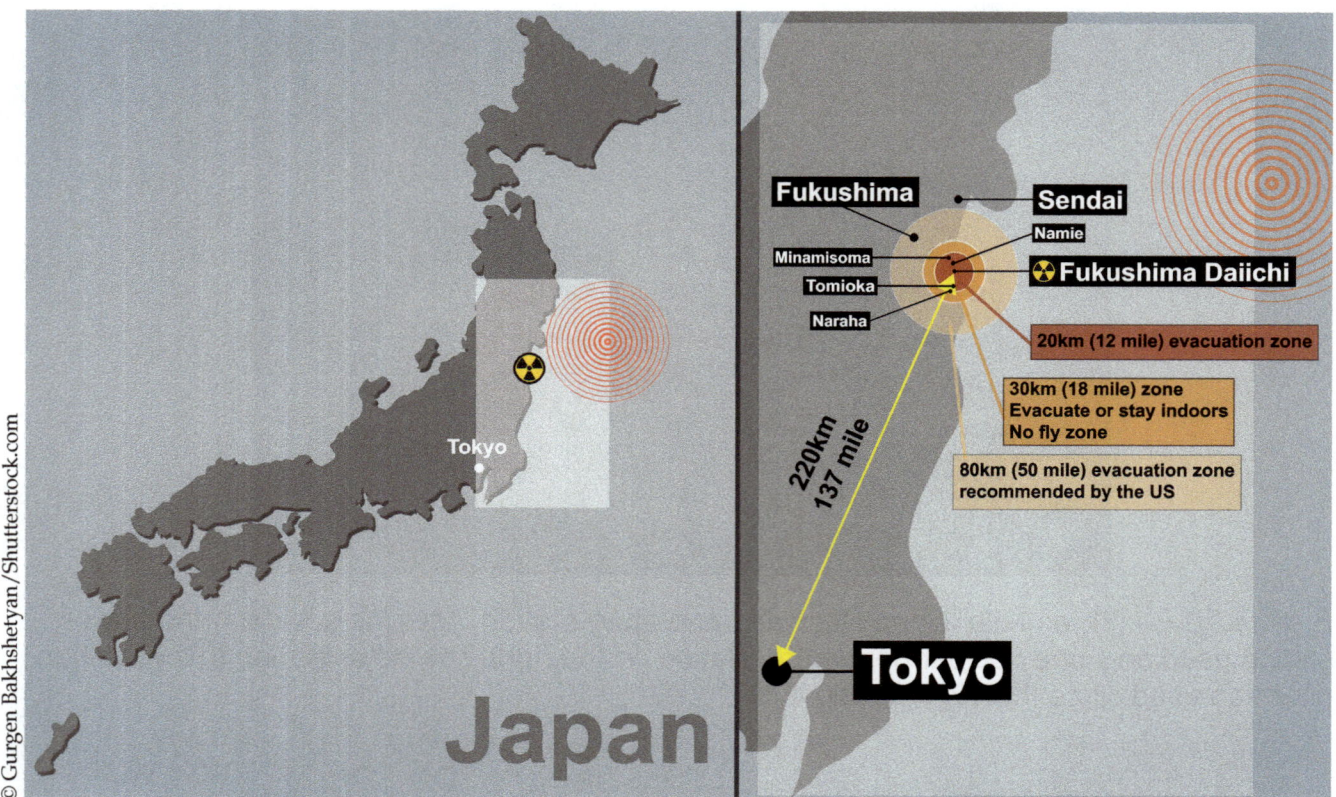

Figure 10.31 Fukushima disaster.

Nuclear power plants have other environmental issues, such as thermal pollution if they use ambient water for cooling instead of cooling towers. Intake points of cooling water could have ecological impacts such as entrainment or entrapment of small fish or fish larvae (similar to that for desalination plant as discussed in Chapter 6). At some places such as California, there are regulations specifically written to control thermal pollution. For example, in California's Ocean Plan, the cooling water should not cause a temperature increase of more than 5°F for most waters, and 2°F (1.1°C) for sensitive waters (State of California, 1975).

One of the key challenges that the nuclear power industry has to deal with is the nuclear waste issue. In the United States, an oversight agency, Nuclear Regulatory Commission (NRC[18]) was established to specifically oversee this important task. Radioactive waste is typically classified as either low-level, intermediate-level, or high-level, dependent, primarily, on its level of radioactivity. Most of the radioactivity associated with nuclear power remains contained in the fuel in which it was produced. This is why used fuel is classified as high-level radioactive waste. After a nuclear fuel unit is spent, it is removed and safely stored until a permanent disposal site becomes available. To put the amount of this high-level waste into perspective, all of the used nuclear

---

[18] NRC was established by U.S. Congress to regulate commercial nuclear power plants and other uses of nuclear materials.

Figure 10.32 A nuclear waste storage facility.

fuel ever produced would cover a football field to a depth of less than 10 yards. This is the same volume of ash that coal-powered plants produce in 1 hour.

Once removed from the reactor, the spent fuel will be stored in a storage pool for 3 to 10 years to allow short-lived radionuclides to decay and for the fuel to cool down sufficiently to be transported safely. Then it will be put into dry casks for storage and transportation to a consolidated interim storage facility (such as the one shown in Figure 10.32) before a permanent disposal facility is available. Such a disposal facility is the Yucca Mountain Nuclear Waste Repository, a proposed deep geological repository storage facility within Yucca Mountain in Nevada, USA. However, despite being approved by the U.S. Congress in 2002, the project encountered many difficulties and strong challenges from the public and politicians and was suspended in 2011. Since then, work (construction, permitting, and politicking) has continued off-and-on to this day.

Knowledge Box

WIPP: The Waste Isolation Pilot Plant (WIPP) in New Mexico is a remarkable place. 660 meters below ground inside an ancient salt formation formed some 250 million years ago, the site was identified by a group of scientists in the 1960s to be an ideal place to store defense-related nuclear waste because this 600-m-thick salt bed is free of water and geologically stable. The most important property of this site is that once the nuclear waste is deposited in the underground vault made entirely of salt, natural processes will take care of the spent fuel—the salt will gradually seal all the cracks and permanently close all openings by itself over time. It took 20 years from approval to completion, when the site received its first shipment of nuclear waste. For more information, watch this interesting video produced by the U.S. Department of Energy: https://www.youtube.com/watch?v=3bo36aKc8EY.

Figure 10.33 Washington DC, USA, May 6, 1979 California Governor Jerry Brown addresses the massive crowd of protestors that gathered at the west front steps of the United States Capitol during the "No Nukes" protest after the Three Mile Island nuclear accident on March 28, 1979.

The decommissioning of an old nuclear power plant could be costly and slow as well. For example, the San Onofre Nuclear Generating Station in California, USA started decommissioning in 2013. 7 years later, the dismantling work has not started. The dismantling work itself will last 8 to 10 years by a team of 600 workers[19] at a cost of between $280 and $612 million. Much of the cost is due to the need to carefully handle the low radioactive waste and find an appropriate site for safe disposal. The cost is so significant that most of the nuclear power plants have to prepay a deposit, purchase sufficient insurance, and set aside a reserve fund during its operating period in order to be able to pay such a large sum for decommissioning.

For the above complicated reasons, the general public has a strong opposition to nuclear power despite its numerous advantages. In the United States, the Three Mile Island incident is a watershed moment when the public's sentiment turned negative. Numerous protests took place (e.g., **Figure 10.33**), and few nuclear power plants have been built afterwards.

10.5 Energy and Sustainability

Increasing global concern over climate change, global warming, and general sustainability issues have put the energy industry in the spotlight. With an ever-increasing population (expected to reach 8.5 billion in 2030) and steadily increasing per capita energy consumption (Figure 10.1), the pressure is increasing for the energy industry to produce larger amounts but also safer and cleaner energy to drive society forward

---

[19] For more information, see: https://www.songscommunity.com/about-decommissioning/decommissioning-san-onofre-nuclear-generating-station

without jeopardizing the very existence of humankind. In the near future, heavy reliance on fossil fuels for energy production will not change. As shown in Figure 10.13, the proportion of renewable energy in the overall energy portfolio will increase and become the largest portion. This is quite encouraging. However, there are also significant issues. To illustrate this point, Dr. Daniel Nocera of Harvard University proposed a formula as shown below:

Global energy used $E = N \times (GDP/N) \times (E/GDP)$,

where

N is the global population,

E is the amount of energy,

GDP is the gross domestic product,

GDP/N is the per capita GDP that represents the average wealth of a person, and

E/GDP is the amount of energy required to generate a unit amount of GDP, hence it illustrates the degree of energy conservation.

The above equation illustrates that if the global population increases without proportionate GDP increase, the average standard of living (as indicated by GDP/N) will drop). If GDP increases without more population or more energy, people's standard of living will improve, but it requires more energy unless we can improve energy efficiency (E/GDP). Currently the world is consuming nearly 20 trillion watts in 2020, and the amount will nearly double by 2050. However, of the 7 billion people in the world, 1.6 billion of them do not use any energy and 3 billion of them use little energy. With an additional 3 billion population by the end of the century, most of them will be born in poor countries and will consume little energy. Therefore 6 billion more people will need energy to get out of poverty and prosper to make our planet truly sustainable. This issue will be revisited in more detail Chapter 12. Take for example the transportation sector, which is one of the most significant consumers of energy (the other two major energy consuming sectors are building and power generation), by 2050, transportationfuels will still be overwhelmingly dominated by fossil fuels, with electricity comprising only ~5% of the transportation energy source.

Of this 5% electricity, most of it will still be produced by fossil fuels (**Figure 10.34**). If we look closer into the transportation industry by focusing on the most important and fastest growing sector, light-duty vehicles, the situation is encouraging but with significant room for improvement (**Figure 10.35**). By 2050, the combined use of "clean energy" such as electricity (regardless of how energy is produced), plug-in hybrid, and hydrogen fuel cell) would still be much less than that for diesel, as shown in Figure 10.34. The only encouraging factor is that the increase of renewable energy appears to accelerate with time.

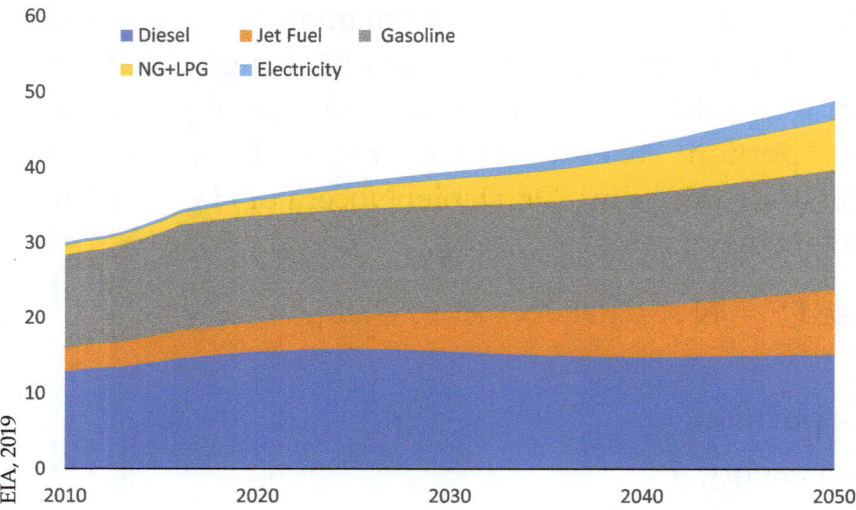

Figure 10.34 The projection of global share of different transportation fuels through 2050. Unit: Quadrillion watt hour.

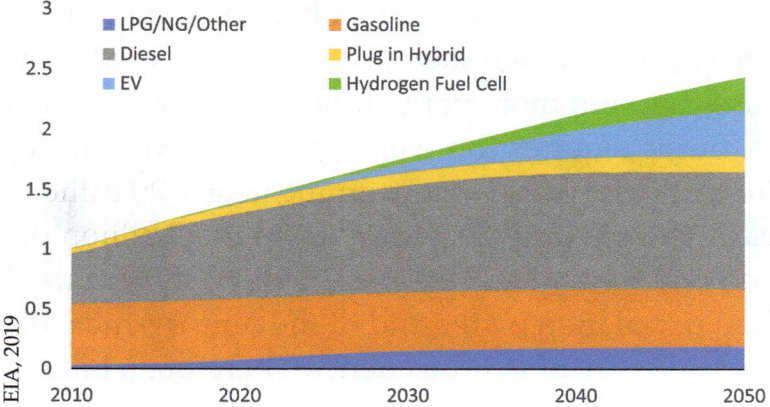

Figure 10.35 Number of light-duty vehicles classified by fuel sources. Unit: billion vehicles.

Therefore, energy and sustainability will still be at odds for decades to come and will be difficult to reconcile. Increased fossil fuel consumption has been factored in the projection of other environmental factors, such as sea level rise and global warming/global climate change. For example, it is currently estimated that sea level will rise by 30 cm by 2050 and 60 cm by 2100 (Bamber et al., 2019). Other researchers came to much higher numbers. These issues will be discussed in Chapter 11. The solution to this global issue will be discussed in Chapter 12.

Further Reading

U.S. Energy Information Administration (EIA), 2019, International Energy Outlook 2019 with Projections to 2050, September 24, 2019. Weblink: https://www.eia.gov/outlooks/ieo/pdf/ieo2019.pdf.

Bamber, Jonathan L., Michael Oppenheimer, Robert E. Kopp, Willy P. Aspinall, and Roger M. Cooke. "Ice Sheet Contributions to Future Sea-Level Rise From Structured Expert Judgment." *Proceedings of the National Academy of Sciences*, 116, no. 23 (May 2019): 11195–200.

McCarthy John. "Facts from Cohen and Others." *Progress and its Sustainability*. Stanford. Archived from the original on 2007-04-10. Retrieved 2006-11-09. Citing: Cohen.

Bernard L. "Breeder Reactors: A Renewable Energy Source." *American Journal of Physics*, 51, no. 1 (January 1983): 75–76.

Energy and the challenge of sustainability (PDF). *United Nations Development Programme and World Energy*.

Solar Energy Perspectives: Executive Summary (PDF). International Energy Agency. 2011. Archived from the original (PDF) on January 13, 2012.

Turcotte, D. L. and G. Schubert. Chapter 4. In *Geodynamics*, 2nd ed. Cambridge, England, UK: Cambridge University Press. pp. 136–137. ISBN 978-0-521-66624-4 (Geothermal energy source: 20% residual; 80% from radionuclide decay).

Peplow, M. "Special Report: Counting the Dead". *Nature*. 440, no. 7087 (1 April 2006): 982–83.

Hore-Lacy, I. Future Energy Demand and Supply. In: *Nuclear Energy in the 21st Century*, 2nd ed., London, UK: WNUP, 2011, ch.1, sec. 6, pp. 9.

Pushker A. Kharecha and James E. Hansen. "Prevented Mortality and Greenhouse Gas Emissions from Historical and Projected Nuclear Power." *Environmental Science and Technology*, 47 (2013): 4889–95.

State of California, 1975. Water Quality Control Plan for Control of Temperature in the Coastal and Interstate Waters and Enclosed Bays and Estuaries of California (California Thermal Plan). Weblink: https://www.waterboards.ca.gov/water_issues/programs/ocean/docs/wqplans/thermpln.pdf. Accessed: March 25, 2020.

Web Resources

World Bank Data for Energy Consumption: https://data.worldbank.org/indicator/EG.USE.COMM.FO.ZS?end=2015&start=1960

U.S. Energy Information Administration: https://www.eia.gov/todayinenergy/detail.php?id=41433

Our World in Data—Global energy portfolio: https://ourworldindata.org/energy

Global Coal Consumption 2019: https://www.iea.org/reports/coal-2019

The Waste Isolation Pilot Project: https://www.youtube.com/watch?v=3bo36aKc8EY

Pew Research Center world population projection: https://www.pewresearch.org/fact-tank/2019/06/17/worlds-population-is-projected-to-nearly-stop-growing-by-the-end-of-the-century/

The World Nuclear Power Reactors and Uranium Requirements, https://www.world-nuclear.org/information-library/facts-and-figures/world-nuclear-power-reactors-and-uranium-requireme.aspx. Access date: March 25, 2020

Chernobyl Disaster: https://en.wikipedia.org/wiki/Chernobyl_disaster#cite_note-16

Information about nuclear waste by World-Nuclear.org: https://www.world-nuclear.org/information-library/nuclear-fuel-cycle/nuclear-wastes/radioactive-waste-management.aspx

United States Nuclear Regulatory Commission, Backgrounder on Radioactive Waste: https://www.nrc.gov/reading-rm/doc-collections/fact-sheets/radwaste.html

Decommissioning San Onofre Nuclear Generating Station: https://www.songscommunity.com/about-decommissioning/decommissioning-san-onofre-nuclear-generating-station

Questions and Exercises

1. Why is storage capability a major issue for solar and wind energy?
2. Other than using huge battery packs for energy storage, design (conceptually) a few other mechanisms to store excess energy (e.g., for solar and wind energy) for later use.
3. Do you think we should use more nuclear energy to combat climate change? Why or why not?
4. If humans were to colonize Mars, what should be the main source(s) of energy? State your reasons.
5. Examine your electricity bill, or check out the local utility company and learn the different sources of its energy portfolio.
6. There are significant government subsidies to renewable energy (such as solar and wind, etc.). Do you think these subsidies should increase or decrease? Why?
7. Visit the website of Carbon Tax Center (www.carbontax.org) and learn about the concept of carbon tax. How could this concept and practice benefit the energy sustainability?

Chapter 11

Climate Change, Sea Level Rise, and Ocean Acidification

	Learning Outcomes	314
	Key Concepts	314
11.1	Climate Change	314
	11.1.1 Greta	314
	11.1.2 Global Climate System	315
	11.1.3 Earth's Energy Budget	320
	11.1.4 Greenhouse Gases and Global Warming	320
	11.1.5 Climate Variability and ENSO	327
11.2	Sea Level Rise	332
	11.2.1 The Concept	332
	11.2.2 Causes of Sea Level Rise	332
	11.2.3 Sea Level Rise in Different Scenarios	333
11.3	Ocean Acidification	335
	11.3.1 What is Ocean Acidification?	335
	11.3.2 The Mechanism of Ocean Acidification	336
	11.3.3 The Impact of Ocean Acidification	336
11.4	Mitigation Measures	338
	11.4.1 The Global Climate Feedback Loop	338
	11.4.2 The Carbon Problem	338
	11.4.3 Potential Mitigation Measures	339
	Further Reading	340
	Web Resources	341
	Questions and Exercises	341

Learning Outcomes

- Understanding of the concept of global climate change
- Knowledge of the nature and cause of ocean acidification
- Understanding of the mechanisms of global climate change
- Understanding of Earth's heat balance and greenhouse effect
- Ability to distinguish the difference between global warming and climate change
- Knowledge of common weather-related natural disasters and their linkage to climate change and global warming
- Understanding of how the interactions between the atmosphere and the oceans determined the global climate system

Key Concepts

Climate change; global warming; sea level rise; ocean acidification; greenhouse gases; climate feedback loop; El Niño; La Niña; El Niño-Southern Oscillation (ENSO); ocean conveyor belt; Paris Protocol

11.1 Climate Change

11.1.1 Greta

Greta Thunberg is a 16-year-old Swedish girl who in 2019 to 2020 became an international icon for promoting awareness of **climate change** by bringing her message *"Skolstrejk för klimatet" (School Strike for Climate)* to millions of people around the globe (**Figure 11.1**). She led school strikes in climate protest, hosted speeches, and attended the United Nations Climate Summit in New York on September 23, 2019. She directly and sharply criticized global leaders on their inaction on the climate change issue. Below are her words at the Summit:

> *"This is all wrong. I shouldn't be up here. I should be back in school on the other side of the ocean. Yet you all come to us young people for hope? How dare you! You have stolen my dreams and my childhood with your empty words. And yet I'm one of the lucky ones. People are suffering. People are dying. Entire ecosystems are collapsing. We are in the beginning of a mass extinction. And all you can talk about is money and fairy tales of eternal economic growth. How dare you!"*

Such a personality or chain of events could have been unimaginable if Greta had not had millions of people, young or old, powerful or weak, share her view and

Figure 11.1 Greta Thunberg joins more than 10,000 school students and activists in a march on the first anniversary of Fridays for Future school climate strike at Lausanne, Switzerland on January 17, 2020.

support her. Greta's story truly reflected the significance of the issue of climate change for most of the people around the world.

Knowledge Box

Climate Emergency Declaration: In December 2016, a climate emergency was declared and since then, over 1,400 government agencies in 28 countries have made such a declaration as of February 2020. This action is somewhat symbolic because many nations do not yet have legal framework to take meaningful actions upon the declaration. However, the message was clear. Humanity is facing a climate crisis like never before. Rhetorically, such declarations acknowledge that climate change exists and that the measures taken up to this point are not enough. Governments should set priorities and take measures to stop human-caused climate change and global warming.

11.1.2 Global Climate System

A basic understanding of the global climate system is needed before the issue of climate change can be discussed. The global climate system consists of atmosphere, hydrosphere, and biosphere with the sun as the main driver and provider of the energy for nearly all processes of the climate on Earth. Earth's lithosphere plays a part in the climate system under certain circumstances (see the Knowledge Box). In this book, the focus will be on the interactions between atmosphere and the ocean, two most important components of the climate system. Lithosphere, biosphere and human impacts will be discussed in the context of their relationship with the atmosphere and the ocean.

> ### Knowledge Box
>
> **Lithosphere and Climate System:** Lithosphere is the solid Earth such as continents, mountains, and ocean floors that provide a 'stage' where the climate system works. In the time scale in which current climate and climate change is concerned, this construct will not change significantly or fast enough. The exception is sudden, large-scale events such as volcanic eruption and meteorite strike. There is strong evidence that both types of events have impacted the global climate system and other processes significantly in the past. Over time, plate tectonics and associated changes in ocean-land distributions, orogeny,[1] and weathering will have a significant impact on long-term climate patterns.

A climate system has physical, chemical, and biological components. In the physical component, land-ocean distribution affects the differential heating and cooling of the surface of the Earth and affects wind pattern, water vapor generation and its distribution. Mountain ranges block air flow and create regional climate; regions with different latitudes receive different amounts of solar radiation. These physical components drive atmospheric circulation globally in a complex pattern, as shown in **Figure 11.2**. The sun heats up the air near the equator the most (because the equator has the smallest angle toward the sun) and causes the strongest uplift of air mass[2] to higher elevation and goes to higher latitude toward both northern and southern hemispheres. When it cools down enough, it sinks to a lower elevation and near the ground surface. The sinking air then splits into two directions, one heading north, one heading south. Such a process goes on from the equator to polar regions and forms distinct three-dimensional air circulation cells called the Hadley cell, Ferrel cell, and polar cell, respectively. The patterns of wind in the lower elevation are affected by the Coriolis Force[3] and form trade winds, westlies, and polar easterlies.

---

[1] The mountain-building process.

[2] When air is heated, it expands and becomes lighter and rises up.

[3] The Coriolis Force is not a force, it is an artifact due to the fact that the Earth rotates and things on Earth appear to veer off a straight line due to the change in reference frame. For more information, consult the web reference at the end of this chapter.

GLOBAL ATMOSPHERIC CIRCULATION

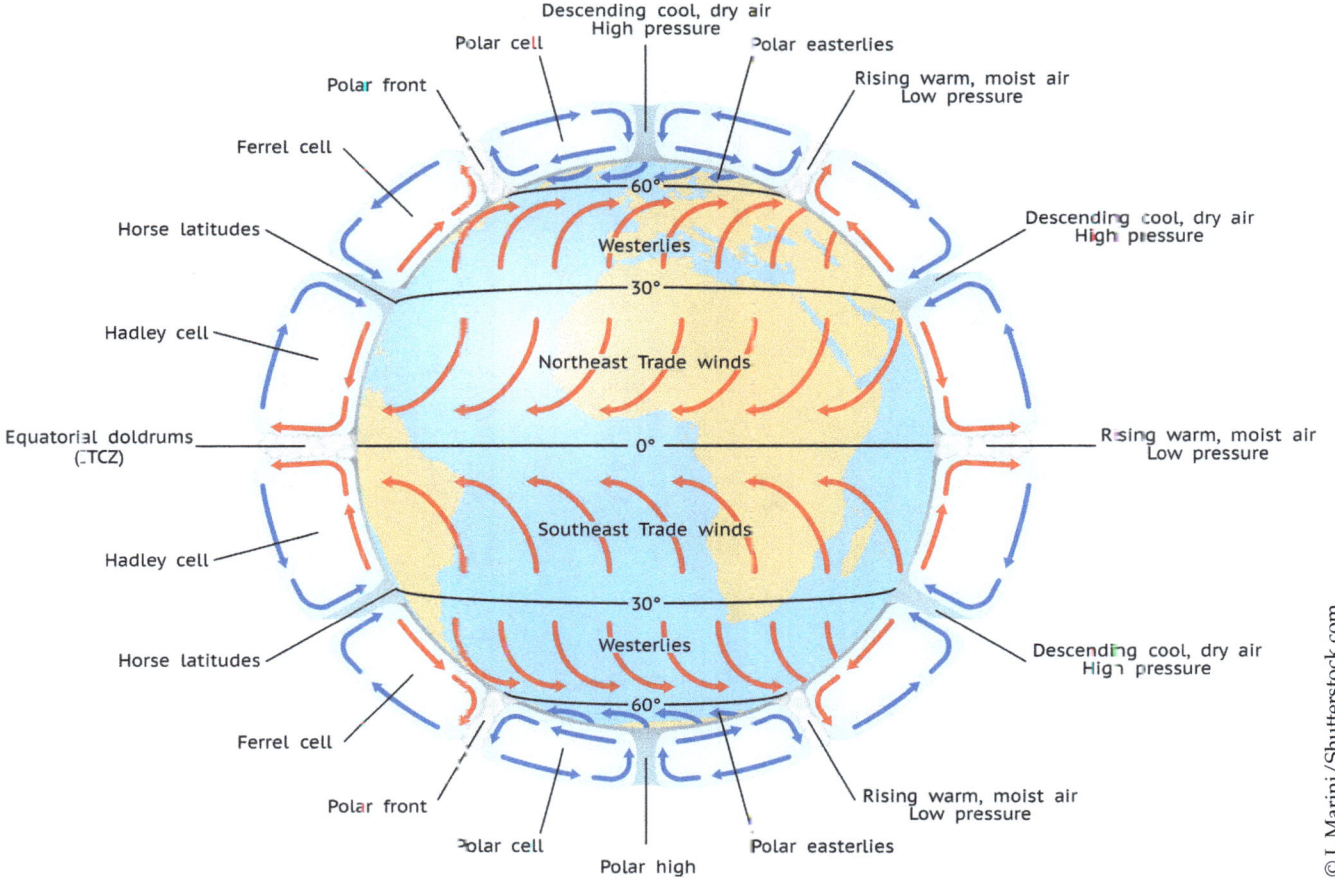

Figure 11.2 A simplified representation of global atmospheric circulation, a critical component of the Earth's climate system.

Atmospheric circulation in turn drives global ocean circulation, as shown in **Figure 11.3**. Essentially, the strong trade winds driven by the Hadley cell and Coriolis force from mid-latitudes of both hemispheres will converge near the equator, driving strong westerly surface currents along the tropical Pacific Ocean, Atlantic Ocean, and Indian Ocean. These currents then split at the western side of the oceans to higher latitudes, creating five large ocean gyres that are clockwise in the northern hemisphere and counterclockwise in southern hemisphere (Figure 11.3). Over time, the ocean and air circulation patterns reach a dynamic equilibrium and they feed into each other. The best example is at the equatorial Pacific Ocean. Under normal conditions, strong trade winds will blow westward, driving the surface current from the west coast of South America toward western Pacific Ocean near Indonesia. This causes the upwelling of cold, deep ocean water in the eastern equatorial Pacific Ocean, and warm water will

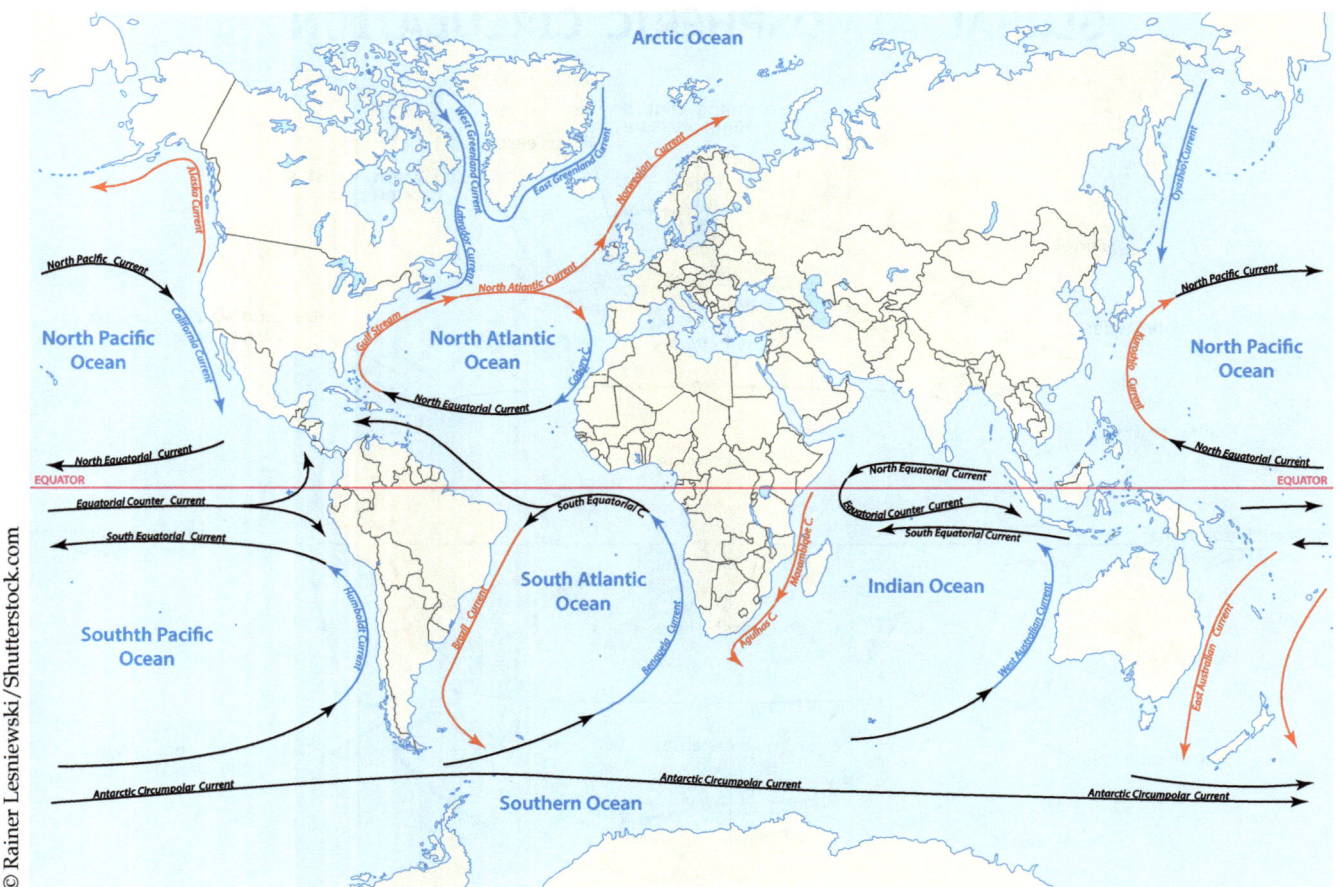

Figure 11.3 Global ocean circulation. Compared with Figure 11.2, it can be seen that ocean circulation is driving by atmospheric circulation with regular patterns. Oceanic gyres in the northern hemisphere are all clockwise and those in the southern hemispheres are all counterclockwise because they are driven by trade wind near the equator and westerlies at higher latitude. Combined with the Coriolis effect, these gyres form regular circulation patterns.

"pile up"[4] in the western equatorial Pacific Ocean. In turn, the cold sea surface temperature in the east will form a higher-pressure zone than that in the west, sustaining the easterly (i.e. blowing from east to west) trade winds and continued ocean–atmospheric circulation.

The above processes have been simulated by numerous global atmosphere-ocean circulation models (IPCC, 1994). A climate model is a representation of the physical, chemical, and sometimes biological processes that affect the climate system. Currently, there are dozens of models developed and used by many different institutions around the world. Most of the leading models use physics laws to drive the atmospheric and ocean circulation based on conservation of mass and momentum, while chemical and

---

[4] This "piling-up" is not an exaggeration. In fact, under normal condition, the sea surface elevation at the western equatorial Pacific Ocean can be more than one meter higher than that in the eastern equatorial Pacific Ocean.

other processes are modeled with parameterized equations and empirical models. These models are forced externally by solar energy and initial conditions such as the observed wind pattern, ocean circulation, and temperatures in the air and the ocean. Based on a complicated observation network around the globe, these models are periodically calibrated, verified, and improved. Many such models have been improved to such a degree that they can be used to recreate climatic events in the past and/or predict future scenarios. Some of these models are used to provide a global framework for local models to conduct finer modeling, including local and regional weather forecasting. A schematic of such a model is shown in **Figure 11.4**. In this schematic, which shows the atmospheric model cells in more detail, the atmosphere is separated vertically and horizontally into many cells of equal or differential sizes (usually the closer to the land/ocean surface, the smaller the sizes are in order to simulate in higher resolution). With solar radiation as the driving force, and oceanic conditions given by a corresponding global ocean model, the model will take time to "spin up," or reach a dynamic equilibrium. The model can be calibrated by observations made on the ground or in the air, or from space by satellites. Due to a large number of complicated processes and physical equations that govern the movement of air masses, such a model requires extremely fast computers to run. In fact, most of the supercomputers around the world are built specifically to run climate models.

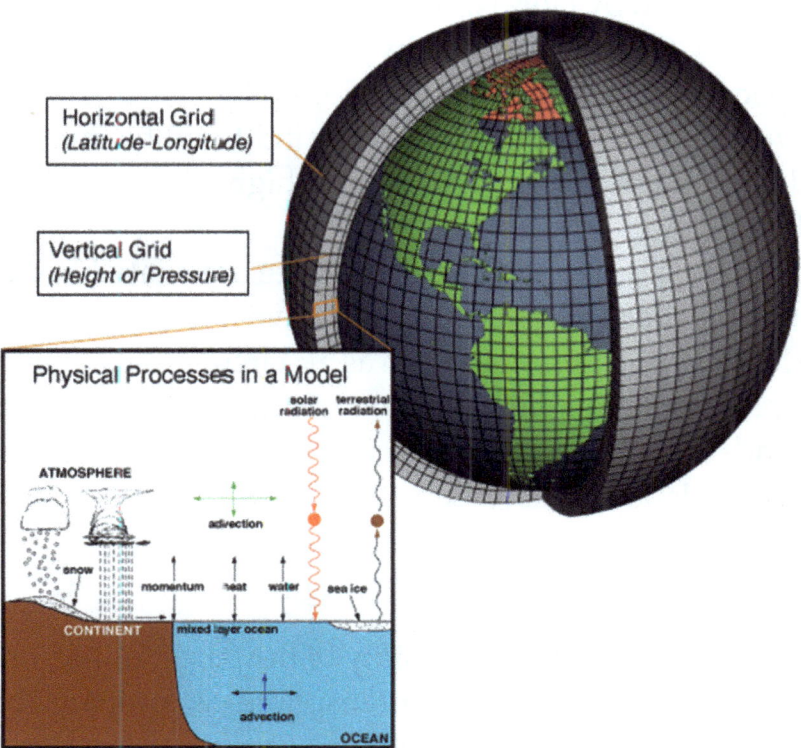

Figure 11.4 The schematic of a climate model (Image source: NOAA. https://www.climate.gov/file/atmosphericmodelschematicpng).

A fully coupled ocean-atmospheric model will require even more computing resources. Therefore, global climate models are usually run on relatively coarse grids, each cell may be hundreds of kilometers in size. These models will be regularly calibrated by observations, and provided to users around the world. The users can then use the modeling results as boundary condition or initial condition to run models on regional levels with higher spatial and temporal resolutions. For example, the weather forecast you would see on TV or on your smartphone is most likely based on a local or regional climate model that is based on and nested in a global or larger scale model as boundary conditions. The local/regional model will then run for up to 10 days, and the modeling results will be provided as the weather forecast you can see. Depending on how often the larger model(s) provide updated boundary conditions, the local/regional model will be updated periodically to provide increasingly accurate forecasts.

11.1.3 Earth's Energy Budget

Since Earth's climate system is ultimately driven by solar energy, it is important to look into Earth's energy balance to understand the climate. Incoming solar energy is in the form of radiation that has a broad spectrum, including visible light. Infrared and ultraviolet are not visible but carry a significant amount of solar energy too. Once it reaches Earth, some will be reflected by clouds, some absorbed by the atmosphere and clouds. About half of the energy will be absorbed by the Earth's surface, with a small portion reflected back to space. The portion of light reflected by clouds and by Earth's surface is called albedo. The higher the albedo, the more light will be reflected back to space. Clouds, snow and other light-colored objects have high albedo, and oceans, forest, and soil have lower albedo.

After the atmosphere, clouds, and Earth's surface are heated by sunlight, they will dissipate most of the heat back to space via long-wave radiation such as infrared. Some of the heat will be lost to drive processes such as air circulation and evaporation. Under normal conditions, all of the incoming and outgoing energy for the Earth as a whole will be balanced. Under this condition, Earth will not get warmer or colder. **Figure 11.5** shows Earth's energy balance as discussed.

11.1.4 Greenhouse Gases and Global Warming

As discussed in Chapter 1 about the habitability of the Earth for life, one of the key beneficial features for the Earth is that it has an atmosphere (as compared to other terrestrial planets such as Mars and Venus), and there are **greenhouse gases** (carbon dioxide, water, methane, etc.) with high enough concentrations to trap heat from the sun from escaping at night. During Earth's history, the concentrations of carbon dioxide have

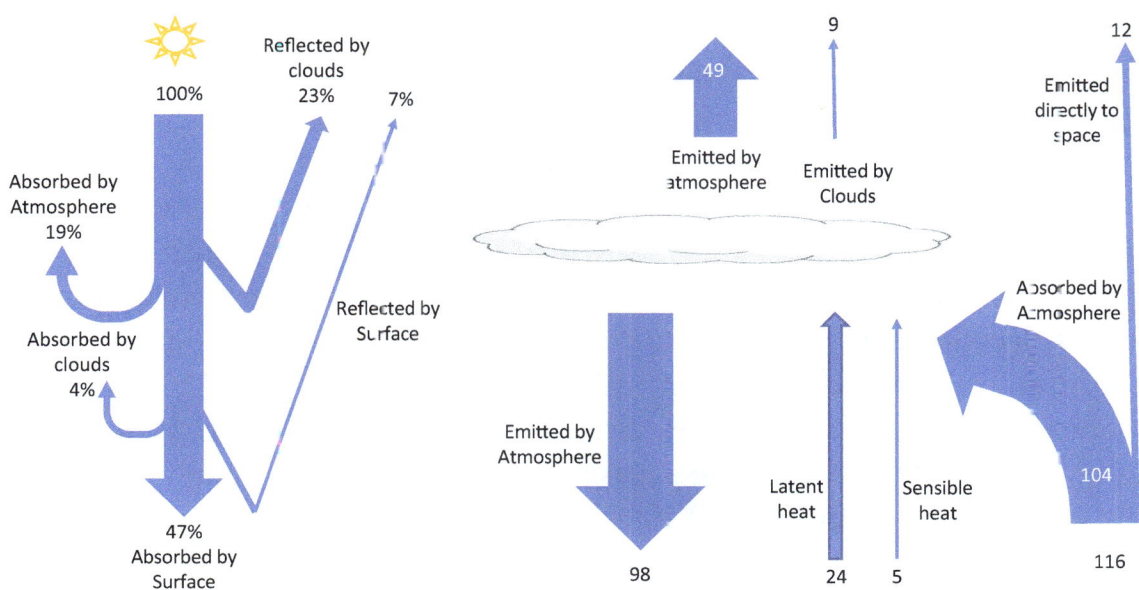

Figure 11.5 Earth's energy budget (source: weather.gov). All numbers are in percentages and solar radiation is set to be 100% as a reference (the actual number is 341 watt/m²). Due to the greenhouse effect, the energy transfer between the ground surface and clouds/atmosphere can exceed the incoming solar energy.

Source: Created by Jian Peng

varied significantly over time. In fact, the atmosphere on early Earth did not have oxygen until 3 to 4 billion years ago. As shown in Figure 11.5, however, there is a fine balance for Earth's energy budget. If there are insufficient greenhouse gases, Earth will cool. When there is too much, there will be **global warming**. Figure 11.6 shows

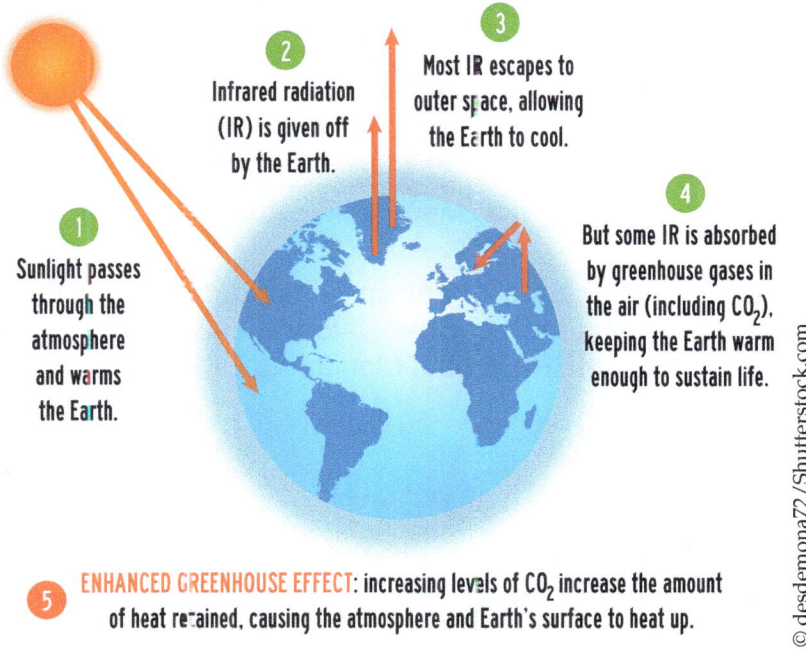

Figure 11.6 The greenhouse effect.

such a scenario. As a greenhouse gas, excess atmospheric carbon dioxide will trap the infrared heat dissipated by Earth back to space. If the concentration of carbon dioxide is too high for too long and other conditions being equal (e.g., albedo; cloud cover; etc.), the average temperature around the globe will increase.

As discussed in detail in Chapter 10, since the 1970s, global consumption of fossil fuel has increased significantly (see Figures 10.1 and 10.5). This can be attributable to increasing population, but only to some extent because per capita consumption of fossil fuels has also been ramping up (Figure 10.1), causing steady and significant increase in atmospheric carbon dioxide concentration (**Figure 11.7**). As a result, during the same period, the global temperature has increased by about 1 °C (**Figure 11.8**). To quantify the degree in which the greenhouse gases trap the heat, climate scientists designed the Annual Greenhouse Gas Index (AGGI) to be a measure of the capacity of Earth's atmosphere to trap heat as a result of the presence of long-lived greenhouse gases such as carbon dioxide, methane, chlorofluorocarbons (CFCs), and other halogenated gases,[5] as shown in **Figure 11.9**.

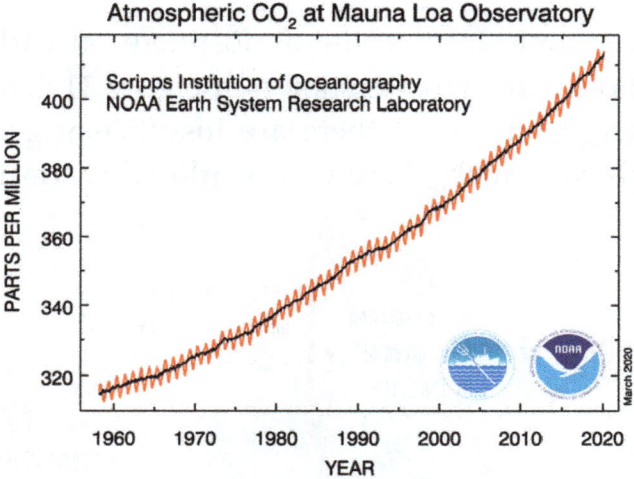

Figure 11.7 The global trend of atmospheric carbon dioxide concentration (parts per million) measured at NOAA's Mauna Loa Observatory in Hawai'i, USA, from 1958 to 2020. Data/Image provided by NOAA ESRL Global Monitoring Division, Boulder, Colorado, USA (http://esrl.noaa.gov/gmd/).

---

[5] Note that there are many other greenhouse gases, and many are more potent than carbon dioxide (e.g., methane is about 25 times more potent as a greenhouse gas than carbon dioxide). Water vapor actually is a greenhouse gas as well, but it is not a "long-lived greenhouse gas" because it is constantly cycled and varies much both temporally and spatially.

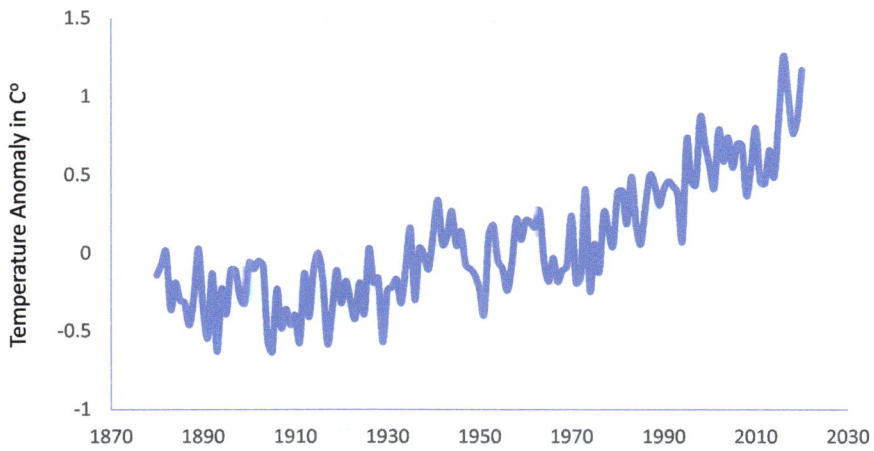

Figure 11.8 Global temperature anomaly from 1880 to 2020. Data/Image provided by NOAA ESRL Global Monitoring Division, Boulder, Colorado, USA (http://esrl.noaa.gov.gmd/).

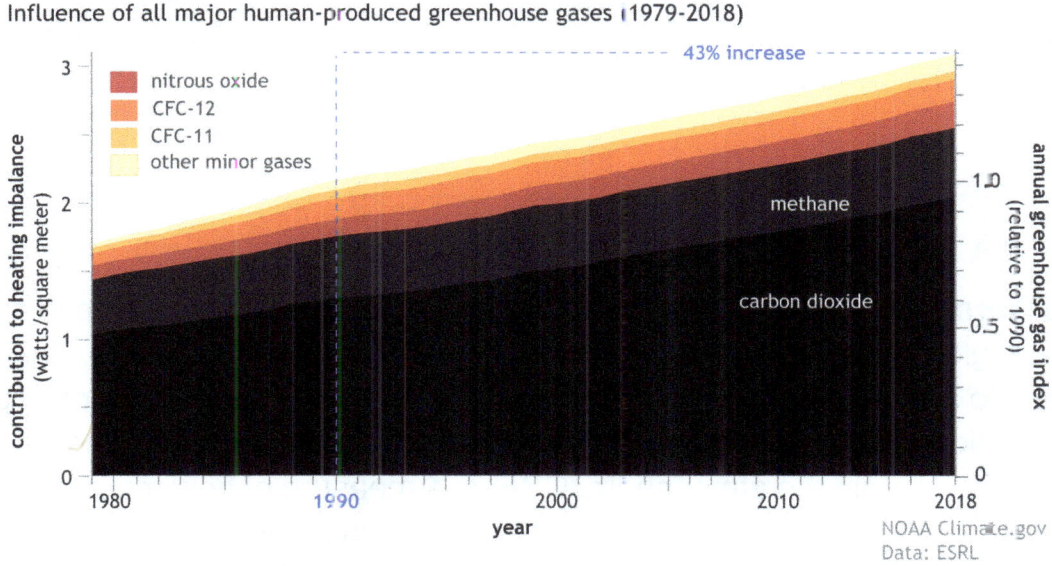

Figure 11.9 Influence of all major human-produced greenhouse gases. The year 1990 is used as the reference point. The curve shows that the Greenhouse Gas Index has increased 46% since 1990. Data/Image provided by NOAA ESRL Global Monitoring Division, Boulder, Colorado, USA (http://esrl.noaa.gov.gmd/).

To better understand greenhouse effects, it is also important to understand the potential, or power, of different gases under similar conditions. **Table 11.1** lists several important greenhouse gases and compared their global warming potential (GWP) in 100 years compared to carbon dioxide, which is used as a reference with a GWP of 1.

Table 11.1 The greenhouse potential (GHP) of different gases over the 100-year lifetime using carbon dioxide as reference.

Greenhouse Gas	Global Warming Potential for 100 years
CO_2	1
CH_4	23
N_2O	296
HFC-23	12,000
HFC-134a	1,300
SF_6	22,200
CFC-12	7,000

Source: IPCC Third Assessment Report (2001); for CFC-12, data are based on UNEP (2020)

It can be seen that methane (CH_4) is a much more potent greenhouse gases (23 times stronger), while nitrous oxide (NO_2) is nearly 300 times stronger, and some hydrofluorocarbon (HFC) compounds thousands of times stronger. Chlorofluorocarbons (CFCs) are roughly as strong as HFC as greenhouse gases.

> ### Knowledge Box
>
> **Global Warming Potential (GWP):** Different greenhouse gases can have different effects on global warming and the difference depends on two factors. The first is the ability to absorb energy (recall that greenhouse gases work by absorbing long-wave radiation that ordinarily would be emitted back to space.), and the second is how long they stay in the atmosphere. GWP is a measure of how much energy the emissions of unit mass of a gas will absorb over a given period of time, relative to the emissions of the same amount of carbon dioxide. The larger the GWP, the more that a given gas warms the Earth compared to CO_2 over that time period. The time period usually used for GWPs is 100 years. GWPs can be added together to calculate the cumulative effect of a mixture of gases.

Despite its relative "weakness" as a greenhouse gas, carbon dioxide is still the most important (Figure 11.9) because it has much higher concentration than the other gases. Methane is the second most important greenhouse gas. Unknown to many, water vapor is also a greenhouse gas (it actually accounts for about 60% of the greenhouse effect), but it is not treated as one simply because the average atmospheric water vapor

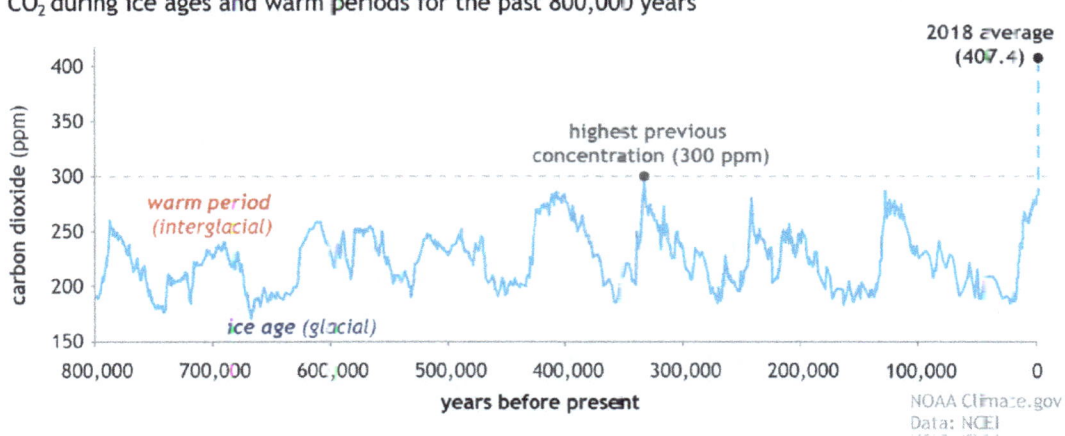

Figure 11.10 Global atmospheric carbon dioxide (CO_2) concentrations in parts per million (ppm) for the past 800,000 years. The peaks and valleys track ice ages (low CO_2) and warmer interglacials (higher CO_2). During these cycles, CO_2 was never higher than 300 ppm. In 2018, it reached 407.4 ppm. On the geologic time scale, the increase (blue dashed line) looks virtually instantaneous. NOAA Climate.gov, based on EPICA Dome C data (Lüthi D., et al., 2008) provided by NOAA NCEI Paleoclimatology Program.

level is largely constant and is considered the "baseline" condition. The influence of carbon dioxide concentration on climate is manifested in **Figure 11.10**, which depicts Earth's climate history for the past 800,000 years. The cold periods (so-called glacial periods; the "valleys" in the graph) are characterized by low carbon dioxide levels, and vice versa for the warm periods (so-called interglacial periods). If the same feedback mechanism holds true, we are in a very precarious situation with a carbon dioxide level higher than at any time in the past 800 millennia.

Due to the significance of these greenhouse gases (especially carbon dioxide) to global warming, it is important to look at the reservoir and fluxes of carbon dioxide on Earth, see **Figure 11.11**. Of the 750 gigatons of carbon in the atmosphere, it interacts with surface ocean (via gas exchange and net primary production), plants on land though net primary production, and with soils through microbial processes. It could be seen that the atmospheric carbon reservoir is not balanced due to the dynamic nature of many processes in different time scales. The other terms such as river input, peat, and deep sediment are either too small or have very long turnover time. Not shown here is the weathering of marine and terrestrial sediment, which put carbon back to the atmosphere. It could be seen that deforestation and fossil fuel combustion contribute to the net increase of atmospheric carbon by 3.2 gigaton per year, with the rest of the carbon entering into the surface ocean. Therefore, less than half of the net increase of anthropogenic carbon is being offset, suggesting that the current situation is not sustainable, and more needs to be done to curb the excess carbon dioxide input in order to sustain this delicate balance (and to avoid uncontrollable global warming).

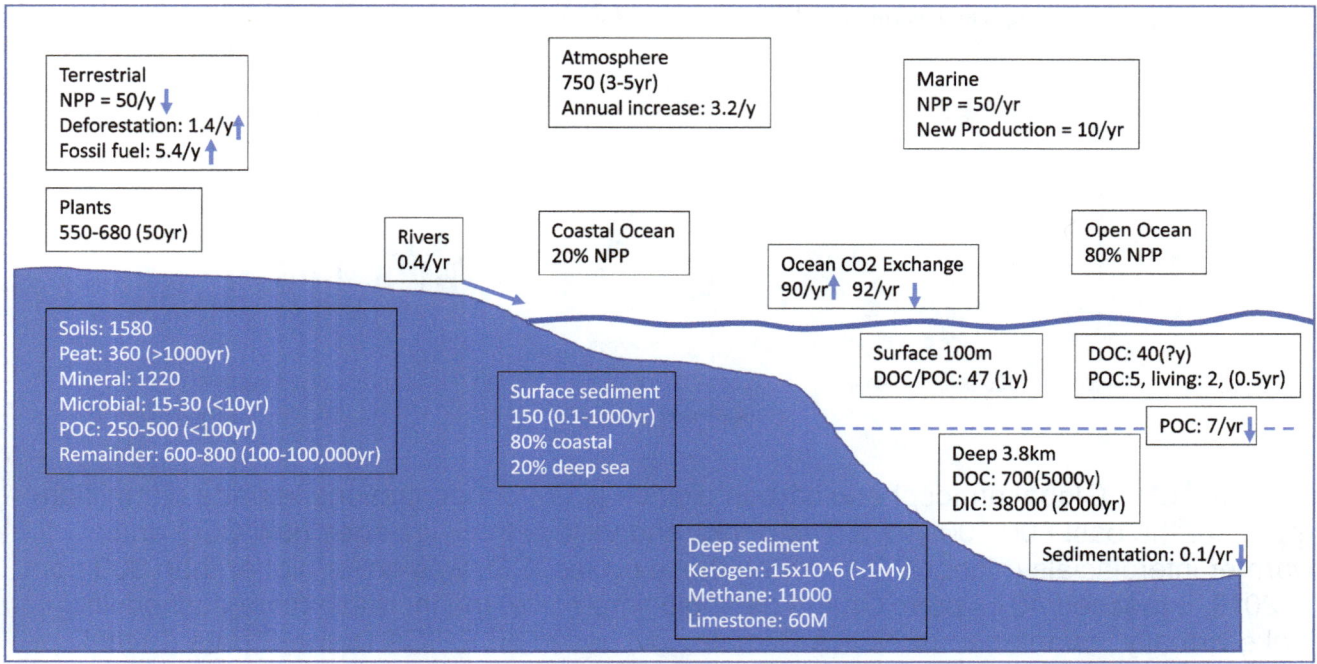

Figure 11.11 Carbon reservoir and turnover rates (recreated based on Siegenthaler and Samiento, 1993). Unit is gigaton in carbon (10^9 tons). Turnover times are expressed as the number of years. Fluxes are expressed as gigaton carbon/year.
Source: Created by Jian Peng

Figure 11.11 also shows that the "balance" among these carbon reservoirs is fairly precarious. For example, there is a very large reservoir of methane in the ocean in the form of methane clathrate (see the Knowledge Box) that is many times larger than the atmospheric reservoir. These clathrates are not stable and could be disturbed by events such as earthquakes or global warming itself because the clathrate becomes unstable in warmer temperatures. For example, in 2008, a research team in Russia's Siberian Arctic found that millions of tons of methane was released with concentrations at some regions more than 100 times of normal level. In 2014, a multi-agency team of scientists led by the United States Geological Survey found significant methane emission from coastal environments around the United States. On the other hand, a large volcanic eruption could input a significant amount of carbon dioxide in the atmosphere, potentially causing a catastrophic consequence.

Another interesting observation that can be made on Figure 11.11 is the impact of fossil fuel on atmospheric carbon dioxide balance. From the standpoint of global biogeochemical cycles, fossil fuel burning (5.4 gigaton per year) is a tiny fraction of the carbon reservoir in the soil and deep sediment (1580 and 15×10^6 gigaton, respectively). However, the issue is that the burning of fossil fuel is taking place at a rate that produces extra carbon dioxide such that it cannot be readily absorbed by terrestrial plants or by the ocean, while the other larger reservoirs are in equilibrium with negligible net carbon dioxide import or export.

> ## Knowledge Box
>
> **Methane Clathrate:** Methane clathrate is also called methane hydrate where methane is trapped within a crystal structure of water, forming a solid similar to ice. It has a nominal chemical formula of $4CH_4 \cdot 23H_2O$. It is usually formed in polar areas where temperature is low, or in the ocean floor where methane seeps out at favorable conditions (low temperature, high pressure).

In his book "Six Degrees: Our Future on a Hotter Planet," author Mark Lynas looked at scenarios where the planet would warm 1 °C to up to 6 °C (the lower and upper bounds of global warming based on a number of climate models reviewed by the International Panel on Climate Change or IPCC). The picture he painted was fairly grim. The snow of Kilimanjaro would melt; boulders of the Matterhorn would fall, the atoll nations in the Pacific would disappear inch by inch under the sea. More coral reefs would die off. There would be mass extinction. Back in 2008, Lynas predicted that humankind had only a decade to fix these apocalyptic problems. Luckily by 2020, it appears there is a little additional time. Global warming could be a highly contentious issue for many people. There are, to this day, some people who do not believe global warming is taking place. From a scientific point of view, "global warming" is a neutral term about the fact that the global average temperature has indeed been increasing. In terms of what is to blame for global warming, the preponderance of scientific evidence does point to the significant increase in atmospheric carbon dioxide and its undoubted linkage to fossil fuel, as shown in Figures 11.7 to 11.9 above.

11.1.5 Climate Variability and ENSO

Every adult likely has a vivid memory of some extreme weather conditions he/she has experienced before. If you pay close attention to the weather, each year is like a chocolate box and is different from most of the years you have experienced. Even with the most sophisticated climate model, it is nearly impossible to predict with certainty what kind of a year ahead it will be[6]. Clearly, there are inherent variabilities in the global climate system.

To elucidate the complex mechanism behind these variabilities, two important phenomena are introduced here: **El Niño** and **La Niña**. These are Spanish words for

---

[6] Nowadays NOAA has put increasing resources on long-term forecast and can make rough predictions on general weather conditions such as El Niño/La Niña conditions, or hurricane severity and frequency, etc., with a heavy dose of salt. With increasing sophistication of the climate models scientists use, predictions will get better with time.

"the boy" and "the girl," with the term El Niño gaining its fame first because it usually arrives around Christmas. As shown in Figures 11.2 and 11.3, the ocean and atmosphere systems are closely coupled and dynamically linked. They have a dynamic and delicate equilibrium in such a way that a perturbation in one place could sometimes propagate far and wide, causing much greater disruption. El Niño is such a prime example. **Figure 11.12** explains the El Niño condition. Under normal condition, strong and consistent easterly (because they blow from the east) trade winds will blow along the equator, causing upwelling of the cold, deep water on the eastern equatorial Pacific Ocean, and piling up warm water off Indonesia on the western equatorial Pacific Ocean. The high pressure on the east and low pressure on the west maintain the trade winds[7] and ensure that the dynamic equilibrium is maintained. Usually there will be little rain to the east along the south American coasts, and plenty of rain in Indonesia and the western equatorial Pacific Ocean. However, such an equilibrium is also delicate. Every 2 to 5 years, the trade winds would somehow weaken, and the warm water accumulated at

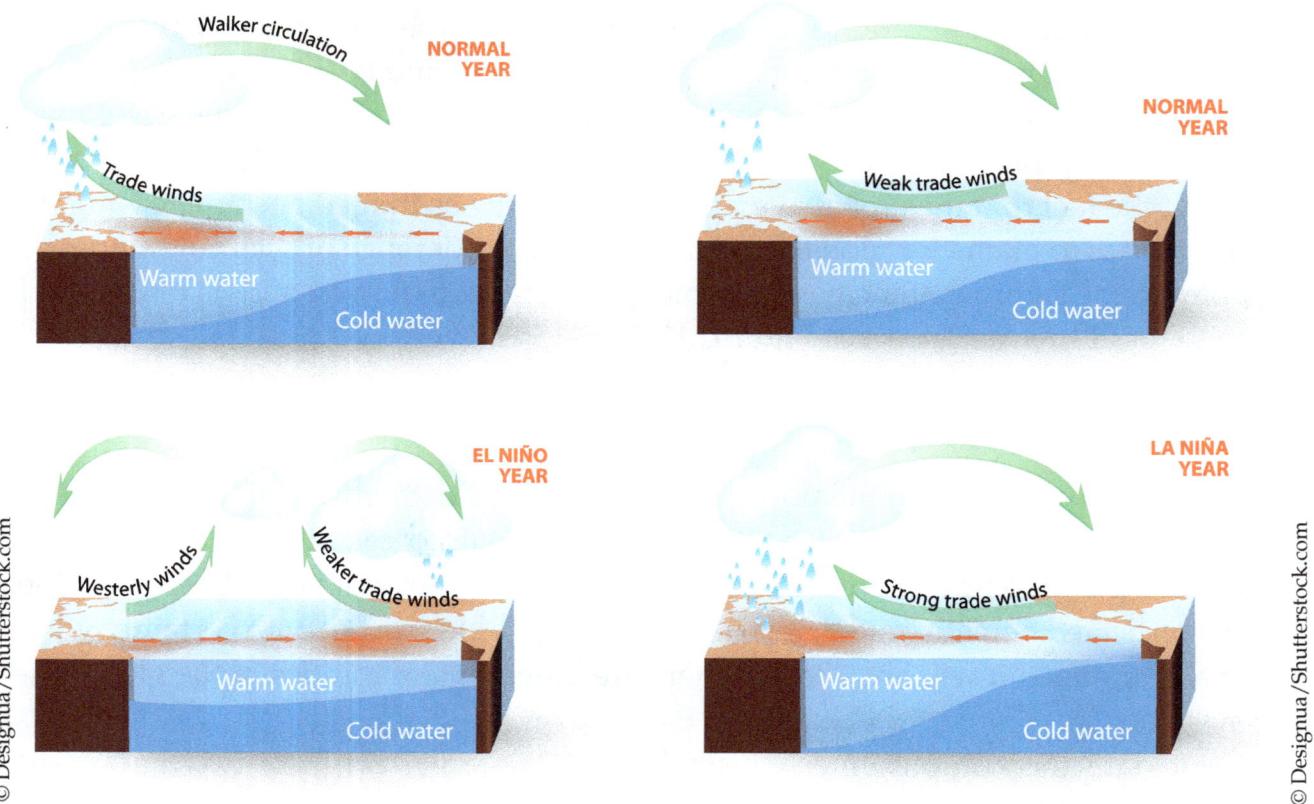

Figure 11.12 Schematics of El Niño and La Niña conditions.

---

[7] Note that winds blow from high to low pressure. High pressure to the east and low pressure to the west promote the normal trade winds, which blows from east to west.

the western equatorial Pacific Ocean would flow eastwards, cutting off the upwelling and bringing rain along the south American coast. The local fisheries, usually very productive because the upwelling water is nutrient-rich, will collapse. The weakened trade winds bring a chain reaction to many other parts of the world (see Figure 11.2 and related discussion for the interactions and connections of the global climate system), including drier western equatorial Pacific ocean and Australia; wetter southern United States, warmer southern Asia during winter time, and more disruptions at other places in the ensuing summer months.

La Niña is almost the opposite from El Niño. During La Niña conditions, the trade winds would strengthen, enhancing the upwelling of the eastern equatorial Pacific Ocean and bringing large volumes of warm water to the west. Under these conditions, there are perturbations to other parts of the world as well and these perturbations are usually quite different from those during El Niño years. Since both El Niño and La Niña conditions can be seen as an oscillation of high-low pressure zones at or near the southern Pacific Ocean near the equator, these phenomena are often termed **El Niño-Southern Oscillation**, or **ENSO**.

> ### Knowledge Box
>
> **ITCZ:** The true location where the trade winds meet is actually not right on the equator. It is a little to the south of the equator (about 5° south) due to the differences in the land mass–ocean distribution between Northern and Southern Hemisphere. This convergence zone is called Intertropic Convergence Zone, or ITCZ. This is the epicenter where a lot of important climate events take place, including El Niño and La Niña.

There are a number of theories about the origin of ENSO. Without going into too many technical details, the phenomena could be explained by an analogy of the Greek myth of Sisyphus, the king of Ephyra (**Figure 11.13**). Being punished for his sins, he was forced to roll a large boulder up a hill, only for it to roll down every time when he was close to the top. The trade winds are like Sisyphus, pushing the warm surface water from east to west along the equatorial Pacific Ocean. Whenever Sisyphus becomes a little tired (and the trade winds slack slightly), the rock rolls back and the warm water flushes back from west to east, further slackening the trade winds and bringing more warm water eastward. This positive feedback loop of weakened upwelling to the east and less warm water to the west eventually initiates the onset of an ENSO event, and poor Sisyphus finds himself and the rock at the bottom of the slope.

Figure 11.13 The Sisyphus myth.

Coupled with surface ocean circulation and ocean-atmospheric interaction is the deep ocean circulation called the **"Ocean Conveyor Belt"** coined by the famed geochemist Wally Broecker (see Chapter 1). Using ingenious methods with radionuclides to measure the age of ocean currents, he identified the pattern of deep ocean circulation, as shown in **Figure 11.14**. The Conveyor Belt, though a closed loop, could be considered to start from the north Atlantic Ocean, where high salinity water from the Gulf Stream is cooled enough and heavy enough to sink to the bottom of the ocean and travel near the seafloor of the Atlantic Ocean all the way to the Southern Ocean around Antarctica. It then turns eastbound and splits into two branches. The smaller branch would upwell in the Indian Ocean, and the larger branch would upwell from the Pacific Ocean. After upwelling, the Conveyor Belt becomes part of the surface ocean circulation, which finds its way back again to the north Atlantic Ocean. The increasing "age" of waters from north Atlantic Ocean, to the Southern Ocean, and to the north Pacific Ocean, as measured by radiocarbon, is strong evidence that the Conveyor Belt is indeed operating, with a round trip taking about 4,000 years. Because the Conveyor Belt is critical to surface circulation as well as global heat balance, any anomalies would bring profound disruption to the global climate.[8]

---

[8] According to Dr. Broecker, one such disruption could be a sudden release of ice melt from the Labrador Sea. The water with low salinity is lighter and will stay afloat on the north Atlantic Ocean, cutting off the global Ocean Conveyor Belt, precipitating a series of global climatic changes. This theory has since been challenged by other scientists, but it opened the door to vigorous scientific research projects that have advanced our understanding of the global climate system and other triggers that could cause significant changes.

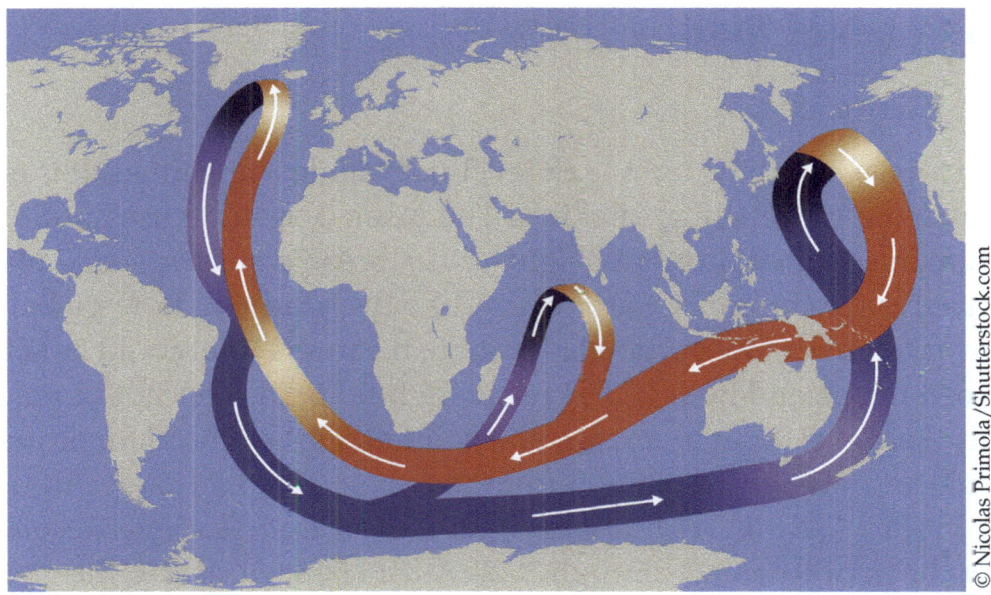

Figure 11.14 The global Ocean Conveyor Belt conceptualized by Dr. Wally Broecker.

Knowledge Box

The Father of Global Warming—Dr. Wally Broecker (1931–2019) of Columbia University's Lamont-Doherty Earth Observatory was arguably the best geochemist in his generation.[9] Using a toolbox of radionuclides such as carbon-14, uranium series, and stable isotopes, Dr. Broecker discovered many important geochemical and oceanic processes, including the Conveyor Belt, and its profound impact on global climate. He also paid special attention to the north Atlantic Ocean, where he believed holds the key to global climate change (and he was correct, for the most part). He helped coin the term "global warming" in the 1970s. In his own words, he would figure something out every six months, and then he would write about it and encourage research on it. Amazingly, he suffered from dyslexia and did not know how to type or use a personal computer. All of his work was done with a pencil and notepad. He was an ardent and most vocal advocate of climate change research.

---

[9] Dr. Broecker was the author's academic grandfather (see Chapter 1).

11.2 Sea Level Rise

11.2.1 The Concept

You probably have heard the term "**sea level rise**" quite a few times. It is measured by NASA around the globe using satellites. The Jason-3 satellite uses radio waves and other instruments to measure the height of the ocean's surface—also known as sea level. It does this for the entire Earth every 10 days, studying how global sea level is changing over time. For any particular location, mean sea level could be set at the midpoint between the mean low tide and mean high tide. The concept of sea level rise is that once the sea level is set at a reference point at a given location, or multiple locations around the globe, any deviation from the reference point can be measured fairly accurately. Figure 11.15 is such an accurately measured sea level record from four satellites for the period of 1993 through 2020. As can be seen clearly, the global sea level is rising steadily at a rate of about 3.3 mm per year.

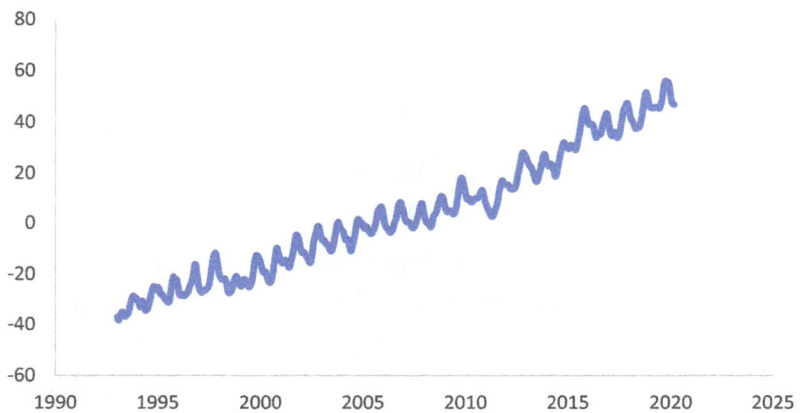

Figure 11.15 Global average sea level from 1993 to 2020 computed at the NASA Goddard Space Flight Center under the NASA MEASUREs program, with data averaged from 4 different satellites.

Data source: GSFC. 2017. https://podaac-tools.jpl.nasa.gov/drive/files/allData/merged_alt/L2/TP_J1_OSTM/global_mean_sea_level/GMSL_TPJAOS_4.2_19909_202002.txt. Created by Jian Peng

11.2.2 Causes of Sea Level Rise

There are two major causes of sea level rise. First, when the global temperature (both air and surface ocean) rises, ocean water will expand, causing sea level rise. Second, the melting of land-based ice, both sea ice (Figure 11.16) and mountain glaciers (Figure 11.17) above the normal amount will increase the volume of ocean water.

Figure 11.16 Ice calving off the Perito Moreno glacier in Argentina.

Figure 11.17 Location and date marker for glacier in Jasper National Park in Canada.

For example, Greenland's ice loss rate during 2012 to 2016 was 247 billion tons a year compared to a mere 34 billion tons a year in 1992 to 2001. Since global warming affects both processes, one can safely say that global warming is the ultimate cause of sea level rise.

11.2.3 Sea Level Rise in Different Scenarios

Since 40% of the global population lives within a short distance from the coast, sea level rise could bring significant issues around the globe. The most common issue is coastal erosion and flooding (Figure 11.18). For infrastructure planning purposes, it is important to predict future sea level rise so preventive measures such as flood control and coastal erosion facilities could be built. However, if sea level rise is not curbed, many coastal communities will become uninhabitable. For example, the low-lying

Figure 11.18 Ho Chi Minh City, Vietnam, October 18, 2016. The street was flooded due to a storm surge and exacerbated by sea level rise.

Figure 11.19 Flooded St. Mark's Square in Venice, Italy. Few places can feel the pain of sea level rise more acutely than Venice, a city already in the water.

areas in Miami, Florida, USA will be severely impacted if sea level rises significantly. Even if sea walls could be built to prevent constant flooding, the beaches will disappear, and access to the ocean will be impacted, making this area quite undesirable to live in. For cities such as Venice, Italy, sea level rise is a lot worse and entire city blocks could be flooded for days (**Figure 11.19**).

Figure 11.20 A Ridgway's Rail, an endangered species, forages in a bird sanctuary in Orange County, California. Its habitat is being threatened by sea level rise.

Sea level rise will also bring ecological issues, especially loss of coastal habitat for many ecologically sensitive species. For example, rising sea levels in recent years has impacted the breeding ground for the endangered Ridgway's rail (an acuatic bird) in Bolsa Chica wetland on Southern California's coast. During high tides, many nests were flooded, impacting the reproduction of this endangered species (Figure 11.20).

Sea level has risen and fallen many times, sometimes drastically, in the Earth's history. During the last glacial maximum around 115,000 to 11,700 years ago, the sea level was more than 125 meters lower than today. The sea level has been roughly constant for the first 2,500 years, but recently has been rising rapidly due to human-induced global warming. Therefore, real actions are needed for nations around the globe to curb this global disaster slowly in the making.

11.3 Ocean Acidification

11.3.1 What is Ocean Acidification?

Ocean acidification, as evidenced by recent decrease in the average pH of seawater around the global oceans, is linked to elevated atmospheric carbon dioxide level, as shown in Figure 11.7. As mentioned earlier in the book, the ocean is highly buffered and basic with a pH of 8.1 to 8.2. Ocean acidification is not a process where the seawater will turn acidic. Rather, the seawater will become less basic. Since the industrial revolution began about 200 years ago, excessive carbon dioxide in the atmosphere has driven the pH of surface ocean waters down by 0.1 pH units. This might not sound like a lot, but the pH scale is logarithmic,[10] so this change represents approximately a 30% increase in acidity.

11.3.2 The Mechanism of Ocean Acidification

The connection between the level of carbon dioxide in the air and ocean acidification can be expressed by the following chemical reactions:

1. $CO_2 + H_2O \rightarrow H_2CO_3$ (carbonic acid)
2. $H_2CO_3 \rightarrow H^+ + HCO_3^-$ (dissolution of carbonic acid)
3. $CO_3^{2-} + H^+ \rightarrow HCO_3^-$ (formation of bicarbonate)

These reactions will gradually drive the pH toward a lower level due to a perturbation of the carbonate system in seawater. **Figure 11.21**, which overlays the atmospheric and oceanic CO_2 levels with pH values measured for seawater in Hawaii, USA, shows a clear correlation as described above.

11.3.3 The Impact of Ocean Acidification

The most important (and known) impacts from ocean acidification are coral bleaching (**Figure 11.22**) and shell thinning (**Figure 11.23**). The acidification makes it more

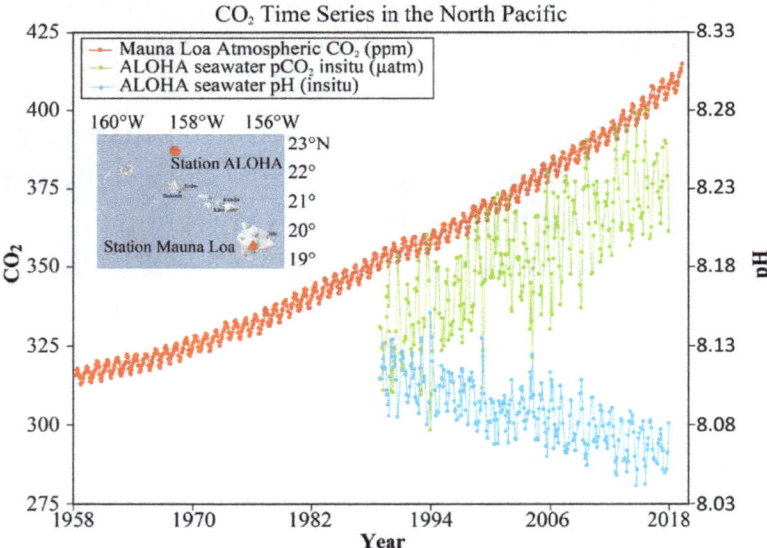

Figure 11.21 Carbon dioxide level and ocean acidification in the North Pacific Ocean. Red line on top: CO_2 levels measured at Mauna Loa; green line in the middle: CO_2 levels measured at the surface ocean ALOHA site; blue line at the bottom: pH measured at the surface ocean ALOHA site. Data source: NOAA.

---

[10] pH is calculated by taking the negative logarithm of the activity/strength of hydrogen ion (H+). Therefore, the higher the acidity is for a solution, the lower its pH will be.

Figure 11.22 Coral bleaching at the Great Barrier Reef, Port Douglas, Queensland, Australia.

Figure 11.23 Shell thinning/pitting due to ocean acidification.

difficult for marine calcifying organisms, such as coral and some plankton, to form biogenic calcium carbonate, and such structures become vulnerable to dissolution. While ongoing ocean acidification is at least partially anthropogenic in origin, it has occurred previously in Earth's history. The most notable example is the Paleocene-Eocene Thermal Maximum, which occurred approximately 56 million years ago when massive amounts of carbon entered the ocean and atmosphere, and led to the dissolution of carbonate sediments in all ocean basins. Ocean acidification has been compared to anthropogenic climate change and called the "evil twin of global warming" and "the other CO_2 problem." Therefore, curbing anthropogenic carbon emission is likely the key to solving this problem.

11.4 Mitigation Measures

11.4.1 The Global Climate Feedback Loop

Global warming, climate change, sea level rise, and ocean acidification are like a freight train carrying the Earth and its inhabitants slowly but surely toward a disaster. Clearly, mitigation measured are needed, and these measures should be guided by solid science. The global climate system is complex and dynamic, with many checks and balances to keep the global climate largely stable. However, certain conditions may drive the Earth into climate chaos due to positive **climate feedback loops**. The first scenario is called "runaway global warming," (Section 4.5, see Figure 4.11) in which excessive carbon dioxide causes stronger greenhouse effect and global warming, rising sea surface temperature forces the release of more carbon dioxide (solubility of carbon dioxide is inversely proportional to temperature), which results in more greenhouse effect. Global warming will also cause sea ice, snow pack, and glaciers to melt. These materials have very high albedo (i.e., they are light-colored and can reflect most solar energy back to space). Decrease in the areas for snow and ice will result in more absorption of heat from the sun, exacerbating global warming.

An opposite mechanism could drive the Earth into an icy condition, and this is called "runaway global cooling," which has happened many times in the past.[11] A slight cooling will allow the ocean to absorb more carbon dioxide, further weakening the greenhouse effect. Similarly, an increase in albedo due to increase in the areas of snow/ice will reflect more solar energy back to space, decreasing the global temperature even further. The preponderance of evidence, however, clearly suggests that runaway global warming is the much more likely scenario and should be the focus of any mitigation measures.

11.4.2 The Carbon Problem

Putting what we have learned together, it is not difficult to find that the root cause of greenhouse effect (and the ensuing global warming) and ocean acidification is excessive carbon dioxide in the atmosphere. From the analysis of our energy consumption, however, we can clearly see that the carbon problem will get worse for the next few

---

[11] In fact, based on Earth's recent history when glacial and interglacial periods alternated on a semi-regular pattern, we are currently overdue for a glacial period. In a sense, anthropogenic global warming may have delayed or stopped the glaciation.

decades before it gets better (hopefully), and the anticipated improvement is in part due to the diminishing amount of available fossil fuels, not because we become smart enough to make right decisions.

With rising temperature in both atmosphere and the oceans, rising sea levels, and more and more extreme weather events around the world, it is dangerous to delay the fight against excessive carbon emission. In view of the dire consequences of a runaway global warming scenario as mentioned above, it is imperative that all nations work together to avoid this classic tragedy of the commons (see Chapter 3). The focus of mitigation measures should be the carbon emission problem.

11.4.3 Potential Mitigation Measures

Global warming and climate change are almost certainly linked to anthropogenic activities, especially fossil fuel burning that disturbed the delicate balance of the global carbon reservoir for both the atmosphere and the oceans. It is clearly extremely challenging to curtail the global consumption of fossil fuels (see Figures 10.1 and 10.5). While we can still hope that more Greta Thunbergs will emerge, and human ingenuity will drive more clean and green energy solutions in the near future, there have been some attempts to alleviate the sharp conflict between the energy industry and global warming. On the energy generation side, as discussed in Chapter 10, there are many options other than the fossil fuel options for energy generation. These options include renewable energy such as solar, wind, hydropower, biomass, etc., or "clean" energy such as geothermal, or non-carbon options such as nuclear power.

Besides the above options, the emission of carbon from fossil fuel power plants could be captured directly. The options include collecting carbon dioxide for reuse or disposal. For larger facilities, some even used pure oxygen instead of air for the process to avoid the dead volume of nitrogen in the air. As a result, the combustion product will be highly purified carbon dioxide that can be pressurized and injected to deep geological formations.

There are also ways to extract carbon dioxide from the atmosphere using a number of ways. They include ocean fertilization, seaweed farming, tree planting, genetically engineered deep-rooted crops; biochar production, etc., to promote biological carbon-fixing. There are also ways to increase the albedo of the earth, including sulfur aerosols dispersion; ocean misting (cloud making), genetically engineered crops to increase albedo; high-albedo roads and buildings; and so on, so that Earth's surface is more reflective and more heat can be reflected back to space.

To combat global warming and curb carbon dioxide production on a global scale and in a coordinated manner, there have been numerous attempts by the United Nations and member nations to implement a series of protocols and frameworks to resolve the "Tragedy of the Commons" issue. These conventions include the 1987 Montreal Protocol on Substances that Deplete the Ozone Layer; the "Earth Summit" in Rio de Janeiro in 1992; the Kyoto Protocol in 1997; COP15, COP17, COP18, and COP19 (in 2009, 2011, 2012, and 2013, respectively) and the most recent **Paris Protocol** of 2015. However, implementing these protocols has proven to be difficult, with the United States pulling out of the Paris Protocol due to political reasons.

Carbon emission trading is also a tool for different nations as well as entities within a nation to manage carbon emission holistically. To meet their obligations specified by the Kyoto Protocol for the reduction of carbon emissions, a country or a polluter having more emissions of carbon is able to purchase the right to emit more and the country or entity having fewer emissions sells the right to emit carbon to other countries or entities. This way the trading market results in the most cost-effective carbon reduction methods being exploited first. Such examples include the Cap-and-Trade Program (in California, USA), and international emissions trading (by the United Nations) scheme. These efforts are still in their fledgling stage, but they offer a glimmer of hope for mankind to deal with the carbon problem, which is the root cause of the global issues discussed in this chapter.

At this point, the reader is encouraged to look around and inside to consider various ways to reduce carbon emissions. The actions one could take include using public transportations and electric cars; avoiding air travel; reducing household wastes by reusing and recycling; promoting sustainability at work and to people around you, among many others. Many such actions will be discussed in the next chapter. Many of us have a Greta Thunberg inside us—we just need to let her speak out.

Further Reading

Ripple, William, Christopher Wolf, Thomas Newsome, Phoebe Barnard, and William Moomaw. World Scientists' Warning of a Climate Emergency. BioScience, Biz088, American Institute of Biological Science (Oxford Academic; Oxford University Press). Retrieved 2019-12-14.

History of Local Government Declarations. *CACE*. Council and Community Action in the Climate Emergency. Retrieved 2020-01-06.

United Nations Environmental Programme, 2020. Introduction to Climate Change. Weblink: http://www.inforse.org/europe/dieret/Climate/climate%20graphics/05.htm. Accessed: June 6, 2020.

Six Degrees: Our Future on a Hotter Planet, the author Mark Lynas.

Intergovernmental Panel on Climate Change (IPCC). Climate Change 1994: Radiative forcing of climate change and an evaluation of the IPCC IS92 Emission Scenarios, Houghton, J.T. et al (eds.), New York, NY: Cambridge University Press, 1994.

C.D. Keeling, R.B. Bacastow, A.E. Bainbridge, C.A. Ekdahl, P.R. Guenther, and L.S. Waterman. Atmospheric carbon dioxide variations at Mauna Loa Observatory, Hawaii, *Tellus*, 28 (1976): 538–51.

Shakhova, N., I. Semiletov, A. Salyuk, D. Kosmach, and N. Bel'cheva. "Methane Release on the Arctic East Siberian Shelf" (PDF). *Geophysical Research Abstracts*, 9 (2007): 01071.

Skarke, A., C. Ruppel, M. Kodis, D. Brothers, and E. Lobecker. "Widespread Methane Leakage from the Sea Floor on the Northern US Atlantic Margin." *Nature Geoscience*. 7, no. 9 (21 July 2014): 657–61.

GSFC. Global Mean Sea Level Trend from Integrated Multi-Mission Ocean Altimeters TOPEX/Poseidon, Jason-1, OSTM/Jason-2 Version 4.2 Ver. 4.2 PO.DAAC, CA, USA. Dataset accessed [6/6/2020], 2017, at http://dx.doi.org/10.5067/GMSLM-TJ42.

Web Resources

Obituary of Wally Broecker: https://blogs.ei.columbia.edu/2019/02/19/wallace-broecker-early-prophet-of-climate-change/
Wikipedia entry for Greta Thunberg: https://en.wikipedia.org/wiki/Greta_Thunberg
NOAA's Climate Website: https://www.climate.gov/
The Coriolis Force: https://en.wikipedia.org/wiki/Coriolis_force
Climate engineering: https://en.wikipedia.org/wiki/Climate_engineering
The melting ice: the future of the Arctic: https://www.youtube.com/watch?v=U0aNeY-ZL8jY

Questions and Exercises

1. Observe Figure 11.5 and ensure that the energy is conserved in the global energy budget.
2. Show mathematically that a drop of 0.1 pH unit is approximately 30% increase of acidity.
3. Observe Figure 11.10. If the feedback loops in the "runaway global warming" and "runaway global cooling" actually works, which feedback loop works more strongly? Why?

4. The carbon problem could be dealt with through two major types of climate engineering methods. The first type is to capture carbon. The second type is to alter the Earth's albedo. Think about possible solutions in both approaches and consult Wikipedia's Climate Engineering page (https://en.wikipedia.org/wiki/Climate_engineering).
5. In Chapter 1, a mechanism that could dampen the carbon problem is the chemical weathering of silicate rocks. Can it be used as a climate engineering option? Why or why not?
6. It is conceivable that in the next few decades the consumption of fossil fuels will decrease significantly due to the exhaustion of these resources. Should we simply allow the fossil fuels to run their paths without intervention? Why or why not?
7. In the positive feedback loops as discussed in Section 11.4, what are the potential processes that could break the loops? Give a few examples.

Chapter 12

Environment, Human Development, and Sustainability

Learning Outcomes	344
Key Concepts	344
12.1 The Definition and Concept of Sustainability	344
12.2 Environmental Issues through Economic Lenses	346
12.2.1 The Environmental Industry	346
12.2.2 Environmental Economics	347
12.2.3 Environmental Regulation and Economics	350
12.3 Life Cycle Environmental Impacts and Costs	351
12.3.1 Life Cycle Cost	351
12.3.2 Ecological Footprint	352
12.3.3 Carbon Footprint	352
12.3.4 Life Cycle Environmental Cost	355
12.4 Sustainability Issues	359
12.4.1 Environmental Mitigation	359
12.4.2 Nature's Intrinsic Value	359
12.4.3 Deep Ecology and Sustainability	360
12.4.4 Resource Depletion	362
12.5 Global Sustainable Development	366
12.5.1 Global Sustainable Development Goals	366
12.5.2 The Earthrise	370
Further Reading	373
Web Resources	374
Questions and Exercises	375

Learning Outcomes

- Thorough understanding of the term "sustainability".
- The ability to dissect sustainability into at least three aspects: environmental, social, and economical.
- Knowledge of the linkage between sustainability issues with the environmental issues we have learned before, such as water, air, soil, solid waste, and energy.
- Appreciation of the issues that impact the sustainability of mankind, including resource depletion, climate change, natural disasters, etc.
- Understanding of the concept of life cycle cost, water footprint, and carbon footprint
- Ability to analyze yourself and assess the sustainability of your lifestyle

Key Concepts

Sustainability; triple bottom line; three pillars of sustainable; carbon footprint; ecological footprint; life cycle environmental cost; intrinsic value; deep ecology; sustainable development goals; Basel Convention; environmental mitigation; 5-Rs (refuse, reduce, reuse, recycle, rot)

12.1 The Definition and Concept of Sustainability

In Chapter 1, we discussed some big-picture issues such as the genesis of the universe and the Earth, the Earth's history and its future. Our understanding of the past should ideally help us better plan for the future, which unfortunately is getting increasingly uncertain because of growing concerns about many environmental issues, global warming/climate change, resource depletion, and so on. While environmental pollution issues (for water, air, soil, and biota) or depletion of natural resources are worrisome, they are largely not severe enough to impact the survival of the human race, at least in the near future. Global climate change, on the other hand, has the potential to render the Earth an uninhabitable or hostile environment. As discussed in Chapters 10 and 11, we are doing especially poorly on controlling our consumption of fossil fuels. Many environmental pollution issues could have been better managed if we improved source controls and stopped the pollution before it took place. Therefore, it is fitting that this book culminates at the **sustainability** issue to encourage you to apply what you have learned in this book and consider ways to steer our planet toward a path to sustainability.

This chapter will focus on sustainability and its relationship with environmental and human development issues. The term 'sustainability' has quite a few different meanings. Lexico/Oxford defines sustainability as "*avoidance of the depletion of natural resources in order to maintain an ecological balance.*" Environmentalscience.org defines it as a science on "*how natural systems function, remain diverse, and produce everything it needs for the ecology to remain in balance.*" Investopedia, a website dedicated to finance and investment, has a surprisingly sound definition, that "*sustainability focuses on meeting the needs of the present without compromising the ability of future generations to meet their needs.*"

In 2005, the World Summit on Social Development identified **three pillars of sustainable** development: economic, social, and environmental (**Figure 12.1**). This is similar to the so-called **"triple bottom line,"** where consideration of any environmental efforts should consider all of these three factors. In Figure 12.1, sustainability is at the intersection of all three pillars, and it also takes into consideration the factors interfacing any two of the three pillars. For example, sustainability at the intersection between environment and economy should be "viable." If a solution, even though environmentally beneficial, is not viable in economic or financial senses, it will not be sustainable. **Figure 12.2** shows another way to understand these three pillars, where economy has to operate within the bounds of the society, and both need to operate within the bounds of the environment. Recently, the so-called "circle of sustainability" carries these three pillars one more step by splitting the social portion of sustainability into politics and culture. Thomas (2016) further separated sustainability issues into seven modalities/modules, including: economy, community, occupational groups, government, environment, culture, and physiology. Again, this approach focuses even more on the social aspects of sustainability.

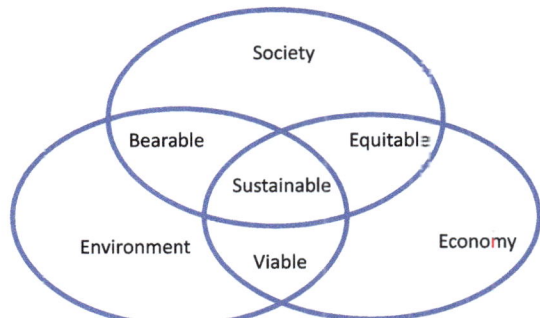

Figure 12.1 Three pillars of sustainability and their interfaces (Adams, 2006).
Source: Created by Jian Peng

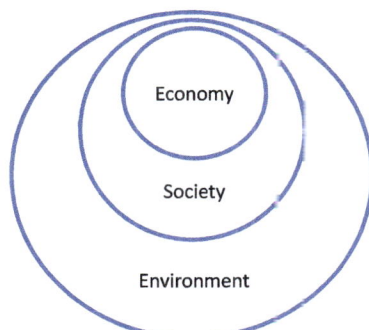

Figure 12.2 The relationship between the three pillars of sustainability.
Source: Cato, 2009. Created by Jian Peng

USEPA defines sustainability as a principle: *everything that we need for our survival and well-being depends, either directly or indirectly, on our natural environment. To pursue sustainability is to create and maintain the conditions under which humans and nature can exist in productive harmony to support present and future generations.*

Wikipedia defines sustainability literally as "the ability to exist constantly," and then dissects this term in many different ways, including most of the aspects discussed above. Its sustainability entry (https://en.wikipedia.org/wiki/Sustainability) is a good resource to get an overview of this concept from various angles.

12.2 Environmental Issues through Economic Lenses

12.2.1 The Environmental Industry

Environmental issues are typical "tragedy of the commons" issues and cannot usually be resolved effectively by individuals in a society. As discussed in Chapter 3, when environmental impacts start to harm enough members in a society, or 'the commons', environmental regulations will be set up and enforced by government agencies. These regulations drive the protection of the environment through increasingly stringent and prescriptive requirements. As a result, environmental compliance, similar to health and safety requirements, is now an integral part of nearly all businesses (mostly as the Environmental Health and Safety or EHS department or division within a business). There are also industry segments dedicated to environmental compliance work. Generally, environmental industry can be segmented into environmental services, environmental equipment, and environmental resources categories, each with several sub-segments, as follows:

- Environmental services category: environmental testing and analytical services; wastewater treatment works; solid waste management; hazardous waste management; remediation and industrial services; and environmental consulting and engineering.
- Environmental equipment category: water and wastewater equipment and chemicals; instruments and information systems; air pollution control equipment; and waste management equipment
- Environmental resources category: water utilities; resource recovery; clean energy power and systems.

The environmental industry in the United States had a revenue of $370 billion in 2016, representing nearly 3% of the gross domestic product (GDP). The industry also provided 1.7 million jobs in 2016 (EBI, 2020). This does not include government agencies and environmental organizations, which are important players in the environmental field. For example, there are many federal agencies that carry out environmental regulation responsibilities led by USEPA, which has about 15,000 permanent staff and an $8.1 billion annual budget. Other agencies such as the United States Geological Survey (USGS), United States Fish and Wildlife Service (USFWS), National Oceanic and Atmospheric Agency (NOAA), and even the Department of Energy (DOE), have sizable staff and budget for environmental work. On state and local levels, environmental staffing and budget are even more significant. For example, in the State of California, the California Environmental Protection Agency (CalEPA) employs about 5,000 permanent staff with an annual budget of $1.5 billion. There are many more staff on local governmental levels.

There were 26,548 environmental nonprofit organizations (usually called nongovernmental organizations or NGO) in 2005 but two thirds of them were very small organizations with revenues less than $25,000 a year. Large environmental NGOs such as the Nature Conservancy could have significant financial power, with over 200 NGOs having annual revenues over $5 million in 2005. Altogether, the environmental NGOs have a combined revenue of $8.24 billion dollars.

These industry sectors are critical to the environmental quality and sustainability of our society. As shown in Figure 12.1, to make environmental compliance work, any environmental improvement projects have to be economically viable. For example, to clean up a contaminated water body, there are available technologies and environmental services firms that are capable of doing the work. However, funding will be needed for the cleanup operations, be it from the government (or from taxpayer), or from the responsible party that caused the pollution. Even for environmental NGOs that use volunteers to do much of the work, substantial costs will also be incurred for their work on monitoring, advocating, and public outreach work that requires staff and funding to be carried out. Therefore, environmental work always comes at a price.

12.2.2 Environmental Economics

In a classic business model, all businesses are set up with the ultimate goal of making money, and the best financial strategy is the one that maximizes the financial return for their owners and stockholders. Under this business model, when a company's executives look at the company's financial balance sheet and income statements, the environmental costs are a simple number and often buried by other 'administrative' costs.[1] Many corporations deal with environmental compliance from purely economic considerations rather than treating environmental protection as their intrinsic responsibility as corporate citizens. To maximize their profitability and minimize the cost including those for environmental compliance, most corporations adopt such common practices as contracting; outsourcing; moving the operation to elsewhere with less stringent environmental regulations; or investing just enough money to meet the bare minimum environmental regulatory requirements. Few would go above and beyond the regulatory requirements

---

[1] Often called "SG&A," which stands for sales, general, and administratives costs.

even though a few extra steps could bring significant environmental benefits (see the Knowledge Box). Since environmental regulations have a long lag time (see Chapter 3), this approach could often result in irrevocable environmental damage.

> ### Knowledge Box
>
> **Export of Hazardous Waste**—if you recall, in Chapter 9 and Section 9.5.2, **Basel Convention** was discussed. This convention was set up to specifically manage the transfer of hazardous waste between rich and poor countries due to increasing handling and treatment costs of hazardous wastes in developed countries. Some developed countries and companies found that it was much cheaper economically to simply export the hazardous wastes to some poor countries in Africa and Asia even they knew that these wastes would not be handled properly and would cause significant environmental and human health damages. Therefore, stronger regulation was needed.

Environmental field is quite unique in that it is almost entirely driven by regulatory requirements. As mentioned previously, each firm has EHS personnel (a person, a division, or even a department, depending on the size of the firm and the nature of the business) to ensure that the firm complies with all environmental regulations. They hire environmental consultants and engineers, they purchase environmental cleanup equipment, conduct environmental monitoring, etc. Local governmental agencies and municipalities are subject to environmental regulations as well and they have staff and budget for environmental compliance. This brings an interesting phenomenon where the EHS department and environmental divisions of regulated governmental agencies are often considered a liability or nuisance inside the firm or agency. On the firm's balance sheet, environmental regulations and obligations for compliance are usually cited as potential liabilities that the firm has to disclose to its shareholders. Some firms even have environmental liability reserve funds or insurance in case they violate environmental regulations and need funding to pay for the penalty or cleanup costs. In many firms and agencies, the EHS function is a nonproductive function and is viewed as a nuisance.

Due to increasing public attention to environmental and sustainability issues, more and more firms, especially large firms, rely on green marketing or public relations campaigns and tout their environmental and sustainability efforts on their publications and websites. When a firm touts its efforts and culture on "green" products, environmental stewardship, and its support for environmental and sustainability causes in its marketing campaign, it is called green marketing. One can easily find such information on the websites of virtually every large firm. In Mark Lefko's book "Global Sustainability: 21 Leading CEOs Show How To Do Well by Doing Good," CEOs talked eloquently about their firms' position on environmental and sustainability efforts. Some of these firms are large firms with worldwide operations. Therefore, to see that they adopt sustainability measures is encouraging.

However, talking about sustainability is different from practicing it, especially if it may impact a firm's bottom line. When a firm tries to exaggerate its effort on environmental and sustainability issues, it is called greenwashing, which is also a fairly common (but not easily discernible) phenomenon (see the Knowledge Box). Therefore, governmental or society interventions are needed to make sure that the public is informed and that society truly benefits from sustainability efforts from all individuals and firms.[2]

Knowledge Box

Environmental expenditures in large firms: According to Chevron's 2019 annual report (10-K), the firm spent about $2 billion on "environmental spending." This is a significant amount, but a small portion compared to its 2019 annual revenue of $146.5 billion. Included in these expenditures were approximately $0.6 billion of environmental capital expenditures and $1.4 billion of costs associated with the prevention, control, abatement or elimination of hazardous substances and pollutants from operating, closed or divested sites, and the decommissioning and restoration of sites. In comparison, other firms such as Apple Inc., environmental-related cost is buried in the Sales, General and Administrative (SG&A) costs and is a negligible portion in its vast annual revenue of $260 billion in 2019.

Knowledge Box

Green Marketing and Greenwashing—Green, green everywhere. Greenwashing is a compound word inspired by the word whitewash. It is a form of false marketing to make the firm or product appear more environmentally friendly than it actually is. This technique is most commonly used for food and medicine, but is often used in other industries as well. Greenwashing could include using a product name that suggests natural, green or healthy benefits. A firm could also greenwash itself by claiming to be an environmentally friendly entity when its actions and practices show otherwise. At some point, the boundary could be blurred. For example, a hotel nowadays will likely tell you to skip room service and/or reuse your towels to support its environmental cause. That might be the case, but it is also quite possible that the financial benefits from such practice is the true underlying driver (lower costs for laundry and room service). We can all blame Richard Thaler[3] for this subtle messaging.

---

[2] One such example is Hove Social Good Intelligence, Inc., which calculates and validates social goods and sustainability impacts of firms and products (disclaimer—the mentioning of this firm does not imply endorsement).

[3] In his best-selling books "Nudge" and "Misbehaving," the Nobel-winning economist offered a lot of ammunition for corporations suspected of greenwashing to "nudge" their customers in certain ways to make their marketing strategies more effective.

12.2.3 Environmental Regulation and Economics

One key consideration for environmental regulation and guidance to improve environmental protection and sustainability is that the regulation should use financial incentives/disincentives to ensure that the regulations are followed. Under the Clean Water Act, the calculation of environmental damage and compensation/penalties depends on four factors: seriousness (the amount and degree of damage caused by the responsible party), culpability (degree of liability or fault on the part of the responsible party vs. human error or circumstances), mitigation (whether and how much the damage is mitigated), and history of prior violations (whether there is a pattern of negligence or wrongdoing). Considerations for each factor (and subcategories within each factor) will result in significant differences (up to 500%) in penalties, up to $25,000 per day. The actual amount will also depend on economic factors such as economic impact of the penalty, as well as the economic benefit of the violation to the responsible party, so that the penalty will ensure that the responsible party will never benefit from committing the violation. The World Bank provided calculation methodology for environmental damages, such as for land plot; earth bowels,[4] ecologo-economic damages to soil, water, forest, plant cover, wildlife, and so on.

Enforcement and penalty-based environmental regulatory schemes are an important deterrent but rarely used mechanism for environmental regulation. More often than not, incentive-based schemes are used. For example, California's cap and trade program was one of the first and largest in the world to regulate carbon emission with a carbon credit system, essentially allowing financial leverage to regulate this difficult issue. Under this framework, polluters pay for the carbon they emit so they are motivated to lower the emissions on their own using a variety of tools, including open-market trading. Open-market trading has the advantage of identifying the most efficient and cost-effective way of meeting specific goals.

On the broader pollutant trading policy, the majority (78%) of economists agree with the notion that "effluent taxes and marketable pollution permits represent a better approach to pollution control than the imposition of pollution ceilings" (Mankiw, 2018). On the face of it, this statement is quite stark, inhumane, and anti-environment. However, in the mind of an economist, everything makes sense. Effluent taxes impose a monetary burden upon a company so that the more pollution it causes, the more taxes it pays. Marketable pollution permits allow the dischargers to trade the pollutant credits. For example, in a large watershed with many wastewater-treatment plants, the regulators could set a ceiling of overall pollutant discharge without specifying the limits of individual dischargers. If the pollutant market is completely open, the treatment plants could trade amongst themselves and find ways to control the overall discharge amount within the limit in the most economical way. Lastly, the hard pollution ceiling for an individual discharger is

---

[4] The Earth bowels include everything beneath the land surface, such as mineral resources.

very difficult to set correctly or implement rationally, and it hampers the ability for the dischargers to collaborate and achieve the best and most cost-effective solutions.

Of course, the fact that one quarter of economists do not agree with the above notion is hardly surprising, because many could argue that the damage to the environment is hard to measure with monetary terms; some may even argue, albeit naively, that there should not be pollutant discharges after all. However, some of the disagreement could be fixed by certain arrangements, for example, a nonlinear scale to calculate pollution tax that allows pollution tax to escalate quickly into a strong economic disincentive for the discharger to pollute. This way, the overall impact to society may not be negative.

12.3 Life Cycle Environmental Impacts and Costs

12.3.1 Life Cycle Cost

Life cycle cost is originally an engineering term about the total cost of purchasing, owning, operating, maintaining, and disposing of an object, a process, or a project. This is a better parameter than just the capital cost or capital plus operation and maintenance cost, because the engineering project eventually needs to be scrapped or rebuilt at the end of its life. For example, if a firm or a family wants to purchase and install solar panels to offset its electricity bills, a common mistake would be to only consider the initial installation cost. Even though the initial investment for the solar panels and their installation appears to be the largest cost, the operation and maintenance; electricity generated and offset; resultant cost savings; disposal/replacement/repair cost at the end of the useful life of these solar panels should all be considered in order to compare with other potential options such as rent or third-party ownership. If one wants to be exact, these values should all be converted to net present value or NPV to make these options comparable (see knowledge box on NPV).

Knowledge Box

NPV—Net Present Value (NPV) is commonly used in accounting and corporate finance and is usually the gold standard to evaluate different options of business decisions. To make NPV work, all future cash flows, expenditures, and other income/costs should be converted to present value by considering the amount, the timing of cash flows, as well as the discount rate. This is important because a dollar at present is not the same as a dollar 10 years from now due to inflation and uncertainty. The discount rate is the expected/desired rate of return if you were to invest the money elsewhere. For more information, please consult a simple tutorial, such as: https://en.wikipedia.org/wiki/Net_present_value.

12.3.2 Ecological Footprint

Ecological footprint is a concept promoted by the Global Footprint Network (www.footprintnetwork.org) to measure "how much nature we have and how much nature we use". It tracks the use of six categories of ecological resources measured by productive surface areas: cropland, grazing land, fishing grounds, built-up land, forest area, and carbon demand on land. The ecological footprint could be calculated for a given population by how much such resources are consumed. The concept is similar to the water footprint discussed in Section 6.8 where water resources, instead of the productive land, are used to calculate the footprint. **Figure 12.3** shows water footprints for some common food items (also refer to Table 6.1). As can be seen, the water footprint could be as much as 50-fold difference between different food items. Ecological footprints for different products, countries, or lifestyles could vary greatly as well. Globally, mankind is consuming ecological resources at a rate that significantly exceeds the capacity of the planet. This issue will be discussed in more details in subsequent sections.

12.3.3 Carbon Footprint

Carbon footprint, or energy footprint, is analogous in concept to water footprint. It is the total amount of carbon dioxide (or greenhouse gases in general) emissions caused by a person, a product, or a process. The unit for carbon footprint is "carbon dioxide equivalent." Similar to water footprint, assumptions and simplifications are commonly used. For example, according to Cnet.com, an iPhone has a carbon footprint of 50 to 100 kg, depending on models. The same concept could be used to evaluate

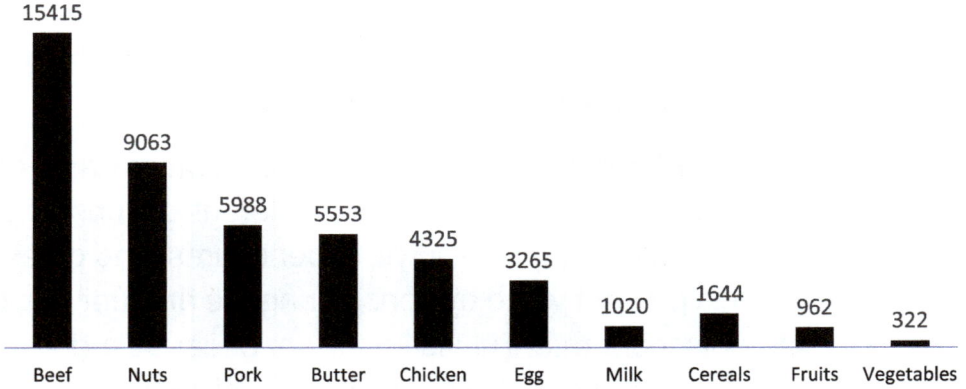

Figure 12.3 Water footprint (liter per kilogram of product) for various food items.
Source: Mekonnen and Hoekstra (2010). Created by Jian Peng

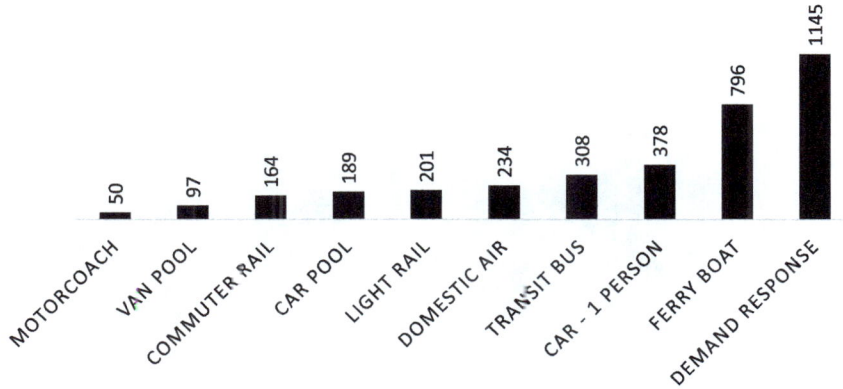

Figure 12.4 Carbon footprint for various transportation modes per capita in grams of carbon dioxide per mile.
Source: Bradley and Associates (2008). Created by Jian Peng

which transportation mode is the most environmentally friendly (Figure 12.4). Based on the data, it appears that long-distance motorcoach (such as GreyHound) has less than one twentieth the carbon footprint as demand response (online ride-hailing such as Uber or Lyft). There are a few factors that control the carbon footprint for transportation. First, shared services have lower carbon footprint. The reason why demand response vehicles have the highest carbon footprint is because the vehicles have significant amount of idle time, and the trip to pick up the passenger consumes additional fuel. Therefore, from the standpoint of carbon footprint, public transportation is the best way toward sustainability. Air travel has a very high carbon footprint too, and recently there have been calls for reduction in air travel to curb global carbon emission. For the same reason, high-density communities (apartment buildings and condominiums) with public transportation have lower per capita carbon footprint than the low-density community (e.g., single family homes) that relies mostly on passenger cars.

Considering the absolute amount of carbon dioxide emissions, light-duty passenger cars have the largest share of 59% due to the sheer number of this type of vehicle around the world, followed by medium and heavy duty trucks at 23%, followed by aircraft, rail, and ships. Clearly, as discussed in Chapter 10, light-duty passenger cars hold the key to transportation-related carbon emissions. To tackle this important source, USEPA set an ambitious goal of increasingly higher gas mileage targets, and the industry responded. From 2004 to 2017, the average gas mileage for passenger cars in the United States improved by an impressive 29%, reaching 24.9 miles per gallon.

Figure 12.5 Condensation trails (contrails) left by airplanes flying at high altitudes. Air travel could have a significant impact on global warming and pollution.

> ### Knowledge Box
>
> **Carbon Footprint in the Sky**—Air travel has significant environmental impact (**Figure 12.5**): aircraft engines produce greenhouse gases, heat, noises, and particulates at high altitudes (in the stratosphere). Due to global trade and communications, air travel has become more and more common over the years, with a 100% increase from 1999 to 2014, when 8.3 million people flew on a daily basis, producing about 900 million tons of carbon dioxide, or 2.4% of global carbon dioxide emissions. Because the emission takes place at high altitude and there are other pollutants, it is estimated that the global warming potential for aviation is 4.9% of all sources, doubling the global warming impact if the same amount of carbon dioxide were emitted on the ground level. This was perhaps the reason why Greta Thunberg traveled around the world for her *Skolstrejk för klimatet* campaign (see Chapter 11) on a solar-powered sail boat, not by air.

Similar calculation could be carried out for electricity production efficiency in terms of carbon footprint, as shown in **Figure 12.6**. It could be seen that fossil fuel sources (coal, oil, and natural gas) have much higher carbon footprint than those for renewable sources (solar, wind, and hydropower). Biomass has minimal impact on greenhouse gas emissions because it is renewable, even though efficiency is not high and its carbon footprint is not ideal. For energy sources with no actual carbon emission (such as solar), the assigned carbon footprint is calculated by estimating the carbon emission during

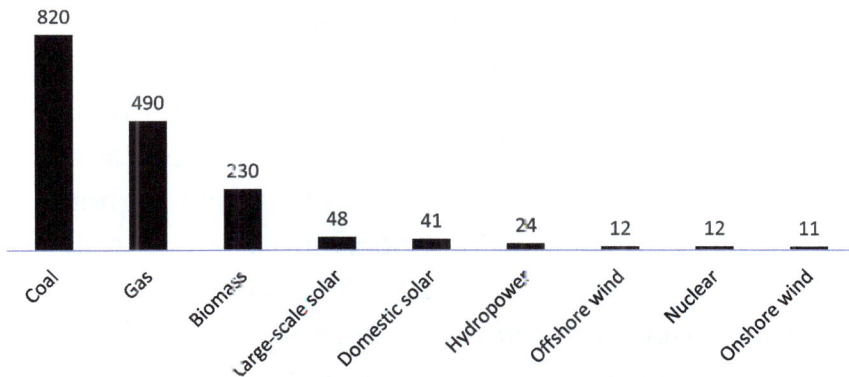

Figure 12.6 Carbon footprint for different energy sources in grams of carbon dioxide per kilowatt hour.
Data source: Sims et al. 2007 (IPCC). Created by Jian Peng

its manufacture, installation, maintenance, and dismantling/disposal after reaching the end of its useful life.

> ### Knowledge Box
>
> **The 2000-Watt Society:** The 2000-watt society is an environmental vision introduced in 1998 by the Swiss Federal Institute of Technology in Zürich for people to reduce their overall average primary energy usage to no more than 2,000 watts by the year 2050 without lowering their standard of living. At the time when this was proposed in 1998, the average power usage was 5,100 watts, while 2,000 watts was the global average. For reference, the energy consumption in the United States was 6,000 watts and 300 watts in Bangladesh.

12.3.4 Life Cycle Environmental Cost

During our daily life, we see people use many "cheap" items, coffee cups (no cost); straws (no cost); disposable food containers (no cost); plastic grocery bags (no or low cost[5]); cheap plastic toys, etc. However, we also see that a lot of these items, for one reason or another, end up in the environment and cause a series of environmental impacts and damages (Chapter 5 and 9). Cleaning up this garbage requires a significant amount of expenditure as well. Sometimes the cleanup cost may far exceed the manufacturing cost itself. One could argue, therefore, that the total environmental cost for these types of items should be much higher, and the manufacturers of these items

---

[5] At many places, grocery stores are starting to charge people a small amount of money if they want to use disposable plastic shopping bags. This is most often because of local laws aimed at preventing plastic pollution.

should bear these additional costs. Other items such as tires for automobiles, glass bottles, polystyrene foam, various disposal plastic packing materials, and many other items that either cannot be recycled or will pose other environmental issues after their use, should be considered to have high **life cycle environmental costs** as well. In this case, the absolute cost of these 'cheap items' will be much higher, making them less competitive than the other items such as paper cups, reusable items, etc. Life cycle environmental cost could be used in two ways. First, it could be used similar to carbon and water footprint as discussed before, taking into consideration other ecological impacts in addition to water and carbon, such as biodiversity; pollution reduction; solid waste reduction; and so on. For example, in work by Amienyo and Azapagic (2016) on the life cycle environmental cost for beers in the United Kingdom, the total cost included costs of raw materials, production, packaging, transportation of all the materials, and post-consumer waste disposal. Then, for each category, detailed breakdown and analyses were carried out. For example, there are nine raw materials (barley, hops, yeast, diatomaceous earth, sodium hydroxide, sulfuric acid, phosphoric acid, carbon dioxide, and light fuel oil); five packing materials (glass bottles, steel caps, cardboard, aluminum cans, and steel cans); six processes for production and filling that also require water and electricity, and so on. In the end, all of the costs for these materials and processes are added up to calculate that the total life cycle environmental costs for beers in the United Kingdom is 553 million pounds. From this exercise it was found that packaging, despite having a small portion (13%) of the cost, had a significantly larger portion (19–46%) of the environmental impacts.

The second way to consider total life cycle environmental cost is to evaluate different lifestyles. From an economic standpoint, financial well-being and quality of life are measured by the amount of money spent on everyday necessities and luxuries. The more a country's citizens spend (measured using the same currency), the better financial well-being the citizens will enjoy compared to those in other countries. Using the life cycle environmental cost methodology, it is possible to calculate total environmental impacts for different lifestyles. To make the result meaningful, the ecological footprint can be shown as the number of "Earths,"[6] with the global average ecological footprint of 1.7 Earths,[7] meaning that we are already overloading the Earth by 70% at the present time. The results for average citizens for countries around the globe are shown in **Figure 12.7**.

---

[6] This sustainability unit represents how many Earths would be needed to support your lifestyle in a sustainable manner if everyone on Earth lives like you. So, if your lifestyle scores 2 Earths, it would take another mother Earth to support 7.8 billion wasteful individuals like you.

[7] This is odd but true—our current collective lifestyle is not sustainable and we are depleting the Earth's resources much faster than we could replenish them. We are leaving behind a planet more stressed to our future generations than the one we inherited from our ancestors.

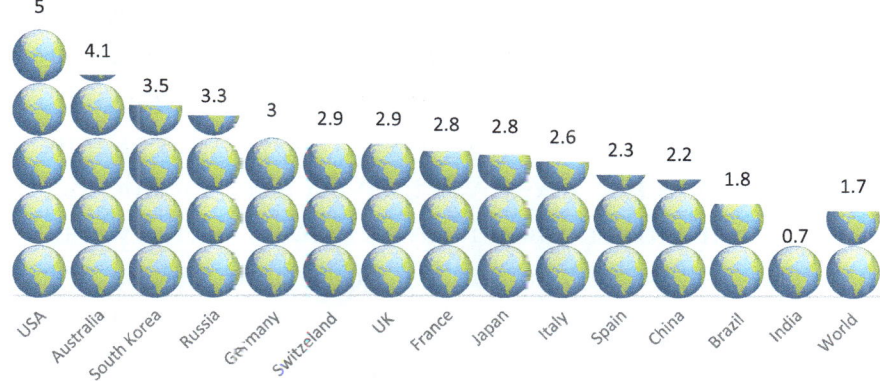

Figure 12.7 Ecological footprint of selected countries. Unit: number of Earths (needed to sustain the lifestyles).

Source: Created by Jian Peng using data from https://www.footprintnetwork.org

> ### Knowledge Box
>
> **Ecological Footprint Measured in Earths:** You can calculate your own ecological footprint by using this link: http://www.footprintcalculator.org/. Once you are done, you may be surprised to see the results. Think about what factor(s) may have affected your scores. You can also tweak some inputs and see how they affect your score. Hint: you may want to reduce your air travel in the future.

> ### Knowledge Box
>
> **Fast Fashion and Sustainability:** In September 2019, a fashion icon, Forever21, filed for bankruptcy. There are many reasons why such a once-popular fashion outlet went out of business, but a deeper dive in the underlying factors found that sustainability concerns of the younger generation played an important role in its demise. The so-called "'fast fashion" fad, as exemplified by Forever 21, Zara, H&M, and other smaller brands, promote cheap, stylish clothing for younger generations. They focus on fashion over quality and expect the consumer to wear the clothing for only a few times before discarding it. Similar to Forever 21, Zara is seeing declining sales in recent years. At the same time, the second-hand apparel market, which allows mostly young people to exchange used clothing, is expected to increase from $24 billion to $64 billion in 2028 (Global Data, 2018). Clearly, sustainability concerns are taking root in society and most encouragingly, in the minds of the young generation. There is hope.

To make this concept of life cycle environmental cost work, environmental impacts will have to be evaluated comprehensively and holistically. Manufacturers of these high environmental cost items could be made to pay for the post-consumption cleanup, disposal, and recycling of these commodities. There also needs to be a regulatory framework in place to enable true consideration of life cycle environmental costs, which should be reflected in the price of these commodities as surcharges, fees, and taxes. These surcharges, fees and taxes will then act as economic incentives and disincentives to encourage firms to produce environmentally friendly commodities because they represent better values for consumers. Sometimes, these environmental costs could be reflected on the prices of these items to act as disincentives for consumers to purchase them, and to bear part of the environmental costs. For example:

- Paper versus plastic straw or paper bags versus plastic bags: surcharges can be levied for plastic straws and plastic bags to encourage consumers to use paper straws, bring their own reusable straws, and use reusable grocery bags. In most of the developed countries and many developing countries, disposable plastic items are being phased out. Many passed legislation to prohibit the production of these items.
- Electric vehicle versus conventional internal combustion engine (ICE) vehicles: if carbon emission can be considered, conventional ICE vehicles could be levied a carbon emission surcharge. Or, cash or tax incentives can be granted to electric vehicles. In most countries, the latter has been implemented, resulting in increasing shares of EV on the global auto market.
- Organic versus modern farming: organic farm products could charge a premium over conventional or modern farming products. This has been implemented in many places, and organic farm products, which are becoming increasingly popular and commanding a clear premium over nonorganic products, have to be certified and inspected for sale.
- Conventional versus renewable energy sources: similar to EV versus ICE, renewable energy sources could be granted incentives; or regulation could mandate that each utility company produce a certain percentage of renewable energy in its energy portfolio. In fact, the latter has been implemented in many places, including California.

For many products, the environmental impact could be before (i.e., raw material sourcing) or after the useful life of the product (i.e., costs for recycling, disposal, etc.), such as the case for nuclear power, where the disposal after decommissioning could be a significant cost, and there are considerable uncertainties in the calculation of "true" life cycle costs due to regulations, public perception, and lifespan of nuclear plants.

Similar to above, life cycle cost benefit analysis could be beneficial to make the best business decisions. If environmental costs and benefits are considered fully, cost–benefit analysis could be a tool that promotes environmentally sustainable business solutions. For example, the author participated in a cost–benefit analysis study to identify the most cost-effective way to reduce illnesses incurred by swimmers who surf in the coastal ocean in Southern California. The current environmental regulation mandates limitations on fecal indicator bacteria (see discussions on Chapter 5), while surfer illnesses are caused by pathogens, not by fecal indicator bacteria. The cost–benefit analysis suggested that mitigation measures should focus on human pathogens instead of fecal indicator bacteria because the cost saving would be very significant. For the same amount of investment, management options that focus on reducing pathogens could yield up to 20 times more reduction in illnesses compared to options focused on fecal indicator bacteria.

12.4 Sustainability Issues

12.4.1 Environmental Mitigation

Here the concept of **environmental mitigation** will be constrained to its narrow sense, where the potential damages to the environment are mitigated or compensated by the responsible party that proposes a project or other activities that will cause these damages. Under both National Environmental Policy Act (NEPA) and California Environmental Quality Act (CEQA), the responsible party will need to create new habitats, purchase credits to such a habitat, enhance a habitat, or preserve/maintain existing (but unrelated to project) habitats in order to receive approval for a project that will take out or otherwise impact a habitat. In California, the requirements for environmental mitigation can sometimes significantly delay or terminate very large projects. Depending on the quality of habitat, the "mitigation ratio,"[8] could vary from 1:1 up to 10:1. Most such mitigation projects need to be maintained in perpetuity, presenting a significant compliance obligation for the project proponent. Therefore, many project proponents resort to mitigation banking, where a third party maintains and operates a "bank" of different habitats that are made available for project proponents to purchase for a fee, and the owner of the mitigation bank takes over the compliance obligation. In California, such a mitigation bank could be worth more than a million dollars an acre. This economic incentive has supported an industry of mitigation banks.[9] The mitigation requirements, even though burdensome, have contributed to the protection and restoration of the nation's wetlands and other valuable habitats.

12.4.2 Nature's Intrinsic Value

Discussion about sustainability would not make sense without examining the concept of nature. Coates (1995) in his book "Nature—Western Attitudes since Ancient Times" describes nature as having five layers of meanings: a physical environment without

---

[8] Mitigation ratio is the amount of habitat that the responsible party must create, enhance, or maintain, versus the habitat that will be taken out by the responsible for a project, such as a roadway through a wetland.

[9] See, for example, the Ecological Restoration Business Association, https://ecologicalrestoration.org/

human modification or under the threat of human activities; a collective phenomena of the world or universe, with or without humans; nature as an essence, quality and/or principle that informs the workings of the world or universe; nature as an inspiration and guide for people and source of authority governing human affairs; and; finally, nature as the conceptual opposite of culture. Therefore, sustainability could be explained as a state in which humans and nature coexist in harmony into perpetuity. Since human activities inevitably will interact with, sometimes harm, our natural environment, and nature as an essence cannot speak for itself, one may ask—does nature have **intrinsic value**? Should nature be given legal rights to be protected? There are many factors associated with this issue, which is a critical aspect in the environmental policy making process and in the center of debate among different interest groups. The answers to these two questions will underpin a policy making process and the agenda of an interest group. The answers to these questions have implications on issues like the right of humans to extract resources, turning habitats into urban uses, and polluting the environment in the process. They also influence opinion on issues such as genetic engineering that seemingly interferes with natural processes. With human's understanding and appreciation of Nature's intrinsic values changing with time, our attitude toward Nature and sustainability issues will evolve as well.

12.4.3 Deep Ecology and Sustainability

Deep ecology, as discussed first at Section 8.5.3, is a movement founded by Norwegian philosopher and environmentalist Arne Naess that establishes principles for the well-being of all life on Earth and for the richness and diversity of life forms. In this section, the implications of deep ecology on sustainability will be focused. Deep ecology promotes the inherent worth of living beings regardless of their instrumental utility to human needs. The movement advocates, among other things, a substantial decrease in human population and consumption along with the reduction of human interference with the natural world. To achieve this, deep ecologists advocate policies for basic economic, technological, and ideological structures that will improve quality of life rather than standard of living. Therefore, the concept of deep ecology requires a fundamental change in one's world views, personal values and ideology. Deep ecology principles also require its followers to make the necessary change happen by restructuring modern human societies, including the so-called billion year Sustainocene.[10]

As can be seen from the above, deep ecology respects nature's intrinsic value and does not treat humans as no superior beings over the natural world, including

---

[10] Sustainocene is a newly coined term following the same logic as "Anthropocene" at the end of the Holocene that started around 11,000 years ago. While the Anthropocene represents a period humans dominate the Earth in more or less a negative way, the Sustainocene could be considered a future stage of the Earth's history when humans achieve true sustainability and harmony with the natural world.

even the nonliving environment. This ideology is somewhat at odds with the philosophy of humanism, where humans are at the center of religion, literature, ideology, and the natural world. For this reason, humanism received criticism that an overarching and excessively abstract notion of humanity or human nature could lead to imperialism and domination of those less than human, including the natural world.

As a doctrine, deep ecology has the following tenets:

- Human and nonhuman life on Earth has a value independent of its usefulness to humans
- Biodiversity contributes to this value
- Humans have no right to reduce this biodiversity except to satisfy vital human needs
- A substantial percentage of the human population must be eliminated
- Humans interfere with the world too much already, and this activity is worsening
- A new political and economic model must be devised to replace that of present governments
- Individuals must be content with their living situation instead of striving for a higher standard of living
- Deep ecologists have an obligation to implement the above tenets

While most of the above make sense, the fourth tenet "a substantial percentage of the human population must be eliminated" received severe criticism and is deemed misanthropy or even ecofascism. While many agree that there are issues with anthropocentric worldview, which is the counterpart of deep ecology, it is more constructive to improve issues with the former (and there are many ways to do it) than to radically change it. In this sense, deep ecology largely embraces the concept of sustainability but with a somewhat radical twist.

Of many proposals that deep ecology believers promote, one well-known technology is global artificial photosynthesis, where renewable energy can be produced by directly harvesting sunlight using nanotechnology-synthetic biology systems. At the same time, carbon dioxide is consumed and oxygen is produced. In artificial photosynthesis, hydrogen gas instead of water can be produced and used as renewable energy. Such systems could be integrated into roads and buildings and drastically increase the effective areas where solar energy can be harnessed. Currently, this technology is still on research level, but its potential can be significant. In the best-case scenario, it might hold the key to true energy resolution and global sustainability because it will fundamentally change energy production, greenhouse gas emissions, and many environmental and socioeconomic issues.

> **Knowledge Box**
>
> **Arne Dekke Eide Næss and Deep Ecology:** Rachel Carson's Silent Spring has produced not just the 1960s environmental movement and a flurry of environmental regulations such as the Clean Water Act, it also inspired many people to be the leaders in other environmental fields. Naess was one such leader inspired by Rachel Carson. He was an influential Norwegian philosopher who coined the term "deep ecology." Believing in Gandhian nonviolence but taking direct actions as well, Naess believed that most environmental groups had focused on only superficial issues and largely failed to look into the underlying cultural and philosophical roots to these problems. To him, the difference lies in between deep and shallow ecological thinking. So, deep ecology promotes a true understanding of nature that appreciates the value of biological diversity and nature.

12.4.4 Resource Depletion

Nearly two thousand years ago, the Roman author Tertullian became pessimistic about the future of Earth because, in his view, humans had nearly exhausted Earth's resources:

> *"Everything has been visited, everything known, everything exploited. Beaches are plowed, mountains smoothed and swamps cleaned... Everywhere there are buildings, everywhere people, everywhere communities, everywhere life."*

Fast forward to the 1960s, Lake Erie's "death" (oxygen depletion caused by eutrophication) captured public concerns of ecological crisis together with other environmental disasters and nuclear arms race during the Cold War. Bill McKibben, an environmentalist and author, famously retreated from the city to live in the Adirondack mountains of upstate New York to escape from the world in a log cabin.[11] In his book "The End of Nature," he lamented that humans have so thoroughly altered Earth and modified nature that the concept of nature is no longer the one we had in mind.

McKibben's concern is not unfounded. Many of the Earth's resources, including natural lands, are being exploited much faster than the rate at which they are replenished. Some cannot be replenished at all. These resources include fossil fuels, minerals, biodiversity, or simply the wilderness. As mentioned in Chapter 10, the world's fossil

---

[11] The other side of the same story was that he resigned from the New Yorker magazine in protest of the departure of longtime editor William Shawn.

fuel reserves for coal, oil and natural gas will be exhausted within 100 years.[12,13] Other lesser-known resources such as rare-earth elements, phosphorus, copper, etc., are facing even worse fate because they are expected to be exhausted even sooner, and many such resources are vital to the global economy.

> ### Knowledge Box
>
> **Depletion of Phosphorus:** phosphorus is a critical element as chemical fertilizers (82% of its use and 7% for animal feed) as well as other applications. Currently global phosphorus production is about 250 million tons, and it is estimated that this resource, which is mostly from phosphate rocks, will exhaust in 50 to 100 years. New findings of mineral resources alleviated the worry to some extent, but it is clear that the world is facing a phosphorus shortage. Because nearly 90% of phosphorus is responsible for food production, a severe shortage of phosphorus will threaten the well-being of the global population profoundly. Possible solutions to this issue include demand reduction through smarter fertilizer use; recycling of phosphorus-materials (such as biosolids from wastewater treatment plants; see Chapter 5); and exploration of more mining areas or new technology for extraction of phosphorus at lower levels.

Such sustainability issues can manifest themselves in many forms. For example, the oil-rich nation United Arabic Emirates (UAE) is the 7th richest country in the world due mostly to its rich oil reserve, which accounts for more than 85% of its economy (**Figure 12.8**). With this finite resource, it is imperative that UAE diversify its economy. To date, the nation has invested in tourism, finance, manufacturing, aerospace, among others, such that once its oil reserve is exhausted, the nation has other means to survive. It also used its oil wealth to invest in ocean desalination to supply nearly a quarter of its freshwater supply, which is one of the key limiting resources for this desert nation. In contrast, the Micronesian island country Nauru is the third smallest country in the world with rich phosphate mineral reserves. At one point, the phosphate ores from Nauru was a significant portion of the world's phosphorus

---

[12] As alarming as it sounds, this may not be a bad thing because depletion of fossil fuels will passively curb greenhouse gas emission and prompt the world to seek renewable energy. This has been happening (see Chapter 10).

[13] As explained in Chapters 2 and 10, the principle of the conservation of mass determines that these resources are not simply being consumed and disappear, they are being consumed faster than the rate they are replenished naturally. For other elements, depletion means that there is no more exploitable mineral resources and these elements can only be obtained via recycling and reuse.

Figure 12.8 Burj Dubai skyline in 2016. Dubai is the biggest city of UAE and important financial center of the Middle East economy. In recent years, its economy has diversified from one that heavily relies on crude oil.

Figure 12.9 A surface phosphate mine site in Nauru, the third smallest country in the world but once accounted for a significant portion of the world's phosphorus supply. Since the mineral existed on the surface of the island, mining caused considerable destruction of landscape and other natural resources

production, bringing significant wealth to this tiny nation. With its natural resources quickly exhausted in late twentieth century, Nauru tried other means to remain economically viable with little success, and is now largely broke and dependent on Australia for financial assistance. With natural resources severely damaged by mining (**Figure 12.9**) and an economy in shambles, Nauru is a textbook case in sustainability. The Mayans, Rapa Nui (Easter Island) and other extinct civilizations are possible lessons for regional or national sustainability as well.

> **Knowledge Box**
>
> **Depletion of Rare Earth Elements (REEs):** Rare earth elements are seventeen chemical elements in the periodic table (see Figure 1.3). Fifteen of them are lanthanides, plus scandium and yttrium. Even though they are not necessarily the rarest of the elements (cerium is the 25th most abundant element in the Earth's crust), they are mostly dispersed in various minerals and are difficult to mine economically. They also tend to occur together. The production of REEs is quite uneven, and currently 81% of the world's production is from China. Most of the REEs are used for catalysts and magnets. Many are critical components of specialized industrial materials and processes. Similar to other finite resources, overexploitation of REEs will eventually result in resource depletion, but other factors such as geopolitical tensions[14] could affect the supply of REEs as well.

The alarming issue of resource depletion has caused many to rethink the waste management strategy (also see Chapter 9 on solid waste issues). Instead of the conventional 3-Rs (reduce, reuse, recycle), a more appropriate way is to have **5-Rs (refuse, reduce, reuse, recycle,** and **rot;** see **Figure 12.10**) with added "refuse" to prevent the resource from being consumed in the beginning. The "rot" in the end represents composting and other natural processes that return the waste back to the environment in a beneficial way. For waste management in general, traditional waste management should be expanded to waste prevention to reduce waste in the first place. Beyond waste prevention, efficient allocation of resources is more fundamental to ensure that the economy is "circular," i.e., one that generates minimal or no wastes. Beyond this, there are human well-being and ecosystem resilience that are rooted in the green economy (**Figure 12.11**).

Figure 12.10 5-Rs: refuse, reduce, reuse, recycle, rot.

Figure 12.11 The Green Economy related to waste and resources (Middleton, 2019).
Source: Created by Jian Peng

---

[14] For example, China has used REE export ban or quota to other nations due to political reasons.

The traditional value of making the most use of things still makes sense, as reflected by the following quote by Calvin Coolidge:

"Eat it up, wear it out, make it do, or do without!"

12.5 Global Sustainable Development

12.5.1 Global Sustainable Development Goals

Sustainability is a global issue and it makes the most sense to tackle this issue on a global scale. The **Sustainable Development Goals** (SDGs; see **Figure 12.12**) are the United Nations General Assembly's current harmonized set of seventeen future international development targets. SDGs were set in 2015 and aimed to be achieved by 2030. SDGs include the following:[15]

1. No poverty—"End poverty in all its forms everywhere." Currently, 1 in 10 people live on less than $1.25 a day. Elimination of poverty would also require provision of basic services such as healthcare, security, and education.
2. Zero hunger—"End hunger, achieve food security and improved nutrition and promote sustainable agriculture". Globally, 1 in 9 people are undernourished. To end hunger would require doubling agricultural productivity and incomes of small food producers, which may itself bring other environmental issues.
3. Good health and well-being for people—"Ensure healthy lives and promote well-being for all at all ages". There was a 47% increase in the mortality rate for

Figure 12.12 United Nations 17 sustainable development goals.

---

[15] For clarity, these goals are listed in the format of title—"short description," followed by discussion/elaboration of each specific goals.

children under 5, but still 5.6 million such children died in 2016 alone. Clean water, sanitation, and infectious diseases still pose significant health risks to many.

4. Quality education—"Ensure inclusive and equitable quality education and promote lifelong learning opportunities for all." Significant progress has been made and the number of out-of-school children decreased nearly 50% from 1997 to 2014. However, 22 million children still do not finish primary school.
5. Gender equality—"Achieve gender equality and empower all women and girls." There are still 52 countries that do not support gender equality. Child marriage and other gender-related issues persist in many countries to this day.
6. Clean water and sanitation—"Ensure availability and sustainable management of water and sanitation for all." Globally, 6 out of 10 people lack safe sanitation services. 2.6 billion people still use open defecation, posing a severe public health risk.
7. Affordable and clean energy—"Ensure access to affordable, reliable, sustainable and modern energy for all." As discussed in Chapter 10, energy consumption is required for any economic activity, and safe and clean renewable energy is critical to the global well-being. Only about half of the world use clean energy for cooking, missing the target of 95%.
8. Decent work and economic growth—"Promote sustained, inclusive and sustainable economic growth, full and productive employment and decent work for all." This includes long-term economic growth and infrastructure investments. For the poorest countries, an annual growth rate of 7% is desired. This would require innovation, entrepreneurship, and growth of small businesses and sustainable tourism, among others.
9. Industry, innovation, and infrastructure—"Build resilient infrastructure, promote inclusive and sustainable industrialization and foster innovation." This would include manufacturing jobs and income for the poorest countries.
10. Reducing inequalities—"Reduce inequality within and among countries." One of such targets is to sustain income growth of the bottom 40% of the population at a rate higher than the national average, and to reduce the transaction costs for migrant remittances to below 3% (currently it is 6%) to protect the migrant workers, often among the poorest of the population.
11. Sustainable cities and communities—"Make cities and human settlements inclusive, safe, resilient and sustainable." This includes affordable housing instead of slums or informal settlements.
12. Responsible consumption and production—"Ensure sustainable consumption and production patterns." This includes eco-friendly production methods and reduction of waste through recycling. This goal also called for a 10-year Framework of Programmes on Sustainable Consumption and Production for both developed and developing countries. An example of this effort is the "One Planet Network," which publishes related data on sustainable consumption and production. This network

includes six parts: Public Procurement; Consumer Information; Tourism; Lifestyles and Education; Buildings and Construction; Food Systems. These programs pool expertise and resources from hundreds of organizations across civil society, government, academia, and the private sector to help countries achieve their sustainable consumption and production priorities.

13. Climate action—"Take urgent action to combat climate change and its impacts by regulating emissions and promoting developments in renewable energy." This goal crystallized at the COP 21 Climate Change conference in Paris in December 2015 that enabled the nations to reach the sustainable development goals. It is also recognized that economic development and climate change are inextricably linked, particularly around poverty, gender equality, and energy, and public agencies should play a bigger role in minimizing the impacts on the environment. IPCC published a special report "Global Warming of 1.5°C" that outlined the impact of a 1.5°C global warming, and a solution would require that the global net human-caused carbon dioxide emissions to fall 45% from the 2010 levels by 2030 and reach "net zero" by 2050. This would require drastic changes in land, energy, industry, buildings, transportation, and cities.[16]

14. Life below water—"Conserve and sustainably use the oceans, seas and marine resources for sustainable development." The issues involved ocean acidification, marine biodiversity, coastal water quality/pollution, sustainable fishery, among others. As discussed before, ocean acidification is related to excessive fossil fuel consumption; pollution is caused by human activities as well, so marine protection requires many other sustainable actions as well.

15. Life on land—"Protect, restore and promote sustainable use of terrestrial ecosystems, sustainably manage forests, combat desertification, and halt and reverse land degradation and halt biodiversity loss." This goal articulates targets for protection of forest, desert, and mountain ecosystems and the biodiversity therein. The goal is to have a "land-degradation-neutral world," which can be achieved by restoration and preservation, as well as mitigation, as discussed previously.

16. Peace, justice, and strong institutions—"Promote peaceful and inclusive societies for sustainable development, provide access to justice for all and build effective, accountable and inclusive institutions at all levels." This includes reducing violent crime, sex trafficking, forced labor, and child abuse.

17. Partnerships for the goals—"Strengthen the means of implementation and revitalize the global partnership for sustainable development." To achieve the above 16 goals, international cooperation and sharing of knowledge, expertise, technology, and financial support is vital.

---

[16] This initiative was adopted by all but four nations: the United States, Saudi Arabia, Russia, and Kuwait. They however wanted to "note" these goals.

> ## Knowledge Box
>
> **Scientists Warning to Humanity:** in November 2017, 15,364 scientists signed World Scientists' Warning to Humanity: A Second Notice written by William J. Ripple and seven co-authors calling for, among other things, human population planning, and drastically diminishing per capita consumption of fossil fuels, meat, and other resources. The Second Notice followed the first notice issued in 1992 led by Henry W. Kendall, a former chair of the Union of Concerned Scientists (UCS) board of directors. A majority of Nobel Prize laureates in the sciences signed the document; about 1,700 of the world's leading scientists appended their signature. The notice included nine time-series graphs of key indicators, each correlated to a specific issue mentioned in the original 1992 warning, to show that most environmental issues are continuing to trend in the wrong direction, most with no discernible change in rate. The article included 13 specific steps humanity could take to transition to sustainability. Interestingly, an opposite viewpoint was given by Heidelberg Appeal, also signed by numerous scientists and Nobel laureates that criticized "an irrational ideology which is opposed to scientific and industrial progress, and impedes economic and social development." However, Heidelberg Appeal did not cite references or offer actionable recommendations.

> ## Knowledge Box
>
> **2019 Warning on Climate Change:** In November 2019, a group of more than 11,000 scientists from 153 countries published a proclamation in the journal Bioscience, led by William Ripple (lead scientist for the 1992 declaration), named climate change an "emergency" that would lead to "untold human suffering" if no big shifts in action takes place. The warning stated that: "we declare clearly and unequivocally that planet Earth is facing a climate emergency. To secure a sustainable future, we must change how we live. [This] entails major transformations in the ways our global society functions and interacts with natural ecosystems." This proclamation was made 40 years after the First World Climate Conference in Geneva, 1979.

These Sustainable Development Goals were a major step forward for the world to work together toward a better and more sustainable future. However, these goals are a result of difficult negotiations and compromises. With key opponents including the U.S. and Russia, it may not be economically viable especially considering the hefty price tag (trilions of dollars a year; up to $200 trillion by 2030 as estimated by the Rockefeller Foundation). These goals are all ambitious and wide-ranging including social, economic, and ecological goals (some are even competing against each other), hence diluting the significance of certain critical issues. After learning various global environmental challenges covered so far in this book, the reader may think that human existential issues such as climate actions (Goal #13), clean energy (Goal #7), and environmental goals (Goals #14 and 15) should have been given higher priorities. Clearly, more globally coordinated efforts are needed for every nation and all their citizens to march toward the same sustainable future.

12.5.2 The Earthrise

Now that major earthly environmental challenges have been discussed by this book, it may be beneficial to revisit Earth from a perspective initiated in Chapter 1, 'How to Build a Sustainable Planet'. Earth is a rather lonely planet (Figure 12.13) because scientists have not yet found a habitable planet after decades of search for similar planets. On the one hand, we should realize how lucky the Earth is (and we are), but at the same time the fact that we have only one such planet to live on and perish from poses a formidable challenge. Humans will have to ensure that Earth remains habitable for eternity because we have nowhere else to go.[17] To make matters worse, the solar system will not last

Figure 12.13 Earthrise—the Earth was on the horizon in the Mare Smythii region of the moon, captured on July 20, 1969 by Apollo 11.

---

[17] Some people such as Tesla's Elon Musk believes that humans can colonize Mars. It is unclear at this time whether such colonization can be sustainable or not. One thing is certain though, very few people will be able to afford it.

forever either because the sun will eventually burn out in about 5 billion years. By then it will turn into a red giant and will engulf Mercury, Venus, and probably Earth as well.

Within the solar system, Venus is too hot, Mars is too cold and too small, Europa (One of Jupiter's moons) is too cold. These Earth-like celestial bodies are all close to us, but not nearly as friendly as Earth due to size, water availability, temperature, and other environmental factors. Other solar systems within the Milky Way or other galaxies could provide new hope. If Earth is indeed not unique, the inability (so far, at least) to find any habitable planets given the vast number of stars[18] and possible solar systems will require some explanation. Earth has seen civilization for only a few thousand years, which is a blink of an eye for Earth's history and even shorter for the history of the universe. If there exist other civilizations, they must have been in existence for much longer and probably have figured out ways to live sustainably.

Our galaxy has about 400 billion stars. Using complex reasoning and calculation, Langmuir and Broecker (2012; see Chapter 1 for related discussions) postulated that there is 0.8% chance (based on the famous 'Drake Equation') that Earth has a companion within the galaxy. Since we have not found such planets despite extensive search, this may mean that sustainable living is not possible; or these planets exhausted their resources or self-destructed before we could connect with them; or Earth is indeed unique. In any case, we should really cherish our planet and our own existence, and do our best to protect it.

For life in our neighboring galaxies, the situation gets trickier because even the closest galaxy, Andromeda, is 2 million light years away. If we receive a 'friending' signal from one of the intelligent beings there, we would know that the signal came 2 million years ago and our friending signal back will take another 2 million years to reach them. This distressing fact has not deterred humans from reaching out to possible extraterrestrial (ET) intelligent life. In 1974, an interstellar radio message (the Arecibo message) directed at the globular star cluster M13 was sent via a radio telescope with simple representations of numbers, atoms, DNA, human forms, solar system, and so on.[19] M13, unfortunately is 25,000 light years away so we will not hear back from ET there (if there are any) for about 50,000 years. In 1977, the United States launched two Voyager spacecraft (Voyager 1 and Voyager 2, see **Figure 12.14**) to explore outer space. Both have reached interstellar space, and both carried information about life on Earth. To this day, we are still waiting for a response from our intelligent interstellar neighbor (if there is one).

---

[18] There are roughly 100 billion stars in each galaxy, but there may be as many as 1 trillion galaxies in the universe. So, the number of stars in the universe is about one billion trillion, or 10 to the power of 21.

[19] These messages were designed by Frank Drake at Cornell University. Drake created the famous equation (the Drake equation) that calculates the likelihood of existence of another planet Earth or intelligent life that Langmuir and Broecker (2012) cited.

Figure 12.14 An artist's illustration of the Voyager spacecraft flying by Jupiter. Both Voyager spacecrafts have reached interstellar space now.

> ## Knowledge Box
>
> **The Voyager Program:** both Voyager spacecraft (Figure 12.14) were launched in 1977 and are now more than 18 billion kilometers from Earth. They are so far from Earth that it will take light nearly a day to reach them. To this day, they are still transmitting information back to Earth, still probing the final intergalactic frontier. Each spacecraft carries a golden record of Earth sounds, pictures and messages. Designed to last billions of years, these discs may likely outlast Earth itself and be the only traces of human civilization after all is gone on Earth and the Earth itself exists no more. For more information, check NASA's Voyager information page at https://voyager.jpl.nasa.gov/mission/status/. For the fascinating story about the Golden Record, check out the Wikipedia page at https://en.wikipedia.org/wiki/Voyager_Golden_Record.

As discussed in Chapter 1, especially concerning the origin of life on Earth, there are eight factors that are critical to intelligent life, including the materials that are needed to form a solid planet and a volatile budget to form a proper atmosphere. Ocean and climate regulation are critical, and life needs to start. However, life on Earth is rather fragile, and it is even more so for humans. Anthropogenic and natural disasters such as war, famine, disease, resource depletion, destruction of biodiversity, super volcanoes/meteorite/solar flares, attack by hostile extraterrestrial life or robots could all take place to wipe out mankind from the surface of the Earth. Or these unfortunate events have happened on our companion planet somewhere in space, but we do not know (hence we have not heard from them). What we do know is that it is absolutely necessary that we realize the criticality of sustainability to not only correct the wrongs we have committed, but also to create a better environment for our future generations.

Further Reading

Langmuir, C.H. and W. Broecker. How to Build a Habitable Planet: A Story of Earth from the Big Bang to Humankind. Princeton, NJ: Princeton University Press, 2012.

Middleton, N. Global Casino: An Introduction to Environmental Issues, Abingdon, Oxfordshire: Routledge, 2019.

Adams, W.M. The Future of Sustainability: Re-thinking Environment and Development in the Twenty-first Century. Report of the IUCN Renowned Thinkers Meeting, 29–31 January 2006 (PDF). Retrieved 16 February 2009.

Thomas, Steve A. The Nature of Sustainability. Grand Rapids, MI: Chapbook Press, 2006.

Scott Cato, M. Green Economics. London: Earthscan, pp. 36–37. ISBN 978-1-84407-571-3, 2009.

USEPA, 1998. Civil Penalty Policy for Section 311(b)(3) and Section 311(j) of the Clean Water Act. USEPA Office of Enforcement and Compliance Assurance. August, 1998.

World Bank. 2011. Methodology for calculating environmental damage assessment and relevant compensation (English). Washington, DC: World Bank. http://documents.worldbank.org/curated/en/804831463331771041/Methodology-for-calculating-environmental-damage-assessment-and-relevant-compensation.

Lefko, M. Global Sustainability: 21 Leading CEOs Show How to Do Well by Doing Good, New York, NY: Morgan James, 2017.

Bradley and Associates LLC. Updated comparison of energy use and CO_2 emissions from different transportation modes. Technical report submitted to: American Bus Association Foundation. Washington, DC, USA, 2008

Lincoln, D. Alien Universe: Extraterrestrial Life in Our Minds and in the Cosmos. Baltimore, MD: The Johns Hopkins University Press, 2013.

The Environmental Business International: Environmental Industry Overview 2017-2018, https://ebionline.org/product/2018-environmental-industry-overview/

Mekonnen, M. M. and A. Y. Hoekstra. The green, blue and grey water footprint of farm animals and animal products. Volume 1: Main Report. UNESCO-IHE., Institute for Water Education. 50 pp., 2010.

Mekonnen, M. M. and A. Y. Hoekstra. The green, blue and grey water footprint of crops and derived crop products. Volume 2. Appendices main report. Value of Water Research Report Series No. 47. UNESCO-IHE Institute for Water Education. 1196 pp., 2010.

Straughan, B. and T. Pollak. The Broader Movement: Nonprofit environmental and conservation organization, 1989-2005. 2008. The Urban Institute. Weblink: https://www.urban.org/sites/default/files/publication/32186/411797-The-Broader-Movement-Nonprofit-Environmental-and-Conservation-Organizations—.PDF. Accessed: March 29, 2020.

Richard H. Thaler. Misbehaving: The Making of Behavioral Economics, New York, NY: Norton & Company, 2015.

Sims R.E.H., et al. Energy supply. In Climate Change 2007: Mitigation. Contribution of Working Group III to the Fourth Assessment Report of the Intergovernmental Panel on Climate Change [B. Metz, O.R. Davidson, P.R. Bosch, R. Dave, L.A. Meyer (eds)], New York, NY: Cambridge University Press, 2007.

Web Resources

Ecological footprint calculator: http://www.footprintcalculator.org/

Hove Social Good Intelligence: https://hovedata.com/

Water Footprint Calculator: https://www.watercalculator.org/

Carbon Footprint Calculator: https://www.watercalculator.org/ and https://www.nature.org/en-us/get-involved/how-to-help/carbon-footprint-calculator/

Carbon footprint of common products: *co2list.org*. Retrieved 4 October 2019.

iPhone carbon footprint: https://www.cnet.com/news/apple-iphone-x-environmental-report/

Freeman, R. E., 2014, Is profit the purpose of business? Weblink: https://ideas.darden.virginia.edu/is-profit-the-purpose-of-business

Similar to Forever21, H&M' story about cheap clothing and sustainability: https://www.cnn.com/2020/05/03/business/cheap-clothing-fast-fashion-climate-change-intl/index.html

Natural Resources Defense Council website: https://www.nrdc.org

The Sierra Club website: https://www.sierraclub.org/

United States Fish and Wildlife Services website: https://www.fws.gov/

United States Geological Survey website: https://www.usgs.gov/

Footprint calculator: https://www.footprintnetwork.org/resources/footprint-calculator/

Lecture on Sustainocene by Harvard professor Daniel Nocera: https://www.youtube.com/watch?v=u92O8LSkezY

One Planet Network—a window into sustainable consumption and production across the globe. https://www.oneplanetnetwork.org/

NASA's information page on the background and status of the Voyager Program: https://voyager.jpl.nasa.gov/mission/status/

Wikipedia page on Voyager's Golden Record: https://en.wikipedia.org/wiki/Voyager_Golden_Record

Questions and Exercises

1. Imagine that you identified a habitable planet just outside of the solar system. The rocket you can use has a maximum speed of 10,000 miles an hour. The radius of the solar system is about 140 billion kilometers. Think about the challenges you may face before you set out for your journey to colonize this planet.
2. Calculate your water footprint using the tools provided in the Web Resources.
3. Calculate your carbon footprint using the tools provided in the Web Resources.
4. Look at the 17 Sustainable Development Goals established by the UN. Why do you think social and economic goals are also important for a sustainable human society? What if we ignore them and focus only on the "true" environmental issues such as pollution and climate change?
5. List five most important issues you have learned after reading this book.
6. Who do you think should bear the life cycle environmental costs of item such as free disposable items such as plastic cups, packaging, straws, and grocery bags? Manufacturer or consumer? State your reason(s).
7. This book does not cover the overpopulation issue since it is more a socio-economic issue than an environmental issue. Many scholars believe that it is the root cause of most of the environmental issues and is the main threat to the global sustainability. Do you agree? Why or why not?
8. List some actions that you can take toward a more sustainable lifestyle. Share the list with your friends/class and implement them.

Bibliography

1. Abdel-Rahman, A.A. On the emission from internal combustion engines: A review. *International Journal of Energy Research* 22 (1998): 483–513.

2. Adams, W.M. *The Future of Sustainability: Re-thinking Environment and Development in the Twenty-first Century*. Report of the IUCN Renowned Thinkers Meeting, January 29–31, 2006. Retrieved February 16, 2009.

3. Bamber, L. Jonathan, Michael Oppenheimer, Robert E. Kopp, Willy P. Aspinall, and Roger M. Cooke. "Ice Sheet Contributions to Future Sea-Level Rise from Structured Expert Judgment." *Proceedings of the National Academy of Sciences* 116, no. 23 (May 2019): 11195–200.

4. Bar-On, Yinon M., Rob Philips, and Ron Milo. "The Biomass Distribution on Earth." *PNAS*. 115, no. 25 (2018): 6506–11.

5. Berner, Elizabeth Kay, and Robert A. Berner. *The Global Water Cycle*. Upper Saddle River, NJ: Prentice-Hall, 1987.

6. Blasing, T. J. (2013): *Recent Greenhouse Gas Concentrations*. Carbon Dioxide Information Analysis Center (CDIAC), Oak Ridge National Laboratory (ORNL), Oak Ridge, TN (United States). January 1, 2016, doi:10.3334/CDIAC/ATG.032.

7. Blatt, H. *Our Geologic Environment*. Upper Saddle River, NJ: Prentice Hall, 1997.

8. Bond, David P.G., and Paul B. Wignall. "The Role of Sea-Level Change and Marine Anoxia in the Frasnian–Famennian (Late Devonian) Mass Extinction." *Palaeogeography, Palaeoclimatology, Palaeoecology* 263, no. 3–4 (2008): 107–18. doi:10.1016/j.palaeo.2008.02.015.

9. Bradley and Associates LLC. *Updated Comparison of Energy Use and CO_2 Emissions from Different Transportation Modes*. Technical Report. Washington, DC, USA: American Bus Association Foundation, 2008.

10. Brooks, Bryan W., Tara Sabo-Attwood, Kyungho Choi, Sujin Kim, Jakub Kostal, Carlie A. LaLone, Laura M. Langan, Luigi Margiotta-Casaluci, Jing You, and Xiaowei Zhang. "Toxicology Advances for 21st Century Chemical Pollution." *One Earth*, 2, no. 4 (2020): 312–16.

11. Brown, Larry R., Robert H Gray, Robert M Hughes, and Michael R Meador (eds). "Effects of Urbanization on Stream Ecosystems." In *Proceedings of the Symposium*, Quebec City, Quebec, Canada, August 11 and 12, 2003.

12. Browning, Edgar K., and Mark Zupan. *Microeconomic Theory and Applications*, 5th ed., New York, NY: HarperCollins College Publishers, 1996, ISBN 0-673-523810.
13. Buchel, K.H. *Chemistry of Pesticides*. New York, NY: John Wiley and Sons, 1983, ISBN 0-471-05682-0.
14. Byron, P.G. Legal opinion on the court case: RB Jai Alai, LLC v. Secretary of The Florida Department of Transportation, 112 F.Supp.3d 1301, 1307-1308 (M. D. Fla. 2015), 2015.
15. California Department of Water Resources. "California Waterways Map." Accessed May 1, 2020. https://water.ca.gov/What-We-Do/Education/Education-Materials.
16. Cao, Yiping, Mano Sivaganesan, Catherine A. Kelty, Dan Wang, Alexandria B. Boehm, John F. Griffith, Stephen B. Weisberg, and Orin C Shanks. "A Human Fecal Contamination Score for Ranking Recreational Sites Using the HF183/BacR287 Quantitative Real-Time PCR Method." *Water Research* 128 (2018): 148–56.
17. Carson, Rachel. *Silent Spring*. Boston MA: Houghton Mifflin Company, 1962. ISBN 0-618-25305-x.
18. Cato, Scott M. *Green Economics*. London: Earthscan, 36–37, 2009. ISBN 978-1-84407-571-3.
19. Ceballos, Gerardo, Paul R. Ehrlich, Anthony D. Barnosky, Andres Garcia, Robert M. Pringle, and Todd M. Palmer. "Accelerated Modern Human-Induced Species Losses: Entering the Sixth Mass Extinction." *Science Advances* 1, no. 5 (2015): e1400253.
20. Cepero, Almudena, Jorgen Ravoet, Tamara Gómez-Moracho, José Luis Bernal, Maria J. Del Nozal, Carolina Bartolomé, Xulio Maside, Aránzazu Meana, Amelia V. González-Porto, Dirk C. de Graaf, Raquel Martín-Hernández, and Mariano Higes. (15 September 2014). "Holistic Screening of Collapsing Honey Bee Colonies in Spain: A Case Study." *BMC Research Notes* 7: 649. doi:10.1186/1756-0500-7-649.
21. Closs, Gerry, Barbara Downes, and Andrew Boulton. *Freshwater Ecology: A Scientific Introduction*. Hoboken, NJ: Blackwell Publishing, 2004. ISBN-13: 978-0-632-05266-0.
22. Coates, Peter. *Nature—Western Attitudes since Ancient Times*, Berkeley, CA: University of California, 1998.
23. Cohen, Bernard L. "Breeder Reactors: A Renewable Energy Source." *American Journal of Physics* 51, no. 1 (1983): 75–76. doi:10.1119/1.13440.
24. Cohen, Michael D., James G. March, and J. P. Olson. "A Garbage Can Model of Organizational Choice." *Administrative Science Quarterly* 17 (1992): 1–25.
25. Council and Community Action in the Climate Emergency. "History of Local Government Declarations." Retrieved 2020-01-06. https://www.caceonline.org/.

26. Dahl, T.E. "Wetlands Loss Since the Resolution." *National Wetlands Newsletter* 12, no. 6 (1990). Washington DC: Environmental Law Institute.
27. Davis, Mackenzie L., and David A. Cornwell. *Introduction to Environmental Engineering*, 5th ed. New York, NY: McGraw Hill, 2013.
28. DeOreo, William B., Peter Mayer, Benedykt Dziegielewski, and Jack Kiefer. *Residential End Uses of Water*, Version 2. Denver, CO: Water Research Foundation, 2016.
29. DeWolf, Wendy. "Engineering Clean Water." *Yale Scientific Magazine*, April 3, 2011.
30. Drever, James I. *The Geochemistry of Natural Waters: Surface and Groundwater Environments*, 3rd ed. Upper Saddle River, NJ: Prentice-Hall, 1997.
31. Dufour Al, Jamie Bartram, Robert Bos, and Victor Gannon, eds. *Animal Waste, Water Quality and Human Health*, London, UK: International Water Association Publishing, 2012.
32. Einstein, Charles. *The Day New York Went Dry*. New York, NY: Fawcett Publications, 1964.
33. The Environmental Business International: Environmental Industry Overview 2017-2018. https://ebionline.org/product/2018-environmental-industry-overview/
34. European Environmental Agency (EEA). *Air Quality Directive—Air Quality Standards*. 2016. Accessed April 26, 2020. https://www.eea.europa.eu/data-and-maps/figures/air-quality-standards-under-the/table-1.eps/image_large.
35. Faure, Gunter. *Principles and Applications of Geochemistry*, 2nd ed., New Jersey: Prentice Hall, 1998.
36. Feng, Huiyun, Dongxia Wei, Xuan Li, Yuheng Zhao, and Zengliang Yu. "Atmospheric Nitrogen and Phosphorus Deposition in the Chaohu Lake Watershed, Anhui, China: Research and Considerations." In *Proceedings of the Expert Consultation Summit on Integrated Environmental Management of Chaohu Lake*, Hefei, Anhui Province, China, February 4, 2018.
37. Fiorino, Daniel J. *Making Environmental Policy*. Berkeley, CA: University of California, 1995.
38. Flagan, Richard C., and John H. Seinfeld, 1988. *Fundamentals of Air Pollution Engineering*. Englewood Cliffs, NJ: Prentice Hall, 1988.
39. Fletcher, Tim D., William Shuster, William F. Hunt, Richard Ashley, David Butler, Scott Arthur, Sam Trowsdale, Sylvie Barraud, Annette Semadeni-Davies, Jean-Luc Bertrand-Krajewski, Peter Steen Mikkelsen, Gilles Rivard, Mathias Uhl, Danielle Dagenais, and Maria Viklander. "SUDS, LID, BMPs, WSUD and More – The Evolution and Application of Terminology Surrounding Urban Drainage." *Urban Water* 12, no. 7 (2015): 525–42. doi:10.1080/1573062X.2014.916314.

40. Garrels, Robert M., and F.T. McKenzie. *Evolution of Sedimentary Rocks*. New York, NY: W.W. Norton. Chapter 6, 1971.
41. Geyer, Roland, Jenna R. Jambeck, and Kara Lavender Law. "Production, Use, And Fate of All Plastics Ever Made." *Science Advances* 3, no. 7 (2017): e1700782.
42. Goddard Space Flight Center (NASA). 2017. Global Mean Sea Level Trend from Integrated Multi-Mission Ocean Altimeters TOPEX/Poseidon, Jason-1, OSTM/Jason-2 Version 4.2 Ver. 4.2 PO.DAAC, CA, USA. Dataset accessed [6/6/2020] at http://dx.doi.org/10.5067/GMSLM-TJ42.
43. Goldfarb, Theodore D. *Taking Sides: Clashing Views on Controversial Environmental Issues*, 6th ed., New York: The Dushkin Publishing Group, 1995.
44. Haeckel, Ernst. *The History of Creation*. New York: D. Appleton and Company, 1880 (this book is freely available at Project Gutenberg at: http://www.gutenberg.org/files/40472/40472-h/40472-h.htm).
45. Harrop, Martin, and Rod Hague. *Comparative Government and Politics: An Introduction*. Okehampton and Rochdale, UK: Macmillan International Higher Education, 2013.
46. Harb, Moustapha, Phillip Wang, Ali Zarei-Baygi, Megan H. Plumlee, and Adam L. Smith. "Background Antibiotic Resistance and Microbial Communities Dominate Effects of Advanced Purified Water Recharge to an Urban Aquifer." *Environmental Science and Technology* 6, no. 10 (2019): 578–84.
47. Hayden, F. Gregory. "Policymaking Network of the Iron-Triangle Subgovernment for Licensing Hazardous Waste Facilities." *Journal of Economic Issues* 36, no. 2 (2002): 479.
48. Hawking, Stephen W. *A Brief History of Time—From the Big Bang to Black Holes*. London, UK: Bantam Dell Publishing Group, 1988.
49. Hoffmann, Michael, Craig Hilton-Taylor, Ariadne Angulo, Monika Böhm, Thomas M. Brooks, Stuart H. M. Butchart, Kent E. Carpenter, Janice Chanson, Ben Collen, Neil A. Cox, et al. "The Impact of Conservation on the Status of the World's Vertebrates." *Science* 330 (2010): 1503–09.
50. Hore-Lacy, Ian. "Future Energy Demand and Supply." In *Nuclear Energy in the 21st Century*, 2nd ed., London, UK: World Nuclear University, chapter 1, sec. 6, pp. 9, 2011.
51. Humes, Edward. *Garbology: Out Dirty Love Affair with Trash*. New York, NY: The Penguin Group, 2013.
52. International Energy Agency. 2011. Solar Energy Perspectives: Executive Summary (PDF). Archived from the original (PDF) on January 13, 2012.
53. Intergovernmental Panel on Climate Change (IPCC). *Climate Change 1994: Radiative Forcing of Climate Change and an Evaluation of the IPCC IS92 Emission Scenarios*. Houghton, J.T. et al. (eds.), New York: Cambridge University Press, 1994.

54. IUCN, 2016. IUCN's Classification of Direct Threats (v2.0). https://cmp-open-standards.org/library-item/threats-and-actions-taxonomies/. August 23, 2019. Retrieved May 10, 2020.

55. Jambeck, Jenna R., Roland Geyer, Chris Wilcox, Theodore R. Siegler, Miriam Perryman, Anthony Andrady, Ramani Narayan, and Kara Lavender Law. "Plastic Waste Inputs from Land into the Ocean." *Science* 347, no. 6223 (2015): 768–71.

56. The Kaspari Lab. 10 Principles of Ecology. https://michaelkaspari.org/2017/07/17/the-ten-principles-of-ecology/

57. Kaza, Silpa, Lisa Yao, Perinaz Bhada-Tata, and Frank Van Woerden. *What a Waste 2.0: A Global Snapshot of Solid Waste Management to 2050. Urban Development Series*. Washington, DC: World Bank. doi:10.1596/978-1-4648-1329-0. License: Creative Commons Attribution CC BY 3.0 IGO, 2018. https://openknowledge.worldbank.org/handle/10986/2174.

58. Kasting, James. *How to Find a Habitable Planet*. Princeton, NJ: Princeton University Press, 2010.

59. Keeling, Charles D., Robert B. Bacastow, Arnold E. Bainbridge, Carl A. Ekdahl Jr., Peter R. Guenther, Lee S. Waterman, and John F. S. Chin. "Atmospheric Carbon Dioxide Variations at Mauna Loa Observatory, Hawaii." *Tellus* 28 (1976): 538–51.

60. Kharecha, Pushker A., and James E. Hansen. "Prevented Mortality and Greenhouse Gas Emissions from Historical and Projected Nuclear Power." *Environmental Science and Technology* 47 (2013): 4889–95.

61. Kiehl, J. T., and Kevin E. Trenberth. "Earth's Annual Global Mean Energy Budget." *Bulletin of the American Meteorological Society* 78, no. 2 (February 1997): 197–208.

62. King, Anthony. "John W. Kingdon, Agendas, Alternatives, and Public Policies, Boston: Little, Brown." *Journal of Public Policy* 5, no. 2 (1985): 281–87.

63. Langmuir Charles H., and Wallace S. Broecker. *How to Build a Habitable Planet—The Story of Earth from the Big Bang to Humankind*. Princeton, NJ: Princeton University Press, 2012.

64. Leakey, Richard E. *The Origin of Humankind*. New York: Basic Books, 1994.

65. Lefko, Mark. *Global Sustainability: 21 Leading CEOs Show How to do Well by Doing Good*. New York, NY: Morgan James, 2017.

66. Lincoln, Don. *Alien Universe: Extraterrestrial Life in Our Minds and in the Cosmos*. Baltimore, MD: The Johns Hopkins University Press, 2013.

67. Linker, Lewis C., Richard A. Batiuk, Gary W. Shenk, and Carl F. Cerco. "Development of the Chesapeake Bay Watershed Total Maximum Daily Load Allocation." *JAWRA Journal of the American Water Resources Association* 49, no. 5 (2013): 986–1006.

68. Locey, Kenneth J. and Jay T. Lennon. "Scaling Laws Predict Global Microbial Diversity." *PNAS* 113, no. 21 (2016): 5970–75.
69. Lynas, M. *Six Degrees: Our Future on a Hotter Planet*. Washington, DC: The National Geographic Society, 2008.
70. Malone, Linda. *Environmental Law: Emanuel Law Outlines Series*, 2nd ed., New York, NY: Aspen Publishers, Wolters Kluwer Law and Business, 2007.
71. Mankiw, N. Gregory. *Principles of Macroeconomics*, 8th ed., Boston. MA: Cengage Learning, 2018. ISBN-13: 978-1-305-97150-9.
72. Martin, Daniel, Helen McKenna, and Valerie Livina. "The Human Physiological Impact of Global Deoxygenation." *Journal of Physiological Science*, 67, no. 1 (2017): 97–106.
73. Mazor, R. et al, 2015. *Bioassessment of Perennial Streams in Southern California: A Report on the First Five Years of the Stormwater Monitoring Coalition's Regional Stream Survey*. Southern California Coastal Water Research Project. Technical Report 844. Costa Mesa, California, USA.
74. McCarthy, J. "Facts from Cohen and Others". Progress and its Sustainability. Stanford. Archived from the original on 2007-04-10. Retrieved 2006-11-09. (2006).
75. McConnel, Campbell R., and Stanley L. Brue. *Economics: Principles, Problems, and Policies*, 12th ed., New York, NY: McGraw-Hill, 2005, ISBN 0-07-112716-x.
76. Mckinney, Michael L. "How do Rare Species Avoid Extinction? A Paleontological View." In *The Biology of Rarity: Causes and Consequences of Rare-Common Differences*, edited by W.E. Kunin, and K.J. Gaston, Netherlands: Springer, 2012.
77. Meacham, Megan, Cibele Queiroz, Albert V. Norström, and Garry D. Peterson. "Social-Ecological Drivers of Multiple Ecosystem Services: What Variables Explain Patterns of Ecosystem Services Across the Norrström Drainage Basin?." *Ecology and Society* 21, no. 1 (2016): 14. doi:10.5751/ES-08077-210114.
78. Mekonnen, Mesfin Mergia, and Arjen Hoekstra. 2010a. *The Green, Blue and Grey Water Footprint of Farm Animals and Animal Products*. Volume 1: Main Report. UNESCO-IHE., Institute for Water Education. 50 pp.
79. Mekonnen, Mesfin Mergia, and Arjen Hoekstra. 2010b. *The Green, Blue and Grey Water Footprint of Crops and Derived Crop Products*. Volume 2: Appendices Main Report. Value of Water Research Report Series No. 47. UNESCO-IHE Institute for Water Education. 1196 pp.
80. Mesler, Bill, and H. James Cleaves II. *A Brief History of Creation*. New York, NY: W.W. Norton and Company, 2016.
81. Middleton, Nick. *Global Casino: An Introduction to Environmental Issues*. Abingdon, Oxfordshire: Routledge, 2019.

82. Miller, R.A. "Analysis of Information Contained in the Completed North American Innovative Remediation Technology Demonstration Projects." Pittsburgh, Pennsylvania, USA: Groundwater Remediation Technologies Analysis Center, TI-97-01, 1997.

83. Molina, M. J. & F.S. Rowland. "Chlorine Atom Catalysed Destruction of Ozone." *Nature* 249 (1974): 810–12. doi:10.1038/249810a0.

84. Muir, Dereck, et al. "Levels and Trends of Poly- And Perfluoroalkyl Substances in the Arctic Environment—An Update." *Emerging Contaminants* 5 (2019): 240–71.

85. Næss, Arne. "The Shallow and the Deep, Long"Range Ecology Movement. A Summary." *Inquiry* 16, no. 1–4 (1973): 95–100. doi:10.1080/00201747303601682.

86. Nicolia, Alessandro, Alberto Manzo, Fabio Veronesi, and Daniele Rosellini. "An Overview of the Last 10 Years of Genetically Engineered Crop Safety Research." *Critical Reviews in Biotechnology* 34, no. 1 (2013): 77–88.

87. O'Connor, Mary I., Matthew W. Pennell, Florian Altermatt, Blake Matthews, Carlos J. Melián, and Andrew Gonzalez. "Principles of Ecology Revisited: Integrating Information and Ecological Theories for a More Unified Science." *Frontiers in Ecology and Evolution*, June 18, 2019.

88. OECD. Guidance Manual For The Implementation Of Council Decision C(2001)107/Final, As Amended, On the Control Of Transboundary Movements Of Wastes Destined For Recovery Operations, 2009.

89. Palyaa, Annie P., Ian S. Buick, and Gray E. Bebouta. "Storage and Mobility of Nitrogen in the Continental Crust: Evidence from Partially Melted Metasedimentary Rocks, Mt. Stafford, Australia." *Chemical Geology* 281, no. 3–4 (2011): 211–26.

90. Peplow, M. "Special Report: Counting the Dead". *Nature* 440, no. 7087 (2006): 982–83.

91. Pirkle, James L., Debra J. Brody, Elaine W. Gunter, Rachel A. Kramer, Daniel C. Paschal, Katherine M. Flegal, and Thomas D. Matte. "The Decline in Blood Lead Levels in the United States. The National Health and Nutrition Examination Surveys." *Journal of the American Medical Association* 272, no . 4 (1994): 234–91.

92. Prata, Joana Correia. "Microplastics in Wastewater: State of the Knowledge on Sources, Fate and Solutions." *Marine Pollution Bulletin* 129, no. 1 (2018): 262–65.

93. Ripple, William, Christopher Wolf, Thomas Newsome, Phoebe Barnard, and William Moomaw. "World Scientists' Warning of a Climate Emergency". *BioScience* 70, no. 1 (2020): 8–12.

94. Ripple, William J., Christopher Wolf, Thomas M. Newsome, Matthew G. Betts, Gerardo Ceballos, Franck Courchamp, Matt W. Hayward, Blaire Van Valkenburgh, Arian D. Wallach, and Boris Worm. "Are We Eating the World's Megafauna to Extinction?" *Conservation Letters* 12, no. 3 (2019): e12627

95. Rong, Y., ed. *Fundamentals of Environmental Site Assessment and Remediation*, Boca Raton, FL: CRC Press, 2018.
96. Rosenbaum, Walter A. *Environmental Politics and Policy.* Washington, DC: CQ Press, 2005.
97. Rowland, F. Sherwood. "Stratospheric Ozone Depletion." *Philosophical Transactions of the Royal Society B* 361, no. 1469 (2006): 769–90. doi:10.1098/rstb.2005.1783.
98. Rubin, Edward S. *Introduction to Engineering and the Environment.* Routledge, Oxford, UK: McGraw-Hill Higher Education, 2001. ISBN 0-07-235467-4.
99. Sato, Toshio, Manzoor Qadir Sadahiro Yamamoto, Tsuneyoshi Endo, and Ahmad Zahoor. "Global, Regional, and Country Level Need for Data on Wastewater Generation, Treatment, And Use." *Agricultural Water Management* 130 (2013): 1–13.
100. Schaefer, Richard T. *Sociology: A Brief Introduction*, 7th ed. New York, NY: McGraw-Hill, 2008, ISBN 0-07-352805-6.
101. Schnoor, Jerald L., and Werner Stumm. "Acidification of Aquatic and Terrestrial Systems." In *Chemical Processes in Lakes*, edited by W. Stumm, 311-338. New York, USA: Wiley Interscience, 1985.
102. The Secretariat of the Basel Convention (SBC). Basel Convention on the control of transboundary movements of hazardous wastes and their disposal. Protocol on Liability and Compensation for damage resulting from transboundary movements of hazardous wastes and their disposal. Texts and Annexes. United Nations Environment Programme, 2018.
103. Shakhova, N., I. Semiletov, A. Salyuk, D. Kosmach, and N. Bel'cheva. "Methane release on the Arctic East Siberian Shelf". *Geophysical Research Abstracts.* 9 (2017): 01071.
104. Seager, Sara, and Drake Deming. "Exoplanet Atmospheres." *Annual Review of Astronomy and Astrophysics* 48 (2010): 631–72.
105. Sims, R.E.H, et al. "Energy Supply." In *Climate Change 2007: Mitigation. Contribution of Working Group III to the Fourth Assessment Report of the Intergovernmental Panel on Climate Change*, edited by B. Metz, O.R. Davidson, P.R. Bosch, R. Dave, L.A. Meyer, Cambridge, United Kingdom and New York, NY, USA: Cambridge University Press, 2007.
106. Skarke, A., C. Ruppel, M. Kodis, D. Brothers, and E. Lobecker. "Widespread Methane Leakage from the Sea Floor on the Northern US Atlantic Margin." *Nature Geoscience* 7, no. 9 (2014): 657–61. doi:10.1038/ngeo2232.
107. State of California. *Water Quality Control Plan for Control of Temperature in the Coastal and Interstate Waters and Enclosed Bays and Estuaries of California (California Thermal Plan).* 1975. Accessed March 25, 2020. https://www.waterboards.ca.gov/water_issues/programs/ocean/docs/wqplans/thermpln.pdf.

108. Straughan, Baird, and Thomas H. Pollak. *The Broader Movement: Nonprofit Environmental and Conservation Organization, 1989-2005*. The Urban Institute, 2008. Accessed March 29, 2020. https://www.urban.org/sites/default/files/publication/32186/411797-The-Broader-Movement-Nonprofit-Environmental-and-Conservation-Organizations—.PDF.
109. Suter II, Glen W., and Michael A. Lewis. *Aquatic Toxicology and Environmental Fate*, 11th Vol., Philadelphia, PA: American Society for Testing and Materials, 1988.
110. Thaler, Richard H. *Misbehaving: The Making of Behavioral Economics*. New York, NY: Norton & Company, 2015.
111. Thomas, Steve A. *The Nature of Sustainability*. Grand Rapids, MI: Chapbook Press, 2016. ISBN 9781943359394.
112. Turcotte, Donald L., and Schubert, Gerald. "Chapter 4". In *Geodynamics*, 2nd ed., 136–137. Cambridge, England, UK: Cambridge University Press, 2002. ISBN 978-0-521-66624-4. (Geothermal energy source: 20% residual; 80% from radionuclide decay)
113. United Nations Environmental Programme. Introduction to Climate Change. 2020. Accessed June 6, 2020. http://www.inforse.org/europe/dieret/Climate/climate%20graphics/05.htm.
114. United Nations Development Programme; United Nations Department of Economic and Social Affairs and World Energy Council. "Energy and the Challenge of Sustainability".
115. United Nations Food and Agriculture Organization. The State of World Fisheries and Aquaculture, 2018.
116. US Energy Information Administration (EIA). International Energy Outlook 2019 with Projections to 2050. September 24, 2019. https://www.eia.gov/outlooks/ieo/pdf/ieo2019.pdf
117. United States Environmental Protection Agency, 1998. Civil Penalty Policy for Section 311(b)(3) and Section 311(j) of the Clean Water Act. USEPA Office of Enforcement and Compliance Assurance. August, 1998.
118. US Environmental Protection Agency (USEPA). Recreational Water Quality Criteria. November 2012.
119. U.S. Environmental Protection Agency (2007) Landfill Bioreactor Performance. Second Interim Report: Outer Loop Recycling & Disposal Facility, Louisville, Kentucky, EPA/600/R-07/060.
120. USEPA. Industrial Stormwater Fact Sheet Series. Sector L: Landfills and Land Application Sites. USEPA Office of Water. EPA-833-F-06-027, 2006.
121. USEPA. Hazardous Waste Listings: A User-Friendly Reference Document. 2012. https://www.epa.gov/sites/production/files/2016-01/documents/hw_listref_sep2012.pdf

122. USEPA. Advancing Sustainable Materials Management: 2015 Fact Sheet. 2018. https://www.epa.gov/sites/production/files/2018-07/documents/2015_smm_msw_factsheet_07242018_fnl_508_002.pdf
123. USEPA. USEPA Priority Pollutant List, 2020. Accessed May 3, 2020. https://www.epa.gov/sites/production/files/2015-09/documents/priority-pollutant-list-epa.pdf.
124. USEPA. *Underground Storage Tanks: Building on the Past to Protect the Future*, USPEA Publication No. EPA 512-R-04-001, Washington, DC, 2004.
125. USEPA. *Superfund Remedy Report*, 15th ed. EPA-542-R-17-001, Washington, DC, USA: USEPA Office of Land and Emergency Management, July 2017.
126. USEPA. *NPDES Permit Writers Manual*. EPA-833-K-10-001. Washington DC, USA, September 2010.
127. USEPA. Menu of Control Measures. 2013. https://www.epa.gov/sites/production/files/2016-02/documents/menuofcontrolmeasures.pdf
128. USEPA. *Air Quality Index—A guide to Air Quality and Your Health*. February 2014, Research Triangle Park, NC, USA. https://www3.epa.gov/airnow/aqi_brochure_02_14.pdf
129. USEPA. *Integrated Science Assessment (ISA) For Oxides of Nitrogen—Health Criteria*. Final Report, EPA/600/R-15/068, 2016. Washington, DC: U.S. Environmental Protection Agency, 2016.
130. USEPA. Integrated Review Plan for the National Ambient Air Quality Standards for Particulate Matter, 2016.
131. United States Geological Survey (USGS). *A Groundwater Primer*. Washington DC: US Government Printing Office, 1963.
132. United States Geological Survey, 2020, Pesticide National Synthesis Project, website: https://water.usgs.gov/nawqa/pnsp/usage/maps/. Access date: February 15, 2020
133. Urban Water Resources Research Council. Pathogens in Urban Stormwater Systems, August 2014.
134. Vörösmarty, C. J., P. B. McIntyre, M. O. Gessner, D. Dudgeon, A. Prusevich, P. Green, S. Glidden, S. E. Bunn, C. A. Sullivan, C. Reidy Liermann, and P. M. Davies. "Global Threats to Human Water Security and River Biodiversity." *Nature* 468, no. 7321 (2010): 334.
135. Wallace, R. John, Goor Sasson, Philip C. Garnsworthy, Ilma Tapio, Emma Gregson, Paolo Bani, Pekka Huhtanen, Ali R. Bayat, Francesco Strozzi, Filippo Biscarini, Timothy J. Snelling, Neil Saunders, et al. "A Heritable Subset of the Core Rumen Microbiome Dictates Diary Cow Productivity and Emissions." *Science Advances* 5, no. 7 (2019): eaav8391.

136. Wang, Jin, Qiuxia Wu, Juan Liu, Hong Yang, Meiling Yin, Shili Chen, Peiyu Guo, Jiamin Ren, Xuwen Luo, Wensheng Linghu, and Qiong Huang. "Vehicle Emission and Atmospheric Pollution in China: Problems, Progress, and Prospects." *PeerJ* 7 (2019): e6932. doi:10.7717/peerj.6932.
137. Wang, Ruo-Nan, Yuan Zhang, Zhen-Hua Cao, Xin-Yu Wang, Ben Ma, Wen-Bin Wu, Nan Hu, Zheng-Yang Huo, and Qing-Bin Yuan. "Occurrence of Super Antibiotic Resistance Genes in the Downstream of the Yangtze River in China: Prevalence and Antibiotic Resistance Profiles." *Science of the Total Environment* 651, Pt 2 (2019): 1946–57.
138. Weber, Karl. *Last Call at the Oasis—The Global Water Crisis and Where We Go from Here*. New York, USA: Public Affairs, 2012.
139. Woese, Carl R., O. Kandler, and M Wheelis. "Towards a Natural System of Organisms: Proposal for the Domains Archaea, Bacteria, and Eucarya." *Proceedings of the National Academy of Sciences of the United States of America* 87, no. 12 (1990): 4576–79.
140. World Bank. The World Bank data; GNI per capita, Atlas method (current US$). 2012. http://data.worldbank.org/indicator/NY.GDP.PCAP.CD.
141. World Bank. *Methodology for Calculating Environmental Damage Assessment and Relevant Compensation* (English). Washington, DC: World Bank, 2011. http://documents.worldbank.org/curated/en/804831468331771041/Methodology-for-calculating-environmental-damage-assessment-and-relevant-compensation
142. World Health Organization. "Diesel and Gasoline Engine Exhausts and Some Nitroarenes." In *IARC Monographs on the Evaluation of Carcinogenic Risks to Humans*, Vol. 46. Lyon, France: IARC, 1989. https://www.ncbi.nlm.nih.gov/books/NBK531303/pdf/Bookshelf_NBK531303.pdf
143. Yang, S.L., J.D. Milliman, P. Lia, and K. Xu. 50,000 Dams Later: Erosion of the Yangtze River and its Delta." *Global and Planetary Change* 75, no. 1–2 (2011): 14–20.
144. Zhao, Xu, Junguo Liu, Hong Yang, Rosa Duarte, Martin R. Tillotson, and Klaus Hubacek. "Burden Shifting of Water Quantity and Quality Stress from Megacity Shanghai." *Water Resources Research* 52, no. 9 (2016): 6916–27. doi:10.1002/2016WR018595.

Knowledge Box Entries

Chapter 1. How to Build a Sustainable Planet

1. Doomsday/Apocalypse
2. Olber's Paradox
3. Redshift
4. Dark Matter and Dark Energy
5. Radioisotope Dating
6. Stromatolite
7. Human Dominance or Self-Destruction?
8. Anthropocene
9. Gaia Hypothesis

Chapter 2. Global Biogeochemical Cycles

1. Paleomagnetism
2. Gondwana
3. Compositions of Natural Waters
4. Weathering process
5. Shellfish farming as a carbon sink
6. Redfield Ratio

Chapter 3. Environmental Regulations

1. Laws and regulations
2. CalEPA
3. Policy Making Process
4. CEQA
5. The Policy Cycle

Chapter 4. Air Pollution and Greenhouse Gases

1. Soot
2. Smog
3. PM2.5
4. Acid Rain
5. Nitrogen aerial deposition

6. Photochemical Smog
7. Air quality challenges in Beijing, China
8. London air quality issues
9. Catalytic Converter
10. Greenhouse Effect
11. Runaway greenhouse effect on Venus
12. Methane and Cows

Chapter 5. Water Pollution

1. Water Quality Standard
2. Point source and nonpoint source
3. BMP
4. TMDL
5. The Ten Tenets for Water Protection of China
6. Natural Freshwater Chemistry
7. Natural Marine Water chemistry
8. HNLC Waters
9. Toxaphene
10. PCBs
11. Selenium
12. The Rational Method Formula
13. Hydrograph
14. Flow Ecology
15. First Flush
16. Stormwater Discharge Permit
17. Combined and Separate Systems
18. BOD
19. The Pretreatment Program
20. Biosolid
21. Microbeads and Microplastics

Chapter 6. Water Resources

1. The Los Angeles Aqueduct
2. Water Quality Challenge of Yangtze River
3. Water Recycling System at the ISS
4. Water purification in the wilderness
5. All Bets are Off
6. The Three Gorges Dam
7. Dam Removal

8. The Three Gorges Dam and the Earth's Rotation
9. California Bay-Delta Issue
10. The Desalination Process
11. The World's Largest Desalination Plant

Chapter 7. Groundwater and Soil Contamination

1. Darcy's Law
2. Well Drawdown
3. MTBE
4. TCE
5. Flint Water Crisis
6. Blue Baby Syndrome
7. Chromium-6
8. Love Canal
9. Valley of Drums
10. UST
11. Conceptual Site Model

Chapter 8. Ecology and Biodiversity

1. Measuring Biodiversity
2. The Panamanian Golden Frog
3. The Catskill Story
4. Bioaccumulation and Biomagnification
5. The Pantanal
6. Sustainable Fishing
7. Sustainable Aquaculture
8. Flavr Savr
9. Baiji
10. Nutria
11. New Pangaea

Chapter 9. Solid Waste

1. Public Relations Management for Landfills
2. Solid waste and sustainability
3. Food waste
4. Sweden Recycling Revolution
5. RDF
6. HazMat
7. The United States and Basel Convention

Chapter 10. Energy and Sustainability

1. Solar Energy Budget
2. Fracking
3. The 1973 Oil Crisis
4. Flare or Release
5. Energy Storage
6. Windmills in Netherland
7. Gansu Wind Farm
8. Tidal Power
9. Future of Nuclear Energy
10. WIPP

Chapter 11. Climate Change, Sea Level Rise, and Ocean Acidification

1. Climate Emergency Declaration
2. Lithosphere and Climate System
3. Global Warming Potential (GWP)
4. Methane Clathrate
5. ITCZ
6. The Father of Global Warming

Chapter 12. Environment, Human Development, and Sustainability

1. Export of Hazardous Waste
2. Environmental expenditures in large firms
3. Green Marketing and Greenwashing
4. NPV
5. Carbon Footprint in the Sky
6. The 2000-Watt Society
7. Ecological Footprint Measured in Earths
8. Fast Fashion and Sustainability
9. Arne Dekke Eide Næss and Deep Ecology
10. Depletion of Phosphorus
11. Depletion of Rare Earth Elements (REEs)
12. Scientists Warning to Humanity
13. 2019 Warning on Climate Change
14. The Voyager Program

List of Acronyms

1. 5-Rs — Refuse, reduce, reuse, recycle, rot
2. BMP — Best Management Practice
3. BOD — Biochemical oxygen demand
4. BETX — Benzene, ethylbenzene, toluene, xylene
5. CAA — Clean Air Act
6. CalEPA — California Environmental Protection Agency
7. CEC — Contaminants of emerging concern
8. CEO — Chief Executive Officer
9. CFC — Chlorofluorocarbon
10. CEQA — California Environmental Quality Act
11. CERCLA — Comprehensive Environmental Response, Compensation, and Liability Act
12. COP — Conference of the Parties
13. CPCB — (Indian) Central Pollution Control Board
14. CSM — Conceptual site model
15. CSEM — Conceptual site exposure model
16. CSO — Combined sewer overflow
17. CWA — Clean Water Act
18. ENSO — El Nino Southern Oscillation
19. ESA — Endangered Species Act
20. FDA — Food and Drug Administration
21. DDT — Dichlorodibenzotrichloroethylene
22. DNA — Deoxyribonucleic Acid
23. DNAPL — Dense non-aqueous phase liquid
24. DOT — (United States) Department of Transportation
25. DPR(1) — (California) Department of Pesticide Regulation
26. DPR(2) — Direct potable reuse
27. EDC — Endocrine Disruption Compounds
28. EEA — European Environmental Agency
29. EHS — Environmental, health and safety
30. EIA — Environmental Impact Assessment, or (United States) Energy Information Administration
31. EIS — Environmental Impact Statement

32. ET — Extraterrestrial (intelligent life)
33. EU — The European Union
34. EV — Electric vehicle
35. FDA — Food and Drug Administration
36. FIB — Fecal indicator bacteria
37. FIFRA — Federal insecticide, Fungicide, and Rodenticide Act
38. FONSI — Finding of no significant impact
39. FWPCA — Federal Water Pollution Control Act
40. GDP — Gross domestic product
41. GHP — Greenhouse potential
42. GI — Green Infrastructure
43. GMO — Genetically modified organisms
44. GWP — Global warming potential
45. HFC — Hydrofluorocarbon
46. HNLC — High nutrient low chlorophyll
47. ICE — Internal combustion engine
48. IDPR — Indirect potable reuse
49. IPCC — Intergovernmental Panel on Climate Change
50. IRWD — Irvine Ranch Water District
51. ISS — International Space Station
52. ISWM — Integrated Stormwater Management
53. ITCZ — Intertropical Convergent Zone
54. IUCN — The International Union for Conservation of Nature
55. KWH — Kilowatt hour
56. LD50 — Lethal dose resulting in 50% death of test organisms
57. LID — Low Impact Development
58. LNAPL — Light non-aqueous phase liquid
59. LOAEL — Low observed adverse effect level
60. MCL — Maximum contamination level
61. MEE — (China) Ministry of Ecology and Environment
62. MNA — Monitored natural attenuation
63. MSW — Municipal solid waste
64. MTBE — Methyl-tert-butylether
65. NAAQS — National Ambient Air Quality Standard
66. NASA — National Aeronautics and Space Administration
67. NDMA — N-Nitrosodimethylamine
68. NEPA — National Environmental Pollution Act
69. NGO — Nongovernmental organization

70.	NIMBY	Not in my backyard
71.	NO2	Nitrogen dioxide
72.	NOAA	National Oceanic and Atmospheric Administration
73.	NOAEL	No observed adverse effect level
74.	NOx	Nitrogen oxide
75.	NPDES	National Pollutant Discharge Elimination System
76.	NPV	Net present value
77.	NRC	Nuclear Regulatory Commission
78.	OCWD	Orange County Water District
79.	OCSD	Orange County Sanitation District
80.	OECD	Organization for Economic Cooperation and Development
81.	OPEC	Organization of Petroleum Exporting Countries
82.	OSHA	Occupational Safety and Health Administration
83.	PCB	Polychlorinated biphenyl
84.	PFAS	Per- and Polyfluoroakyl Acids
85.	PFOA	Perfluorooctanoic acid
86.	PFOS	Perfluorooctane sulfonate
87.	POP	Persistent Organic Pollutant
88.	PPB	Parts per billion
89.	PPCP	Pharmaceuticals and personal care products
90.	PPM	Parts per million
91.	PRB	Permeable reactive barrier
92.	PPT	Parts per trillion
93.	PV	Photovoltaic
94.	RCRA	Resource Conservation and Recovery Act
95.	REEs	Rare earth elements
96.	RIDM	Risk-informed decision making
97.	SC	(China) Sponge City
98.	SDG	Sustainable Development Goals
99.	SG&A	Sales, General and Administration
100.	SUDS	Sustainable Urban Drainage Design
101.	TBEL	Technology-Based Effluent Limitation
102.	TCE	Trichloroethylene
103.	TMDL	Total Maximum Daily Load
104.	TSCA	Toxic Substances Control Act
105.	TSDF	Treatment, storage, and disposal facilities (for hazardous waste)
106.	UAE	United Arabic Emirates
107.	UN	The United Nations

108. USEPA/EPA   United State Environmental Protection Agency
109. USFWS/FWS   United States Fish and Wildlife Services
110. UST   Underground storage tank
111. VOC   Volatile organic compounds
112. WHO   World Health Organization
113. WOTUS   Waters of the United States
114. WQBEL   Water Quality-Based Effluent Limitation
115. WSUD   Water Sensitive Urban Drainage

CPSIA information can be obtained
at www.ICGtesting.com
Printed in the USA
LVHW061643120821
695016LV00001B/1